The Lady in the Ore Bucket

✷ ✷ ✷

A History of Settlement and Industry in the
Tri-Canyon Area of the Wasatch Mountains

Charles L. Keller

The University of Utah Press
Salt Lake City

First paperback printing 2010

The Defiance House Man colophon is a registered trademark
of the University of Utah Press. It is based upon a four-foot-tall,
Ancient Puebloan pictograph (late PIII) near Glen Canyon, Utah.

LIBRARY OF CONGRESS CATALOGING-IN-PUBLICATION DATA

Keller, Charles L.
The lady in the ore bucket : a history of settlement and industry in
the tri-canyon area of the Wasatch Mountains / Charles L. Keller.
p.   cm.
Includes bibliographical references and index.
ISBN: 978-1-60781-021-6 (acid-free paper)
1. Lumber trade—Utah—Wasatch Range (Utah and Idaho)—History.
2. Mineral industries—Utah—Wasatch Range (Utah and Idaho)—History.
3. Water-power—Utah—Wasatch Range (Utah and Idaho)—History.
I. Title.
HD9757.U8 K45   2001
330.9792'2—dc21
00-012732

The Lady in the Ore Bucket

To my "Thursday Evening" hiking friends, who have shared my love for the Wasatch Mountains for more years than any of us care to admit.

Contents

Preface

I, THE AUTHOR OF THIS VOLUME, have had a long and satisfying relationship with the Wasatch Mountains. At first they provided me tremendous pleasure and satisfaction in my discovery of existing trails and my exploration of ridges, drainages, and hidden spots of incredible beauty, not to mention the exhilaration of mounting a high peak and standing there with a view of the whole world sweeping away below. Then they began to unfold other points of interest for me with the discovery of faint trails or roads, eroded mine dumps, remains of cabins, ditches, and breached dams, as well as abandoned structures and machinery from mining, power generation, recreation, or other unimaginable origins. I began to question the sources of these remains as well as the myriad names whose own sources could scarcely be imagined. Inscriptions found on aspen trees and rocks aroused my curiosity, causing me to wonder about those who carved them: their lives, their hopes, their fears. An inscription, "J. B. Hartzog 1890," carved into a smooth sandstone slab is an intriguing example. I could run my finger over the engraved letters and reach back over the years to feel the same indentations in the stone that J. B. Hartzog felt after he carved his name over a century ago. I wondered about him, and why he was there, and where he went after he left his name behind. I felt a kinship with him, however faint, sitting where he once sat and feeling the warmth of the sun on my back, just as he did so many years ago.[1]

In such experiences a new world appeared, begging to be discovered. To gain a better understanding and appreciation of what is there I began to explore the annals of history to supplement my explorations of the mountains. As bits of information from various sources began to merge, stories unfolded, begging to be told. Casual exploration became an obsession, resulting in a wealth of information, some of which is presented in this volume. Perhaps what I have learned will enable others who read these words to better understand, appreciate, and enjoy what they find along the trails in the Wasatch Mountains.

*

In the process of my research I discovered that much of what has been written about events in these mountains is little more than legend or hearsay. And often those accounts that appeared credible could not be substantiated by other sources. As a result, this effort became as much one of producing an authentic and well-documented history as it was one of satisfying my curiosity. If an account was interesting enough for me to include in this narrative but could not be substantiated, that fact is mentioned so the reader can make his or her own choice of whether or not to believe it. Also, if some well-known name or subject is not mentioned at all, it is probably because I could find nothing that I considered to be of suitable validity. The many source notes I have included should allow future researchers to build upon this history without having to duplicate my researches.

In the process of searching for more verifiable sources of information, an unexpected amount of material came to light. My original intent was to write about all the Wasatch Mountain canyons in Salt Lake County—the Central Wasatch—but I discovered there was far more history than could be contained in a single volume. For that reason this narrative is restricted to the tri-canyon area, extending from Mill Creek Canyon on the north to Little Cottonwood Canyon on the south. This area contains the highest elevations in the county as well as some of the most beautiful and challenging terrain. It is also the area that receives the most attention from recreationists, hikers, and climbers.

Another restriction that evolved due to the voluminous material was my decision to limit this narrative to nineteenth-century history, to explore those activities before their traces grow even more faint. But events do not conveniently begin or end just because a century rolls over. Therefore, many stories running well into the twentieth century have been discussed as well. The reader will note an absence of text describing the origins and growth of the ski industry in the Wasatch. There are two reasons for this: it is something that has dominated only the past fifty years, thereby being a fairly modern event, and it is a subject that has been covered very well in recent publications. The reader who has interest in the history of skiing in the Wasatch should read *Skiing in Utah: A History* by Alexis Kelner, and *For the Love of Skiing: A Visual History* by Alan K. Engen.

Historical research can be a very grueling endeavor, with many hours of reading or scanning newspapers, books, journals, letters, reports, and any other form of printed or written documentation. It is often tiring and boring, but each new discovery brings with it an elation that refuels the desire to continue. Unfolding stories also beckon, urging one to continue, to explore new sources, to uncover one more fact. In this way, research becomes very satisfying. As with the search for the Holy Grail, the search itself can provide as great a reward as the object of the search.

This history is also somewhat depressing, however, for it is a story of exploitation and destruction. It begins with the Wasatch Mountains essentially untouched by man, with the possible exception of a few Indian trails. The first exploitation was by the lumbermen who stripped the slopes of their virgin forests. They were followed by miners who took those few trees that remained and left great scars of mine dumps and roads, and the resulting erosion ditches and gullies. Then came the grazers, whose herds stripped the slopes of much of the vegetation that remained and pounded the hillsides to dust. The early twentieth century saw some salvation in the U.S. Forest Service's massive reforestation efforts, the withdrawal from exploitation of remaining public lands in an attempt to preserve Salt Lake City's watersheds, and the decline of the mining industry. Nature was given a reprieve to heal Her wounds, and today we see forests that are resembling what was here when the first settlers arrived. It is easy now to forgive the transgressions of our forebears, but we should not forget the lessons of the past. History repeats itself, it is said, and so it does, for today the Wasatch Mountains face new threats in the demands of recreation and development interests, threats at least as ominous as those of the past. How the mountains will survive these threats remains to be seen, and this may be the subject of explorations by future historians.

*

A word of caution: even though this narrative may describe trails and vanished by-ways, it is not intended to be a hiking guide. While it may add a new dimension to hiking pleasures and may be used as a hiking companion, there are other references that describe trails, their locations, pitfalls and pleasures. One good source is *Hiking the Wasatch* by John Veranth.

Charles L. Keller

White Men Come to the Wasatch

The year 1847 was a major turning point in the history of the Central Wasatch Mountains. It was as if the sun rose to begin the bright day of civilization after a long night of evolution. Or perhaps, considering man's incredibly poor record as custodian of his environment, it might be seen as a twilight, ending a near-eternal day of brightness in which animals and "uncivilized" human beings enjoyed Nature's bounty, and beginning the darkness of settlement, development, exploitation, and destruction. Whatever one's view, the history of the Salt Lake Valley and the Central Wasatch before and after July 1847 is as different as night and day. For the story prior to that time we have to expand our horizons well beyond the Central Wasatch to review the events that pushed men in this direction and the actions that eventually influenced events in our sphere of interest.

Before 1847 the area saw little impact by humans. There were Indians, but they were relatively few in number and were hunters and gatherers, nomadic people who were not inclined to spend time building permanent dwellings or making "improvements" to their lands. The Salt Lake Valley was a buffer zone between the Shoshone tribes to the north and the Utes to the south; while each ventured into the area, they did not remain. They traveled frequently and at times for some distance, often using well-defined trails, as witnessed by later journal entries made by whites who followed Indian trails or heard Indians tell of distant Indian wars or events. However, the impact upon the land by Native Americans was small, and the most notable remnants of their presence are place names. Even those would not have survived had it not been for the occasional passage through the area by white men: explorers, trappers, hunters, and, eventually, immigrants. Some of these people, being literate, recorded

1

impressions of their travels in journals or diaries that have survived for modern study. It is impossible to know how many people may have passed this way; it is only through those who recorded impressions in surviving journals that we are able to surmise what the area was like in those earlier days. These sources also provide an interesting view of how the country unfolded and how its names evolved.

The first known explorers came into the area in 1776, a time when the western part of the present United States was a large blank spot on the map. The Domínguez-Escalante expedition was headed by Franciscan priests seeking a route from Santa Fe, New Mexico, to the missions in California. While Friar Francisco Atanasio Domínguez was the leader of the group, it was Friar Silvestre Velez de Escalante who kept the journals that have provided knowledge of their ventures. These men never saw the Salt Lake Valley, but they did reach the shores of Utah Lake on or about 23 September 1776, arriving there after traveling west through the Uinta Basin and entering Utah Valley through Diamond and Spanish Forks. This particular area was dominated by the Timpanogas branch of the Ute Tribe. The Indians were friendly and receptive to the teachings of the Catholic priests. In fact, the Indians urged the visitors to remain, or at least to return soon so that they might become Christians. Whether this was a wish of the natives or wishful thinking on the part of the Fathers is not known, but certainly there was a much better rapport between the visitors and natives than evolved years later when settlers began to encroach upon the Indians' hunting and gathering grounds.

Through the Indians the Fathers learned there was a larger lake to the north but not how the two lakes were related. On his map of the route and terrain, Don Bernardo Miera y Pacheco, a retired military engineer and the cartographer for the expedition, drew the two lakes adjoining one another, separated only by a narrow channel. The lake was called "Laguna de Las Timpanogas" and the mountains to the east of the valley were labeled "Sierra de las Timpanogas." Miera's map also shows a river flowing into the upper lake from the northeast and another flowing out to the west.[1] The latter was an assumption on Miera's part, who probably expected that the lake had an outlet that flowed to the Pacific Ocean. The river to the northeast was not as fanciful and might have been the Bear. That river may have been as significant to the Indians as it was to the explorers and trappers who used it in later years.

*

The Bear River flows north from its headwaters in the northwestern Uinta Mountains, meanders through Wyoming and Idaho, and crosses through the Wasatch Mountains at the extreme north end of the range. It then veers south through Cache Valley and finally empties into the Great Salt Lake. Along the way it gathers water from numerous tributaries, including the outlet from Bear Lake on the Utah-Idaho border. With all these streams the area became an early favorite for trappers and hunters. In fact, it was an early trapper, Michel Bourden from the Northwest Company, who named the river while trapping there in 1818.[2] As explorers and trappers, and eventually emigrants, came through the area, the Bear River provided a convenient route through the Wasatch and thereby kept travelers well north of the Central Wasatch that is of interest here.

In 1824–25 there was considerable activity in the general area. The Ashley-Henry exploring and trapping party, under William H. Ashley and Major Andrew Henry, came west across Wyoming. In the group were many whose names would remain prominent in western history, including Jedediah Smith, James Bridger, James Clyman, and John H. Weber. In western Wyoming the party split, with Smith and six men heading north while the rest followed the Bear River with Weber. While the latter party wintered in southern Idaho, Jim Bridger followed the Bear River downstream and discovered the Great Salt Lake. He found the water to be salty and returned to report his findings. That spring the entire party explored as far south as the Weber River, to which the leader left his name. In later years the river was also called the Weaver, supposedly after Pauline Weaver, an Arizona frontiersman who was in the area.[3] However, no documented evidence can be found to support this story; it is more likely that Weaver was the result of a phonetic spelling. Since Weber is pronounced Wee-ber, the two words Weber and Weaver sound very much alike.

The Ashley-Henry group under Jedediah Smith discovered a large party from the British Hudson's Bay Company, led by Peter Skene Ogden. As it was heading southward they followed it back into Bear River country. Ogden's party continued down through Cache Valley, then, leaving the Bear, came due south through the mountains until it reached the valley and river that now bear Ogden's name. The group crossed the divide and dropped into the Weber Valley near present-day Mountain Green but didn't get any closer to the Central Wasatch.[4]

Ogden's chief clerk, William Kittson, kept a journal in which he recorded that one of the party, Charles McKay, had climbed to a high spot on 12 May 1825 and had seen the Great Salt Lake. But by this time Jim Bridger had seen (and tasted) it.[5] Kittson also drew a map of his group's travels in the Bear River country. He called the Great Salt Lake "Large Bear Lake," perhaps implying that he knew of the present Bear Lake even though he failed to include it on his map. He drew the Weber River on his map but gave it no name.[6]

Ogden is known to have met Etienne Provost, a Canadian-born mountain man working out of New Mexico, along the Weber River. The previous year Provost (pronounced Provo) was on an expedition that came through the Uinta Basin and into Utah Valley by way of the river that now bears his name. He ran afoul of some Indians and was lucky to escape alive. The next year he returned with a trapping party and met Ogden. How much of the Wasatch he saw or explored is not known, for Provost left no journal.[7]

The first exploration of the Great Salt Lake was conducted in the spring of 1826 when James Clyman and three others paddled their bull boats around the lake for three weeks seeking places to trap beaver. In the process they discovered that the lake had no outlet.[8] This venture has significance for the Central Wasatch history: from their vantage point along the south shore of the lake the men had a good, if somewhat distant, view of the mountains to the east of the Salt Lake Valley. Twenty years later Clyman would leave the first written record of a journey through them.

In 1826 Jedediah Smith, with a group of about sixteen men, set out from Cache Valley to blaze a trail to southern California. He headed south along the Wasatch, through Salt Lake and Utah Valleys, then southwest toward his destination. While he kept a journal of his travels, it gives little detail of his trip along the Wasatch. He wrote, "on the 18th of August [1826] I struck over west on to the Big Salt Lake then south crossing webers fork to the outlet of Uta Lake then up the outlet to the Lake."[9] The "outlet of Uta Lake" is the Jordan River and "the Lake" is Utah Lake. Notice the name used for the Great Salt Lake and the use of "webers fork." John Weber was at this river scarcely a year earlier, but both he and Smith were part of the same Ashley-Henry party.

After wintering along the West Coast, Smith returned the next year over the Sierra Nevada, coming across Nevada, making the first crossing of the salt desert, and coming around the south shore of the Great Salt Lake before heading north to Bear Lake, which he reached on 3 July 1827. His journal mentioned nothing about the Salt Lake Valley or the mountains, only describing briefly some difficulty in crossing the swollen "outlet of Utah Lake."[10]

In 1833 Captain Benjamin L. E. Bonneville sent Joseph Redeford Walker on a trapping expedition via the north end of the Great Salt Lake and the Humboldt River. While his route was a great distance from the Central Wasatch, the information he brought back was reflected in Bonneville's

1837 map, including islands in the Great Salt Lake. It also showed the Bear River flowing in from the north through "Shoshoni Indians" land, "Eutaw" Lake to the east of the larger lake, "Eutaw Mtn." northeast of the smaller lake, and "Eutaw Indians" land to the south.[11] Bonneville took advantage of the situation when he published the map and gave the large lake the name Lake Bonneville, even though he had never personally seen it. The name did not stick; the commonly used names all reflected the lake's salty character, an attribute that could not be denied. Jedediah Smith had used the name "Big Salt Lake" in his journal in 1826, but the following year, on 17 July 1827, in a letter written at "Little Bear Lake" to General William Clark, Superintendent of Indian Affairs, he used the term "Great Salt Lake," which soon became commonly used.[12]

Another trapper, Osborn Russell, arrived at Fort Hall from Independence, Missouri, in 1834. For the next six years he traveled throughout Wyoming and Idaho trapping and hunting and recording his experiences in his journals. He spoke "Canadian French and Indian tongue" and apparently got along well with other trappers, French or English, as well as with the Indians. In the winter of 1839–40 he went as far south as the Salt Lake Valley, traveling alone but living with the area Indians. While he recorded no excursions into the Central Wasatch, his writings during this period give a good insight into Indian life at that time.[13]

Russell left the Salt Lake Valley on 27 March 1840 to return to Fort Hall and points north. He later lamented the diminishing supply of wildlife: "In the year 1836 large bands of Buffaloe could be seen in almost every litle Valley... at this time the only traces which could be seen of them were the scattered bones of those that had been killed. Their trails which had been made in former years deeply indented in the earth were grown with grass and weeds The trappers often remarked to each other as they rode over these lonely plains that it was time for the White man to leave the mountains as Beaver and game had nearly disappeared."[14] The great days of the trappers were ending, and in their place came exploring parties, followed by emigrant trains—in ever increasing numbers.

*

Of the many exploration trips of Captain John C. Frémont, two are of interest here. In 1843–44 he and his party came south along the shores of the Great Salt Lake to Weber's Fork, where they camped and inflated an India rubber boat they had brought, and Frémont and four companions paddled out into the lake to the nearest island. Expecting to find lush vegetation, animal life, and fresh water there, they were disappointed to find nothing but barren rock. Accordingly, Frémont named it Disappointment

Island. Before they returned to shore and headed north again, his cartographer Charles Preuss prepared a map of the lake showing the locations of all its islands. In recognition of Frémont's visit to the island, Captain Howard Stansbury renamed it Fremont Island in 1850.[15]

In 1845 Frémont came through Utah again on what was his most important expedition as far as the Central Wasatch is concerned, for it began a chain of events that placed the first documented white men in those mountains. This time Frémont traveled west through the Uinta Basin to the Provo River, following it down to Utah Lake. He then turned north, followed the Jordan River through the Salt Lake Valley, and in mid-October camped at the site of today's downtown Salt Lake City, remaining there for six days before moving on.

Unfortunately, Fremont's published description of this trip was written many years later, after his journals had been destroyed, so we don't know if his explorations extended into the Wasatch. They probably did not, since the Great Salt Lake offered a much more interesting natural feature than did another range of mountains. Frémont had traveled through mountains since entering the Rockies in Colorado, but the Great Salt Lake was different. Indeed, his recollections dwelt on the terrain and salt incrustations around the lake. He tells about riding to "a large peninsular island, which the Indians informed me could at this low stage of the water be reached on horseback." On the island he and his companions found grass, water, and pronghorn antelope. Several of the animals were killed to replenish the food supply and, wrote Frémont, ". . . in memory of the grateful supply of food they furnished, I have given their name to the island." Thus was Antelope Island given its name in October 1845. This story was corroborated by Kit Carson, who was with Frémont on this trip as well as the one in 1843.

Frémont went on to write that on their return to their camp an Indian was there to tell them he owned the island and the animals. Since he demanded payment for those they had killed, Frémont paid him with some red cloth, a knife, and some tobacco.[16]

From there the party headed west, crossing the salt desert towards Pilot Peak in Nevada and then the Sierra Nevada, reaching Fort Sutter near Sacramento at the very late date of 10 December. On the way Frémont named Pilot Peak and the Humboldt River. Although some writers may dismiss his crossing of the salt desert with only a few words, it was no trivial feat. Only one party was known to have done this previously, that being Jedediah Smith's crossing with two companions from west to east in 1827. When Frémont reached Sutter's Fort, word of his achievement quickly spread. And one young man was especially interested.

Lansford W. Hastings was twenty-three years old in 1842 when he traveled to Oregon with an emigrant train. After spending the winter there, he headed south to California, where he recognized a budding land of opportunity. He returned east by way of Mexico and Texas and published *The Emigrant's Guide to Oregon and California.* The usual route west came across Wyoming to Fort Bridger, went northwest to Fort Hall, north of present-day Pocatello, and then west along the Snake River. There the California travelers left the Oregon Trail and went southwest to the Humboldt River and on to California. While this route took a long detour to Fort Hall, it did bypass the hostile unknown region of the salt desert west of the Great Salt Lake. Hastings recognized the lengthy detour, and with the great experience of one trip to Oregon he boldly described a shorter route in his guide:

> The most direct route for the California emigrants, would be to leave the Oregon route, about two hundred miles east of Fort Hall; thence bearing west southwest, to the Salt Lake; and thence continuing down to the bay of St. Francisco.[17]

Hastings returned to California with a small emigrant party late in 1845, following the traditional route, and arrived at Sutter's Fort a few days after Frémont. The news of Frémont's transit of the salt desert assured him that his recommendation in his guidebook was correct. During the winter he had an opportunity to discuss the trip with Frémont, but the latter's own accounts made the feat appear simple, which was much at variance with the opinions of those who accompanied him. In April Hastings left Sutter's Fort to travel east to Fort Bridger via the new route and guide emigrant trains back the same way. When he left California, his small group included James Clyman, one of the men who had explored the Great Salt Lake in 1826. By 27 May they had reached Pilot Peak and in four more days were at the springs at Grantsville. The next day they rounded the north end of the Oquirrh Mountains and camped in the northwest corner of the Salt Lake Valley, in full view of the Central Wasatch Mountains. Here, Clyman noted in his journal, "several Indians ware seen around us after considerable signing and exertion we got them to camp and they apeared to be friendly."[18]

An interesting sidelight is provided by another entry he made that day: "These Ewtaws as well as we could understand informed us that the snakes and whites ware now at war and that the snakes had killed two white men this news was not the most pleasant as we have to pass through a portion of the snake country." This is another bit of evidence that suggests the Indians traveled long distances or communicated with their neighbors enough to spread news of this sort.

Clyman's journal entry for 2 June 1846 takes them across the Salt Lake Valley and into the Central Wasatch:

> acording to promis our Eutaw guide came this morning and conducted us to the ford on thee Eutaw river which we found Quite full and wetting several packs on our low mules but we all got safely over and out to the rising ground whare we found a fine spring brook and unpacked to dry our wet baggage....
>
> Afternoon took our course E into the Eutaw [Wasatch] mountains and near night we found we had mistaken the Trail and taken one that bore too much to the South camped in a cove of the mountain...."[19]

In his extensive footnote commentary, J. Roderick Korns analyzed the route followed across the Salt Lake Valley. From the northwest corner of the valley the full extent of the Central Wasatch is visible, with the very obvious red sandstone buttes of Red Butte at the north, the wide cut of Emigration Canyon, and farther south the more precise slash of Mill Creek Canyon. Parleys Canyon, the one they wanted to use, is nearly invisible from that vantage point, and so the group headed across the valley toward Mill Creek. Korns's analysis has them crossing the "Eutaw river," which is the Jordan, at present-day 2700 South Street. The "fine spring brook" is a stream rising from a spring at the northeast corner of today's Fairmont Park, its waters flowing southwest to Mill Creek. As the group climbed onto the eastern benches the mouth of Parleys Canyon came into obvious view and they realized they were going too far to the south. The cove of the mountain where they camped is one of the several coves that were more obvious several miles up Parleys Canyon before Interstate 80 altered so much of the scenery.

One question that arises from Clyman's journal entry and Korns's analysis is how they knew they wanted to go up Parleys Canyon and knew that route well enough to recognize they were too far south after Parleys Canyon opened to view. Clyman was the only member of the party who had been in this area before, but that was twenty years earlier and there is no evidence he was in the Salt Lake Valley itself or in the mountains east of it. The only answer that can be given is contained in the first few words of

Clyman's journal entry for the day: the "Eutaw guide" must have known and described this route.

The next day Clyman wrote:

> N.E. up the Brook [Parleys Canyon] into a high ruged mountain [Big Mountain] not verry rocky but awfull brushy with some dificulty we reached the summit and commenced our dissent which was not so steep nor Quite so brushy...this ridge or mountain devides the waters of Eutaw from those [of] Weebers rivers and desended the South branch of Weebers rivir untill it entered a rough Looking Kenyon when we bore away to the East up a small Brook [Dixie Creek] and encamped at the head springs.[20]

This route took them a short way farther up Parleys Canyon to Mountain Dell, then north about five miles before climbing to Big Mountain Pass. On the other side the party turned east to go down Little Emigration Canyon until it met East Canyon, at which point they turned north again. The "rough Looking Kenyon" was the narrows where East Canyon Dam has since been built. The camp was in Dixie Hollow, a short distance northeast of East Canyon Reservoir on Utah Highway 65.

From this point they continued down to the Weber River near Henefer, then up the Weber to Echo Canyon and east to Fort Bridger. There the party broke up. Hastings recruited westbound groups to take his new route, while the others dispersed. Interestingly, Clyman headed east and at Fort Laramie met an old friend, James Reed, who was with the westbound Donner-Reed party and was interested in Hastings's new route. Clyman strongly recommended against it; at the time only three groups had crossed the salt desert, and none of them had taken a wagon.[21]

Hastings soon found four groups who knew about his shortcut through his *Emigrant's Guide* and were pleased to let him guide them over the new route. The first was the Bryant-Russell party, a small mounted group without wagons. Following them was the Harlan-Young party, a large, cumbersome group of at least forty wagons. The third group is known by its scribe, Heinrich Lienhard. This was a small party, mostly Swiss and German emigrants, traveling alone or with other groups as their paths converged. Finally came the Donner-Reed party, another large group that had been advised against taking this route but succumbed to the prospect of cutting several hundred miles off the journey ahead of them.

The first two groups left Fort Bridger the same day, 20 July, but with different leaders. The Bryant-Russell party learned that James M. Hudspeth, who had just come across the salt desert with Hastings and Clyman, was heading west again to scout an alternate route between the Bear River and the Salt Lake Valley. He agreed to guide them as far as the west side of the Great Salt Lake. The route they followed took them along the Bear River to present-day Evanston, west and southwest to Lost Creek, and down that drainage toward the Weber River. At Croyden they turned south-southwest to go over a rise and descend into the Weber Valley, thereby avoiding a narrow, rough section of lower Lost Creek. Hudspeth scouted downstream on the Weber, through the section passing Devil's Slide, and concluded there had to be a better route. Indians confirmed that by telling him of a route up Main Canyon and down East Canyon, a route they in fact followed. Hudspeth had been on the first part of this route in the opposite direction only a few weeks earlier, so that section was familiar.

When they reached East Canyon they turned downstream only to find a narrow canyon choked with fallen boulders, but a small Indian trail went up the side of the mountain to bypass the obstacle. They used the trail, men and mules following one another in single file over an impossible route for wagons, until the canyon opened into the Morgan Valley, which they followed to the Weber River. Taking the main stream westward they had one more obstacle to overcome—the narrow canyon at Devil's Gate. While it required "laborious exertions for several hours," they passed through without incident and camped at the mouth of the canyon, where they remained for two days while Hudspeth went back up the Weber to explore. Apparently he recognized the unsuitability of the route they followed for wagons and wanted to find a better alternative. This time he followed the Weber past Devil's Slide and decided it was not as bad as it had first appeared.

The second party to leave Fort Bridger was under the guidance of Hastings. He took them west by a rather meandering route, finally arriving in Echo Canyon. There he directed his party to continue down the canyon while he went back to scout a more direct route than the one he had followed. In doing so he met Lienhard and his party, who had left Fort Bridger on 26 July. Hastings described to them the new, more direct route to Echo Canyon, then went on ahead to catch his original group. Meanwhile the Harlan-Young wagon train had reached Echo, where they met Hudspeth, who told them of his exploration of the Weber and assured them they could continue down the river to the Great Salt Lake. He then went downstream to join his party, while the wagons followed behind him. What they faced was suggested by Edwin Bryant, who wrote his thoughts in his journal after Hudspeth returned and told him of meeting the wagon train. "The difficulties to be encountered by these emigrants by the new route will commence at that point; and they will, I fear, be serious. Mr. Hudspeth thinks that the passage through the cañon is practicable, by making a road in the bed of the stream at short distances, and cutting out the timber and brush in other places."[22]

By the time Hastings caught up with the Harlan-Young wagons they were well into Weber Canyon. Apparently he was unable to convince them they were going the wrong way; he had intended taking them up over Big Mountain Pass and down Parleys Canyon, but they continued down the Weber. Since there is no journal from that group, we don't know their experiences in Weber Canyon, but years later Jacob Harlan wrote his recollections, saying that they worked for six days building a road and got through on the seventh day.[23] Hastings once again went back to redirect those who were following and met the faster-moving Lienhard party only a short distance up the canyon, near present-day Peterson. He was convincing enough for the group to halt and camp for two nights while he took a few of the party on a reconnaissance, apparently to convince them to take the route he had intended. Or perhaps it was his intention to intercept and warn the Donner-Reed party. In any event, he failed in both, for the last party was still well behind and the Lienhard group chose to follow the wagons that had already descended the Weber. However, Hastings did leave a note for the last group. Taking advantage of the work done by the party ahead of them, the Lienhard party was able to get through the lower canyon in less than one day's time, albeit not without considerable difficulty. Once out of the canyon they were able to follow the leading groups south along the Wasatch Mountains, across the Jordan River, and west around the south end of the Great Salt Lake.

The Donner-Reed party left Fort Bridger on 31 July 1846. By this time the countryside to the west was littered with tracks; a stroke of fate dictated that this last group would follow the longer trail of the Harlan-Young party rather than the more direct Lienhard track, probably slowing them enough that they missed meeting Hastings when he traveled back to Echo. On 6 August they did find his note sticking on top of a sagebrush bush. Hastings asked them to send for him and he would come back to guide them over a shorter and better route than the one through Weber Canyon. The group camped there, in Weber Canyon between Echo and Henefer, for four days while James Reed, accompanied by William Pike and Charles Stanton, rode down the Weber in an attempt to catch Hastings. They rode for two days before finding him with the Lienhard group at Adobe Rock in Tooele Valley. By that time their horses were exhausted. Reed borrowed a fresh horse from the Lienhard party and started the return trip with

Hastings, while his companions stayed behind to allow their mounts to
recuperate. Hastings took Reed across the Salt Lake Valley and up Parleys
Canyon, spending the night in the canyon, probably at the same place he
had camped with Clyman two months earlier. Finding the distance greater
than he had planned, Hastings decided not to go all the way back to the
Donner-Reed camp; instead, he took Reed to Big Mountain Pass, where
they could see the entire route below them. After giving directions, Hastings returned to guide the advanced parties across the salt desert. Reed
made the descent per instructions, blazing and marking the trail as he
went. When he returned to camp on the evening of the tenth, he reported
his opinion that many of their wagons would be destroyed if they attempted
to go through Weber Canyon. He thought his return route to be fair, but
it would take considerable road-building effort.[24]

It took the group six days before they reached Big Mountain Pass, which
they called Reed's Gap.[25] During this time Reed's companions, Pike and
Stanton, returned to report the impracticability of taking wagons through
Parleys Canyon, an observation confirmed by the Mormon pioneer group
the following year. Accordingly, they changed their direction, crossed "Small
mountain" into Emigration Canyon, and spent another three days cutting
trees and brush and making a road almost to the mouth of the canyon.
There they discovered that a short spur extending from the south wall
nearly obstructed the canyon except for a narrow channel for the stream,
thickly overgrown with willows. Another day was spent doubling teams
to pull the heavy wagons up the steep slope to the top of the spur, since
known as Donner Hill. Sunday, 23 August, found the party encamped on
the east bank of "Utah outlett," the Jordan River. The next day they continued west into history, in one of the great, tragic dramas of the western
migrations, one that could be said to begin as they forged one of the very
first roads through the Wasatch Mountains.

*

There is one more page to be written in bringing the feet of white men
into the Central Wasatch. The following year saw the great, and much
celebrated, migration of the Mormon pioneers into the Salt Lake Valley.
Their travels were much eased by the efforts of those who went before
them. They left Fort Bridger on 9 July 1847 and followed the tracks of the
Harlan-Young party from the previous year, but they then took the more
direct path of the Lienhard group west into Echo Canyon. Their track became the favored emigrant trail in the years that followed. After going
down Echo Canyon and a short distance down Weber Canyon from Echo,

they turned to follow the Donner road up Main Canyon, over the Hogs-back and down Dixie Hollow, although in the latter drainage the Don-ners had taken a detour west out of the hollow and over a hill to descend into East Canyon about a half mile below Dixie Hollow, thereby avoiding a great amount of brush cutting. Even though the detour was longer than would be a direct route down Dixie Hollow, it was used for many years before a road was cut directly down the hollow.[26]

The Mormon pioneers followed the Donner-Reed track up East and Little Emigration Canyons and over Big Mountain. There Orson Pratt and John Brown, acting as scouts, continued several miles down Mountain Dell Creek, which they named Browns Creek, and "passed through a very high mountain, where we judged it impossible for wagons to pass." The "high mountain" was the deep gorge of Parleys Canyon below Mountain Dell. They retraced their steps and found the wagon trail that ascended Little Mountain and dropped into Emigration Canyon. The pioneers' pas-sage down the latter canyon was fairly easy, although they observed that the emigrants of the previous year "must have spent a great deal of time cutting a road through the thickly set timber and heavy brush wood." When they got to Donner Hill they chose not to attempt to drag their heavy wagons up the steep slope. Instead they spent about four hours clearing brush and cutting a road along the creek.[27] They continued along Emigration Creek and camped near present-day 500 East and 1700 South Streets on 22 July. The next day they moved north to settle along the south fork of City Creek and immediately began tilling the soil. By the time their leader, Brigham Young, arrived on the official Pioneer Day, 24 July, seeds had been sown and irrigation ditches dug, and a new era of Salt Lake Valley history had begun.

It was immediately apparent that the mountains surrounding the val-ley would be important for the sustenance of the new settlers. In their first day in their new home they discovered that the soil had been baked so hard by the summer sun that it could be plowed only with great diffi-culty. A small dam was built on the stream flowing from the mountains and ditches were dug to carry the water to the fields. Once the soil was soaked the plowing could proceed, but it was obvious that irrigation would be necessary if agricultural pursuits were to have any degree of success. The mountain streams that offered water for culinary and domestic pur-poses now had the much greater task of nurturing the fields. Within a short time they also would be used to power a variety of machinery, es-pecially grist- and sawmills.

Many local features received names during the early days of the new settlement. The peak directly north of town was named only a few days after the pioneers arrived. On Monday, 26 July, a group of at least eight men, including Brigham Young, went on a short exploring expedition north of the camp. William Clayton, one of the party, reported they "began to ascent the mountains, the President signifying a wish to ascend a high peak to the north of us. After some hard toil and time we succeeded in gaining the summit, leaving our horses about two-thirds the way up.... Some of the brethren feel like naming this Ensign Peak."[28]

At a special conference on Sunday, 22 August, shortly before he departed the new settlement to return to Winter Quarters, Brigham Young assigned numerous names to area features. He first moved that they "call this place 'The Great Salt Lake City of the Great Basin of North America,'" thus giving the settlement the name it carried until 1868, when the initial "Great" was dropped. This was not a spontaneous thought on Young's part, for nearly two weeks before those of the Quorum of the Twelve who were in the valley adopted the name to be used for all letters and documents issued from the valley.[29] Heber C. Kimball then moved that the river running west of the settlement be called the "Western Jordan." Before this time it was known as the Utah Outlet, it being the outlet of Utah Lake. Brigham Young then named many of the local creeks: "I move that this creek that we are encamped on be called City Creek. I move that the large creek running a few miles south of here be called Mill Creek, that the little creek a little south of here be called Red Butte Creek, that the next creek south be called Canyon Creek and the next Big Canyon Creek." All motions were seconded and carried unanimously.[30]

Directly east of the settlement, at the edge of the mountains, stood a high butte of red sandstone, one that was as obvious and prominent then as it is today. It was perhaps inevitable that it would be called the Red Butte, and the stream that issued from the mountains at the base of the butte therefore was called by the same name. That the next creek south was to be called Canyon Creek is interesting. Some of the pioneers called it Last Creek as they came down the canyon into the valley, but that name wasn't adopted. When they got near its mouth, where they spent several hours cutting brush and building a road along the stream, William Clayton wrote, "We named this a canyon because of the very high mountains on each side leaving but a few rods of a bottom for the creek to pass through."[31] Perhaps this was the origin of the name; if not, one cannot deny that the creek does flow out of a canyon. Unlike other canyons that assumed the same name as their creek, for years this one did not have a distinctive name. Travelers followed Canyon Creek, without mentioning the canyon itself. Finally, in 1852 it began to be called Emigration Canyon, after the Emigration Road that ran through it.[32]

The next stream to the south also flowed out of a canyon, but this was a bigger one, resulting in the name of Big Canyon Creek. This was the stream that was named Browns Creek by John Brown and Orson Pratt when they saw it in Mountain Dell during their exploration in advance of the pioneers. When the first group camped in the valley they correctly recognized the stream where they camped as the one called Browns Creek, but that name wasn't accepted either. The Big Canyon (usually spelled Kanyon at that time) name was used for many years, although the present name, Parleys Canyon, was used as early as 1853.[33] The Mormon pioneers seemed to like to use simple and descriptive names, like Canyon Creek. When they came to another creek flowing from a canyon, they would often give it the same name. If the two were close together, they would differentiate between them by calling the larger one Big Canyon Creek, as was the case here. This tendency is seen in a number of names, like Big and Little Sandy, Big and Little Mountains, and Big and Little Cottonwood Canyons.

That Mill Creek should have been called by that name is also interesting. It suggests that the stream must have been recognized as suitable to power mills, although it was no better in this regard than a number of others, such as Big Canyon Creek. There were no mills on the creek at the time, nor would there be for another six months.

The two major streams south of Mill Creek, Big and Little Cottonwood, also received their names in these early days, although not by proclamation as did their neighbors to the north. It was inevitable that the two Cottonwood creek bottoms would soon become popular and important to the settlers. They supported a heavy growth of cottonwood trees that had never been disturbed by humans. Cottonwoods grow fast but are easily damaged by wind and storm. In their natural undisturbed state, cottonwood groves have copious amounts of large and small branches littering the ground, often to the point of restricting easy passage through them. A solitary cottonwood tree often will grow irregularly—twisted and deformed—but a group of trees growing together compete for sunlight and thereby grow straight and tall, offering good logs for construction. This area became a wonderful place for the settlers who were in desperate need of wood for fuel, fences, and shelter. And what better name to give the place than the name of the tree that grew there. In January 1848 Lorenzo D. Young noted in his diary, "Briant and John went to the cottenwoods to get out timber for Briant's house."[34] Briant would have been Briant Stringham, one of the original pioneer group, and John was Lorenzo Young's ten-year-old son, John Ray.

Although both streams were collectively referred to as the Cottonwoods, there was a need to differentiate between the two. Little Cottonwood was first called the Further Cottonwood, and South Cottonwood, while the northerly stream was called Big Cottonwood, Near Cottonwood, Great Cottonwood, or simply Cottonwood.[35] The use of "Big" may have referred to the trees that grew along the stream, the stream itself, or simply to the creek in closest proximity to the settlement. The use of the name Big Cottonwood has been found as early as April 1848, while the use of Little Cottonwood has been found as early as January 1851.[36]

*

Other features that received their names during these first days of the Great Salt Lake City settlement were Twin Peaks and Lone Peak. Both names were used as early as 3 August 1847. The Twin Peaks are very obvious from downtown Salt Lake City and the name would come naturally to anyone who looked at them long enough to recognize the two peaks at about the same height with only a shallow depression between them. The first reference for the latter peak spoke of "A Lone peak—S.S.E. from camp."[37]

Twin Peaks received special attention when, on 21 August 1847, four men climbed the peak from the west, three reaching the summit. At half past eight on Friday evening, 20 August, Albert Carrington, John Brown, George Wilson, William Rust, and Alva Calkins started on horseback to ride to the base of Twin Peaks by moonlight. Carrington was one of the pioneer group, while the other four men were members of the Mormon Battalion and were part of the group that arrived in the valley several days after the first pioneers, on 29 July. The five reached the base of the mountain about midnight and encamped for the night. John Brown left an excellent account of their climb in his autobiography:

> Four of us commenced to ascend the mountain at eight AM, leaving a guard [Calkins] with our horses. This was the 21st day of August. After toiling about eight hours and being very much fatigued, three of us reached the summit of the west peak. One man gave out and lay down by a snow bank. We had a barometer, ther-

mometer and compass. We took some observations by which we learned that the peak was 11,219 feet above the sea; temperature 55 above zero at five PM. The same day at noon it was 101 in the valley.

Their descent was interrupted by darkness, forcing them to spend the night on the mountain. At first light they continued down to their camp, then returned to the city, "satisfied with our first attempt at climbing mountains."[38]

Although this climb has been cited as an example of the first recreational use of the Wasatch Mountains, it is not likely that it was done simply for pleasure.[39] On the day the men left the settlement Howard Egan wrote in his journal that the group was going to ascend the peak "in search of coal, etc."[40] While there seemed to be an implicit faith that coal, the domestic fuel of choice, would be available near the new settlement, none was found on the route through the mountains into the valley. Thus, it is quite possible that the search for coal was one of the group's goals. After their return Egan wrote that "they had found no coal, but plenty of black slate."[41] This was certainly a disappointment, for the scarcity of wood made the discovery of coal all the more important; however, when coal finally was found in useful quantities, at Coalville and in Sanpete County, it was so far from the new city that, for all practical purposes, it long remained unavailable. Even though wood was difficult to access, it still was easier to gather and transport into town than was coal. This remained the case for many years until roads were built. Over twenty years later the coming of the railroad finally provided suitable transportation to bring coal to the city.

It is likely the climbing group had another purpose equally as important as a search for coal. Brown noted that they had a barometer, thermometer and compass, and noted that they learned the peak was 11,219 feet above sea level (the height is now accepted as 11,328 and 11,330 feet for the west and east peaks, respectively). George Smith mentioned that Carrington carried the barometer, and after their return he summarized various elevation estimates: "Barometric height of Temple Block by Prof. Orson Pratt 4300 feet, of West Twin Peak by A. Carrington 11,219 feet." Then he added, "Height of peak above Temple Block 6919 feet."[42] In these comments may lie the most important purpose of the climb. After the Temple Block was located, Orson Pratt and Henry Sherwood began to lay out the city around it. On 3 August they chained off three blocks south and three blocks west from the Temple Block. Three days later they went south of the city some six or seven miles, established a base line of 300 rods (4,950 feet) and measured the two highest peaks in the Central Wasatch Range. These points gave them good references for future surveys, but if their elevation estimates could be confirmed, or corrected, the reference points would be that much better. Carrington's barometric readings at the summit of the Twin Peaks did just that.

CHAPTER 1

Civilization's Pioneer Machine:
The Sawmill in Mill Creek and Neffs Canyons

IMMEDIATELY UPON THEIR ARRIVAL in the Salt Lake Valley the settlers looked to the mountains for timber, although there were serious questions about how much would be found. This was a matter of great concern, for a good source of timber was essential for fuel, fencing, and shelter, probably in that order of importance. The first impression of most people entering the valley was the lack of trees. "On our arrival we could behold nothing but one vast waste, scarcely a tree or shrub to be seen," wrote pioneer Levi Jackman.[1] John R. Young later wrote, "From ... City Creek canyon, in 1847, one could see a lone cedar tree on the plain southeast of us, and on the south fork of the creek ... stood seven windswept, scraggy cottonwood trees. On the north side of City Creek stood a large oak tree. No other trees were visible in the valley."[2]

Actually there were many trees along the river bottoms, most of them out of sight at a distance. The first view of the valley, a vast expanse of grassland, much of it turning yellow in the July heat, must have been overwhelming, so impressing the senses with a feeling of emptiness that it would have been easy to overlook the relatively small number of trees. William Clayton, in describing his first view of the valley from the mouth of Emigration Canyon, wrote, "There is but little timber in sight anywhere, and that is mostly in the banks of creeks and streams of water. . . ." Then he added, "Timber is evidently lacking but we have not expected to find a timbered country."[3] The concern about a shortage of wood was manifested immediately upon arrival at the chosen spot near City Creek on 23 July 1847, when George A. Smith recommended that the men gather dead timber, leave the live timber standing, and use as little wood as possible for cooking.[4]

Faced with this timber situation, parties were dispatched to explore the nearby countryside. On Sunday, 25 July, William Clayton reported in his journal, "It is now certain that there is considerable timber in the ravines and valleys between the mountains. . . . There is a mountain lying northeast from here on which is considerably larger timber."[5] The next day he reported that a company of men had gone to build a road into the ravine on the north, this being City Creek Canyon. They must have worked fast, because only two days later he wrote, "A company of brethren have been to the mountains to get more lumber . . . they returned this evening bringing a very handsome pine log about twenty inches through and which, probably, when whole, would measure sixty feet long."[6]

The next day, Wednesday, 28 July, Joseph Hancock and Lewis Barney returned from a two day's tour in the mountains to the east. They reported an abundance of good timber, principally pine, balsam fir, and a little cottonwood; but access to it was very difficult.[7] Unfortunately, neither Hancock nor Barney left an account of their explorations, so we don't know what part of the Wasatch they explored, but it most likely was not Emigration Canyon. That was the route followed to enter the valley and Clayton had observed on arrival that "the mountains through which we have passed have very little [timber] on them."[8] Indeed, that canyon never did become a major source of timber in the years that followed.

While the timber situation now looked more favorable, there was still much concern about availability. At an assembly on Sunday afternoon, 1 August 1847, it was decided to build houses for the winter and to build them in the form of a stockade. There was considerable discussion about the construction materials; it was finally decided to use adobe for the houses. The arguments stressed the scarcity of timber and the difficulty of its access. Willard Richards said, "If wood is put into houses it will be a waste of it. We want all the timber to make floors and roofs."[9] When the stockade, or Old Fort, as it became known, was built, it did have a row of log cabins along the east side, but the rest was built of adobe bricks. Adobe remained the favored building material in the city for many years, and to this day there are adobe houses to be found in the Salt Lake Valley.

In spite of Willard Richards's desire to have wooden floors and roofs, very few of the first houses had either. Most floors were dirt, and the roofs were covered with branches and soil or sod. When it rained, the water soaked through and dripped inside so that it soon was about as wet inside as out. The few who had umbrellas used them to keep dry *inside* their houses. Although logs brought out of the mountains could be used to build cabins, they had to be cut into planks to make good roofs. Pit saws were set up for this purpose. The pit saw could be as simple as two logs that spanned a pit in the ground and supported the log being cut. While this configuration is the source of the name, there were many variations, such as two logs on posts forming trestles to support the log being cut, or the log being simply propped up at an angle to the ground. In any case, two men, one working above and one below, used a two-man whipsaw to cut the log into planks. It was a dirty job for the man below, as all the sawdust fell on him. There may have been more than one pit saw in the new community, but the output would have been very small since it was a slow and laborious process to cut a log into planks. It has been said that two men could cut only 200 to 300 linear feet in a day.[10] As a result, there were few wooden roofs in Great Salt Lake City that first season.

A pit saw used in the new settlement figured in a tragic and poignant event a few months later. Soon after the pioneers arrived Indians began to gather to watch their new neighbors, and for years thereafter Indians could be found camped in the vicinity, often at the very edge of town. On 28 February 1848, a six-year-old boy, the oldest of three sons of Edward Oakey, was killed when a log rolled off a pit saw. Several Indians were seen running from the scene immediately after the accident. One of them had caused the log to roll onto the boy. When he told his chief what had happened, the chief told him to take one of his children, a little boy, and deliver him to the man who lost the child, to be killed or adopted into his family as he saw fit. To the astonishment of Edward Oakey, the Indian came with his son and offered him in sacrifice to atone for his crime. Oakey told him that the settlers had no such custom and that the Indian could take his boy back home.[11]

In 1850 the United States commissioner of patents said, "The ax produces the log hut, but not till the sawmill is introduced, do framed dwellings and villages arise; it is civilization's pioneer machine: the precursor of the carpenter, wheelwright and turner, the painter, joiner and legions of other professions."[12] It did not take long before "civilization's pioneer machine" arrived in the Salt Lake Valley, the first one being the product of brothers Archibald and Robert Gardner.

Eighteen members of the Gardner family, ranging from one to sixty-seven years of age, arrived with the second group of pioneers at the end of September 1847. In 1822–23 the family of Robert Gardner, Sr., had migrated from Scotland to Canada, where members engaged in farming and built several mills. Following his arrival in Utah Archibald Gardner scouted the area north and south of Salt Lake City for a suitable sawmill site. On 13 October, only two weeks after his arrival, Gardner reported his findings to the High Council and asked permission to build a sawmill near the warm springs north of the new settlement. The twelve-member High Council was presided over by President John Smith, uncle of the Mormon Church founder Joseph Smith, and two councilors. It was the governing body for secular and spiritual matters and had control over all development and use of resources. The council granted the request the same day and Gardner went to work constructing his mill.[13]

It might seem strange that Gardner should have chosen the warm springs as his mill site, since it was some distance from any good supply of timber. But at this time there was still concern about how the Indians would react to the settlers' presence. On 4 October, before Gardner went searching for mill sites, Ira Eldredge wanted to build a gristmill on Mill Creek, but the council thought it would be going too far south before they knew the disposition of the area's Indians.[14] Whatever the reason for choosing the warm springs location, the mill was a failure. Due either to the water being too warm or to an insufficient flow, there was not enough power to drive the mill. A biography of Gardner stated that men had to help turn the waterwheel and only three boards were turned out.[15] Robert Gardner later wrote, "We had been used to running mills in Canada with heavy streams and...a fall of from two to eight feet, and we thought a very little water would do, but we had too little there and we could not make lumber."[16]

The Gardner brothers went back to the High Council and on 5 February 1848, by a clear vote, received permission to move their sawmill to Mill Creek.[17] By this time there was less concern about the Indians. Indeed, in the previous month land had been set aside along the Mill Creek stream for farming, and for this reason the Gardner brothers were instructed not to interfere with irrigation. The location of the mill was near the place where Mill Creek crosses present-day Highland Drive, about 3500 South. They built the mill without nails, using mortices and wooden pins. The mill had a muley saw—a long vertical saw guided at both ends—powered by an overshot waterwheel. The shafts, bearings, and cog wheels were made of locally acquired mountain maple.[18] By 6 March 1848 the Gardners' new mill was reported nearly ready for operation, although theirs was not the only sawmill in the area. On this same date Isaac Chase had a sawmill in operation on a spring a short distance from the Pioneer Garden, where Liberty Park is located today. Also, Charles Crismon had a sawmill under construction at or near his small gristmill on City Creek, and another sawmill was under construction about ten miles north of the city.[19]

When the Gardner mill began operation timber was drawn from Mill Creek Canyon. On 1 July 1848 the High Council authorized Archibald Gardner to call on men to work on a road up the canyon, to see that no one wasted or monopolized the timber, and to ensure that those who worked received a reasonable number of logs for their labor.[20] This action gave Archibald Gardner an important presence in the canyon, one that continued for nearly twenty years.

When the provisional State of Deseret was established by the Mormons in March 1849 (to be replaced by the Territory of Utah the following year), its legislative body, the General Assembly, assumed most of the resource management. Unfortunately, there is not a good record of the General Assembly proceedings. It is only through indirect references that its actions relating to control and allocation of resources in the mountains and the canyons during the next few years are known. In February 1852 the Utah Territorial Legislature established county courts and gave them control of all timber and water resources in their respective counties.[21] After that time the minutes of the various county courts provide good documentation for activities in the mountains, but only when people followed the established rules of law. There were surely many who built roads, canals, and mills without bothering a county court for permission, and therefore their activities and achievements remain in obscurity.

*

While the Gardner mill operated with apparent success, it was a long distance from the source of timber that supplied it. After one year of operation the brothers decided it would be better to have the mill closer to the timber, so they returned to the High Council and requested and were granted the privilege of building a sawmill near the forks of Mill Creek, at or near the place where today's Porter Fork joins the main canyon. The council reserved the right-of-way for all persons and teams into and out of the canyon.[22] The new mill was known temporarily as Gardner's upper mill, to differentiate it from the original mill much farther downstream. It was running by the summer of 1850 but was not long in operation before it was destroyed by fire, probably in August 1851 when a fire set by some careless persons raged through the timber in the canyon.[23] The mill was soon rebuilt and operating again. As part of its operations, roads were built to access the timber farther up the canyon. On 7 June 1852 the newly established Salt Lake County Court gave Archibald Gardner permission to build another mill in Mill Creek Canyon, to be located several miles above the first one.[24] The mills continued to operate, with many members of the family joining hired men to do the work. Even some of Gardner's plural wives joined the force. Two of them reportedly hauled logs on ox carts from near the head of the canyon to the mill, where men unloaded the carts, and two others cooked for the mill hands.[25] However, by 1876 the timber reserves were growing slim and the United States Land Office had assumed control of timber on public lands, making lumbering operations increasingly difficult, so Gardner abandoned Mill Creek Canyon in favor of operations he had established farther south in Little Cottonwood Canyon.

As early as 1848 Mill Creek Canyon had become an important source of timber, wood, and poles for the public at large. More and more immigrants were settling on the farmlands south of the city, closer to the canyon. When the Gardner brothers moved their first mill to Mill Creek their road into the canyon made access much easier. Soon the wood along the road disappeared and the scavengers had to go farther up the slopes above the road. Logs dragged down the hillside created slides, which were used repeatedly and which grew longer and deeper as operations moved farther up the slopes. Many of the logging slides still can be seen in the canyon, the most obvious one being at the lower end of the Alexander Basin trail. The trail goes straight up the old slide where logs were brought down from the basin. Another is part of the trail going up Thayne Canyon.

In the winter, when the slopes were covered with snow, the slides were very effective, for a log, once started, would slide swiftly to the bottom of the canyon. Of course, this created a dangerous situation if a person were in the slide or too close to the bottom when a log came rushing down. Robert Gardner found himself in just such a situation, and later wrote about it:

> I went to the mountain on foot to slide some dry timber for firewood. The snow was very deep and the weather very cold. The place of sliding was about five miles from home. The place for sliding was very narrow, and on a steep mountain side, and the snow was about five feet deep. When I reached the sliding place I was not aware that men had gone up in head of me. I had climbed about a quarter of the way up the slide and I was met by a log which was running like an arrow and it struck my right leg below the knee peeling off all the flesh clear to the bone, about four by six inches. On account of the mountain being so very steep my foot gave way in the snow and

did not break the bone. When I looked down and saw the blood and the wound... I took hold of my leg with both hands and raised it and found it was not broken. ... My next thought was: "Get out of here or another log will hit you and take the rest of you." I crawled out of the track... to a high place on my hands and knees where I could see the road below. Two men were coming up the canyon. I hollored and they heard me and came to my relief. I placed one above me to watch for logs which might be coming and give the alarm and the other dragged me down the slide. ...

After a while a team came up with a sleigh without a box or bottom, they rolled a few logs on it and laid me on and started home. ...

When we reached Father Neff's mill, one of Bishop Brinton's sons went to Neff's home to get some liquor for me.[26] Porter Rockwell came back with him and brought me a tumbler of whiskey and molasses.[27] I began to pour it on my wound, but Porter said to pour it inside so I did both. He wanted me to go to his house so he could sew it up for me, but I didn't want to go to any one's house covered with blood as I was, so I asked him to go with me to my home and do the work there.

He placed me down before the fire and washed my leg, and got a handful of fine salt and laid it on the bone and lapped the flesh on it in it's place and commenced to sew it with silk thread. He put in a few stiches [*sic*] and then his heart failed him. He could not do any more, and no one could help me. So they held me up and I sewed it myself. That is, I took the needle through and he tied the threads and we made a good job of it.[28]

With a road into the canyon it was inevitable that others would build sawmills there. There have been several publications listing and even locating mills in the canyon,[29] but all are based on the oral history of Selina O. Stillman published by the Daughters of Utah Pioneers.[30] Unfortunately, neither the order in which the mills were built nor all the people involved, as listed in these documents, can be substantiated by alternate sources, so what will be presented here is only that which has been well documented.

The first mill in the canyon, predating Gardner's upper mill, was built by Samuel Thompson, Joseph Mount, and William W. Willis near the mouth of present-day Thaynes Canyon. The three men had submitted a petition to build a mill and received High Council approval on 21 October 1848.[31] Their mill was in place by March 1849 when the council granted a request by the three men for a strip of land on the south side of the canyon above the mill, provided they keep a road open for the convenience and use of the public.[32] Both Thompson and Willis were members of the Mormon Battalion. Willis, a millwright and a Third Lieutenant in Company A, arrived in Salt Lake City on 29 July 1847 as part of Captain James Brown's detachment. His wife and six children arrived on 25 September 1847. He helped build the mill, then sold his share to Thompson sometime in 1850 and moved his family out of the valley in 1852.

It was Thompson's name that stayed with the mill over the years that followed. He had been a Second Lieutenant in Company C of the Mormon Battalion and went the full distance to southern California, where he was discharged. He then made his way north and was one of the Mormons at Sutter's Mill when gold was first discovered there.[33] The following season he crossed the Sierra Nevada and headed east to arrive in Salt Lake City sometime in 1848. It was a long trip, not without hardship and loss. Two years earlier, his wife, who had suffered the Mormon hardships in Missouri and Illinois, was faced with the prospect of traveling west with her four children and the rest of the pioneers, but without her husband, who was leaving on the trek with the Battalion. This she refused to do. She kept her two youngest children and stayed behind in Iowa. Thompson never saw them again, but his two oldest children did travel west with their aunt to join him in 1848.[34]

Joseph Mount arrived in the Salt Lake Valley with his wife and daughter on 25 September 1847. He worked briefly with Charles Crismon at the latter's gristmill on City Creek, but he fell through some scaffolding and broke several ribs, incapacitating him for some time. He then joined Thompson and Willis to explore the mountains and select a site for a sawmill. Once this was done, they built a cabin and Mount moved into it with his wife and two daughters, one eleven years old and the other a three-week-old newborn. There he and the other workmen spent the winter of 1848–49 building the sawmill. In the spring, when the mill was ready for work, Mount decided to go to California to seek his fortune in the gold fields. He may well have built his enthusiasm for this venture after hearing Thompson's account of the discovery of gold at Sutter's Mill the year before. At any rate, he departed in April 1849, having agreed that Thompson would provide his family with half the proceeds from the mill, while he would send Thompson half of his profits in California.

At the end of the summer Thompson moved Mount's family into the city and about the same time Mount sent money in gold coin, which was split with Thompson. Late in 1850 Mount wrote to his wife instructing her to come to California after having Thompson sell his interest in the mill. Thompson had already bought out Willis's share, so now there were only the two partners. In 1851 Mount's brother, who had been with him in California, came to Utah to fetch Mrs. Mount and the children. The partnership with Thompson was dissolved, but at the last minute Mrs. Mount had second thoughts and refused to leave the city. Mount's brother was furious at this turn of events and left for California without her but with most of her assets. Later that year, 1852, she filed for divorce and custody of the children. Mount returned to Salt Lake City briefly in 1852 but then returned to California, where he died in December 1876.[35]

In July 1850, after the mill was in operation, Thompson alone was granted the privilege of controlling the timber in the canyon south of his mill, provided he build a road and bring the lumber into market as quickly as possible, an indication of the critical need for lumber at that time.[36] He ran the mill for only a year or two after that, for in 1852 he moved his family to Spanish Fork. Although he later moved back to Salt Lake City for a short while, he was not active in the Wasatch Mountains again. His mill, however, was known as Thompson's mill as late as 1859, when it was referred to as "the mill formerly known as Thompsons mill."[37]

*

Next on the scene was Chauncey W. Porter, who operated a mill near the mouth of the South Fork of Mill Creek. Porter arrived in the valley in September 1849 and settled in Mill Creek with his parents, who had arrived in 1847. His daughter, in a personal history written many years later, wrote that he had leased the mill from the Gardner brothers in the spring of 1850.[38] That would seem reasonable, for there would be no reason to build a new mill so close to an existing one when a whole canyon full of trees awaited the woodsman's axe. It also would seem reasonable because in December 1849 the High Council granted the Gardner brothers the privilege of building a flouring mill on Big Cottonwood Creek.[39] As was often the custom, this mill, actually located along the Jordan River near its confluence with Big Cottonwood Creek, was built as a combination saw- and gristmill, but it was best known in the years that followed as the Jordan Grist Mill. With the original Mill Creek mill still operating, the brothers had little spare time to personally run the sawmill in the canyon. What is known is that Porter was operating a mill in January 1851 when Wilford Woodruff wrote that he had ridden horseback to Porter's mill in Mill Creek Canyon.[40]

In March 1852 Peter White was denied the privilege of erecting a lath and shingle mill just below Porter's mill, probably because Porter petitioned for the same right and drew preference due to his mill already operating.[41] With the lath and shingle mills added to his sawmill, Porter took steps to stake his own claim to some of the timber resources. In June 1853 he petitioned the court for permission to build a road up the South Fork and control the sawing timber found there, but he allowed that poles and small timber, as well as use of the road, would be free to the public. His request was granted and the road was built.[42] It was the convention that a canyon or a tributary of a canyon was often named after the person who opened the first road into it, and in this case the South Fork of Mill Creek Canyon became known as Porter's Fork. Curiously, the men who opened the first road into Mill Creek Canyon and eventually to the head of it, Archibald and Robert Gardner, never had any part of the canyon named after them.

Porter's new road was hardly opened before Isaac Bowman petitioned the county court to open a road into the first left-hand tributary above Porter's mill in the South Fork, and to charge twenty-five cents for each load of wood or poles hauled out. His request was granted on 21 September 1853 and he built the road into the area known today as Bowman Fork.[43] Isaac Bowman had been headed for the California gold fields when he arrived in Salt Lake City in July 1850. His party decided to remain in Utah for the winter, but by the next spring Bowman had joined the Mormon Church and become a permanent resident. Like many early residents, he worked at many different jobs. He clerked for several early merchants, taught school, and opened a road in Bowman Fork. He died in Salt Lake City in 1895 at the age of sixty-five.[44]

Isaac Bowman did not build a mill or supply timber to any other mill; his was a business proposition in which he provided a road and charged the public for its use to haul out poles and timber. The seniority enjoyed by Chauncey Porter was emphasized when the court specified that nothing in Bowman's charter could interfere with the rights already granted by the court to Porter. However, Porter did not stay much longer; by 1855 he had moved his families to Centerville, having sold his rights to the South Fork to Archibald Gardner, who in turn held them for more than twelve years before selling them to John A. Hill.[45] Porter thereby disappeared from the Wasatch scene, but his name is still used there to this day.

Also on the scene was Peter White, who had petitioned the court for permission to build a lath and shingle mill near Porter's mill in March 1852. While that request was denied, White did run a mill on Thayne Flat, farther down the canyon. How he came to possess the mill is not known, but there is little doubt that it was the one built by Samuel Thompson. His mill was a recognized fixture in December 1854, when another petitioner used it as a point of reference.[46] And it was still there in October 1855 when the county court proposed moving the toll gate up the canyon near White's mill.[47] Very little is known about Peter White, except that he quietly disappeared from the scene. The mill was taken over, at least briefly, by Enoch Reese, who in June 1859 petitioned the court and was granted "exclusive control of all saw and shingle timber in . . . Little Mill Creek Canyon, running south of the mill formerly known as Thompsons mill."[48] Enoch Reese was an early merchant in Salt Lake City, he and his brother John having opened the third store in the city, which was doing business in 1850. He represented Carson County, in present-day Nevada, in the Utah Territorial Legislature of 1855. He was sent on a church mission to Europe in 1857, leaving with the first missionary handcart company, which, incidentally, also included Robert Gardner among its members. Upon his return he got into the sawmill business in Mill Creek Canyon; but he didn't stay long, for a year later, in June 1860, he transferred the grant to Edmund Ellsworth, a son-in-law of Brigham Young.[49]

Ellsworth had started west with the first group of Mormon pioneers in 1847, but when they reached the upper crossing of the Platte River the water was so high that they were obliged to build a ferry boat. By the time they had ferried all their people across, emigrants for Oregon had begun to arrive. Ellsworth was one of ten men Brigham Young ordered to remain behind and ferry the Oregon-bound emigrants. As a result, he arrived in the Salt Lake Valley with his wife and two small children, who had started the journey with the second group of pioneers, on 12 October 1847.[50] He later went on a mission to England, and on his return was ordered to take charge of the first handcart company to cross the plains, arriving in Salt Lake City on 6 September 1856. In 1859 he was working, or at least keeping books, for Enoch Reese at his sawmill.[51] The following year he received the grant from Reese and was in the sawmill business for himself. He operated the mill for a half-dozen years before moving to Davis County.[52] But he left his name behind, for the tributary formerly known as Little Mill Creek Canyon then became known as Ellsworth Canyon.

The mill was next taken over by J. J. Thayn, who was there in 1866 or early 1867. Selina Stillman claimed that Jack Hill and Daniel Brian ran the mill for Thayn.[53] The first named was John A. Hill, a resident of Mill Creek, who bought the rights to Porter Fork from Archibald Gardner and received county court recognition of that fact in May 1867.[54] Daniel Brian was with Thayn late in 1867 when a newspaper article placed both of them at the sawmill in Mill Creek Canyon. In October of that year the

Deseret News reported that one "brother Hanson," who was hunting for his cattle in Church Canyon (Church Fork), "suddenly came upon an old she bear and cubs. She showed fight, and he being unarmed beat a hasty retreat to the mills in South Mill Creek Canon, from whence brs. Thayne [*sic*] and Daniel G. Brian accompanied him with their guns. After a short chase they ran the old bear and one of the cubs up a tree, and shot them both. They were of the brown or cinnamon bear variety, and the old one weighed 300 lbs., being very fat."[55] Such an encounter with bears was a fairly common occurrence in those days, usually ending disastrously for the bear, which helps explain why bears are not seen in the Wasatch today. The article also tells us that Thayn was well established in the canyon at this time and that Brian was working with him.

In February 1869 Thayn advertised: "I am running my saw mills in South Mill Creek, 12 miles from the Temple Block, and turning out a superior article of Yellow and Red Pine Lumber, From which I am able to fill bills with promptness and at reasonable prices. Slabs in any quantity, for sale at the mill, suitable for fencing and other purposes, and for stove wood as cheap as any in the market."[56] A slab is the first cut taken from the side of a log; it has one flat surface exposing the freshly cut wood, but the other side is the rounded outer surface of the log with the bark intact. Slabs were used primarily for fencing or firewood but sometimes were used as siding for buildings. At this time the purchaser had to travel up Mill Creek Canyon to buy Thayn's lumber, but in later years Thayn established the Utah Lumber Yard in the city, on Third South between East Temple (Main) and First East (State) Streets.[57]

John Johnson Thayn was born in 1825 in Scotland and emigrated to Canada with his parents. He married there and had four children before he moved his family to the United States, where he bought and operated a sawmill in Iowa. Four more children were born there. In 1861 he and his family made the journey to Utah. There were two young ladies who were anxious to make the trip, so Thayn agreed to take one of them. Upon their arrival in Utah the Thayn family settled in Salt Lake City. In December 1865 the county court granted his application for control of "Little Willow Creek Kanyon between the two Cottonwoods" with the privilege of collecting toll at the rate of 75 cents per load, on the condition that he make and maintain a good wagon road into the canyon.[58] However, the following spring Thayn came back to the court and reported that the condition of the canyon would not allow him to make a good wagon road.[59] Shortly after that he became involved in the Mill Creek Canyon sawmill. During this time he took the young woman who had traveled west with his family as his second wife, and subsequently had four children with her. She generally lived in the city while Thayn's first wife and family lived at the mill. After his second wife died in 1875 his first wife reared the woman's children as her own.[60]

During one period Thayn employed a young man by the name of Elisha Jones at the mill. Jones had been traveling to the gold fields in California, but he never got beyond Utah. He went to work at the sawmill, where he met Thayn's daughter Eliza. The two young people soon became close friends, a relationship Thayn discouraged because Elisha was not a member of the Mormon Church. Despite his disapproval, the young couple continued seeing one another, and finally Elisha sent a note to Eliza urging her to run away with him and elope. Now, John J. Thayn was a rather strict parent and was known as a man with quite a temper. Eliza, fearing her sister would see the letter and show it to her father, hid it inside her dress. That evening the letter was no longer there. She searched for it, but to no avail. Although she feared the worst the next day, nothing was said by anyone. But John J. Thayn had found the note, and his reaction showed that he had a heart as well as a temper. He called the two together and told them that if they were so determined to get married, then they could get married and work at the mill, which they did. Elisha worked in the mill while Eliza cooked for the men working there. She was seventeen years old at the time.[61]

In 1875 Thayn took steps to control the harvesting of timber in the canyon above his mill by petitioning the county court for the privilege of "collecting toll in Ellsworth Canon" to the amount of fifty cents for every load hauled out, which request was granted.[62] While Ellsworth's name had survived nearly fifteen years, it soon was replaced by Thayn's after he had control of the canyon. His name continues to be used to this day, albeit now spelled Thayne.

Thayn ran the sawmill for another two years until the night of the Fourth of July in 1877 when he and his wife were awakened by the snapping sounds of fire. Upon arising they discovered the sawmill in flames. The mill, with a small amount of lumber, was completely destroyed, involving a loss of a few hundred dollars.[63] Arson was suspected but never proven. The Thayns then moved to Summit County and later to Carbon County, where John had a continuous involvement with sawmills. John J. Thayn died in Wellington, Utah, on 21 May 1910 at age eighty-four.

In 1854 another logging operation, though not involving a sawmill, took place directly across the canyon from the Peter White mill. At that time Abraham O. Smoot had the task of constructing a building to house

sugar-making machinery, intended to produce sugar from beets. The building, located on Big Kanyon Creek, was known as the Sugar House. Smoot
needed timbers for the structure, so he petitioned the county court for
the right to make a road and control the timber in that branch of Mill
Creek Canyon immediately north of Peter White's mill. Since he was doing this work at the direction of Brigham Young, his petition was made in
behalf of the Trustee in Trust of the LDS Church.[64] As a result, when the
petition was granted, Smoot's name was not attached to the tributary
canyon; instead, it was called the Church Fork. Although the grant had a
duration of only two years, no one else approached the court for control
of the timber or road in that fork, and it continues to be known as Church
Fork to this day.

        Another presence in Mill Creek Canyon that should be mentioned was
that of Alva Alexander and his son. In September 1853 Lorenzo Brown
wrote that he had stopped at the Alexanders' shingle-machine operation
and bargained for 2,500 shingles in exchange for twelve and a half bushels
of onions.[65] This mill was located in the lower canyon, where the Boy
Scout camp is found today. In March 1855 Alexander and his son petitioned the county court for, and were granted, the privilege of making a
slide up the first small canyon on the south side, about a half mile below
White's mill.[66] By this time the Alexanders also had a mill in the upper
canyon, but without having negotiated with the court, probably because
the mill was located within the area of Gardner's grant and any negotiations that may have been carried out would have been with Gardner.
Alexander got his timber for the sawmill out of the basin above, which
today carries his name. In May 1856 Alexander and son dissolved their
partnership, with the father taking "the shingle machine above Gardner's
mills," and the son, Henry Samuel Alexander, taking the lower machine
and canyon.[67]

        During the year 1863 Alva Alexander had responsibility for the canyon
road as an agent of the county, but beyond that time neither he nor his son
was active in the canyon. Alva Alexander died in East Mill Creek on 27
March 1890 at age ninety-one. His son Henry relocated to Wasatch County,
where he continued in the sawmill and lumber business. Selina O. Stillman
claimed that the Alexander mill was later owned by John Osguthorpe,
her father.[68] In October 1873 John Osguthorpe was responsible for the
road from the mouth of the canyon "up to his mill," which suggests that the
Alexanders were gone by that date.[69] As late as 1899 there were two sawmills
with water rights in the canyon, those of Stillman and Osguthorpe.[70] The
Stillman mill was below the mouth of the canyon. The location of the

Osguthorpe mill is not known; John Osguthorpe died in 1884, but it could have been his old mill that was referred to, still in his family.

Selina Stillman also mentions mills belonging to Peter Ranck and Hyrum Rose, both of whom did indeed have mills in the canyon, as they were listed in the 1880 Utah territorial census.[71] Unfortunately, that document does not provide a location for the mills other than that they were in Mill Creek Canyon; but Stillman said that Hyrum Rose had the Gardner mill and Ranck had a shingle mill at the mouth of Alexander Canyon. A biographical sketch of Peter Ranck also claims that he owned a shingle mill.[72] However, before drawing the conclusion that this mill was at Alexander Basin, it should be recalled that Alexander had received permission to make a slide up what is today known as Green Canyon. In the years before that tributary received its present name, it may well have been known as Alexander Canyon, in which case Ranck may have had the shingle mill that belonged to the younger Alexander. Another writer, describing a trip up Mill Creek Canyon in the late fall of 1879, wrote, "The dilapidated lumber mills which along the lower canyon have marked the receding steps of the woodman, now give place to *four* noisy working mills, all within a distance of two miles from the Elbow."[73]

All this underscores the problems encountered in trying to identify these old mills. The source data are extremely sketchy. The mills were small and changed owners often. The winters in the canyons were extreme, with heavy snows, and the spring thaws often brought devastating floods, both elements wreaking havoc upon structures that may have been there. The mills were easily and quickly built, and even more easily damaged or destroyed. Also, mills often assumed the name of their operator, who was not necessarily the builder or owner. Selina Stillman listed twenty mills, ten of which were in the canyon, the others being on the stream below the mouth of the canyon. It could well be that there were only three or four mills in the canyon, however, mills that were operated by different people at different times, or were rebuilt after being damaged or destroyed by natural disaster, yet in each case listed by her as a separate mill. For instance, it has been suggested that Porter's mill was leased from the Gardner brothers, and the implication in this narrative is that the mill, or at least the mill site at Thaynes Canyon, was operated by several different people over the years. Unfortunately, extant source material does not clarify these suppositions, and we are left with a fuzzy picture of exactly what existed in the past.

One additional mill operation in Mill Creek Canyon is worthy of mention, if only for its uniqueness, for its intent was the manufacture of

wooden pails. In mid-July 1856 an immigrant train arrived in Salt Lake
City, having among its members several returning missionaries, including Jeter Clinton and Samuel A. Woolley. Clinton was a somewhat enig-
matic character whose role in life seemed to change repeatedly over the
years. At that time he had established a reputation for being an exceptionally effective missionary, so effective in fact that Brigham Young sin-
gled him out during remarks at the opening of the church's general conference in Salt Lake City in April 1857. Jeter Clinton, Young said, was sent
"to Philadelphia, and when he got there, those who professed Mormonism
were dead, dead, dead; they were withered and twice plucked up by the
roots. Brother Clinton had not been there six months before the Church
numbered a great many more than when he went there. The old members revived, and they began to baptize and have calls from the country,
and when he left he could probably have employed from ten to thirty El-
ders in his field of labour."[74] Clinton was forty-four years of age when
these heady accolades were given, no longer a young man, but held in
such high esteem by President Young that he enjoyed presidential favors
until Young's death twenty years later.

Clinton also enjoyed the title of "Doctor." How he acquired this title is
not known; there is no record of his ever having been a practicing physician, although he was the quarantine agent at the temporary hospital at
the mouth of Emigration Canyon during the 1852 emigration season, one
of several occupations he followed between his many church missions.[75]
But his title predated that service. On the occasion of his homecoming in
1856 he brought with him some 4,200 pounds of machinery for the manufacture of pails, or wooden buckets, and expected to go into business
immediately. He enlisted the aid of Woolley, who had been on a mission
to India but who on his return had spent a year in the Philadelphia mission with Clinton. The two chose a site at an existing sawmill in Mill
Creek Canyon, most certainly the one recently vacated by Peter White.
There they built a new building to hold the machinery; however, by the
time it was ready Clinton was off on another mission and Woolley was
left to manage the installation of the machinery by himself.[76] Brigham
Young took a personal interest in this endeavor. After Clinton's departure
Woolley wrote that he had gone to see Young about Clinton's business
and Young said he could draw on the tithing office for such things as
they had on hand and could do the iron work at the public works shop.[77]
Still, the machinery was not installed until the end of the 1857 season. By
the time the new season began the following spring, the Utah War had
taken place. The decision had been made to vacate Salt Lake City, as well

as other northern communities, and move the people south to Utah County and beyond before federal troops under Col. Albert Sidney Johnston entered the valley. At the direction of Daniel H. Wells, who was in charge of the public works at Temple Square, Woolley removed the newly installed machinery and took it into town. It is not likely the machinery was ever run in Mill Creek Canyon; on three occasions, once in October 1857 and twice in March 1858, Woolley noted that he had tried to start the mill but did not succeed. When the public works department was relocated to Parowan, the pail machinery was taken with it and Samuel Woolley was sent there to set up a new manufactory. He ran it successfully until late 1860, when the machinery was again dismantled and moved back to Salt Lake City. Several years later the machinery was again in use when Edwin D. Woolley, Samuel's older brother, installed it in the old nail factory, not far from the Sugar House on Big Kanyon Creek, where buckets and pails were manufactured for a number of years.[78]

*

In the early days it was the practice to give control of a canyon or tributary to one or more persons, allowing them to charge a toll for the removal of wood or timber in exchange for their construction and maintenance of the road. This was not, as has often been charged, a monopolization of the canyons, but rather a practical way to provide access to the timber that benefited everyone. Occasionally assistance was given to the grantee, as was the case in July 1848 when the High Council authorized Archibald Gardner to call upon men to work on a road up Mill Creek Canyon, but usually the grantee was expected to supply the necessary resources and was allowed to charge tolls to defray the expenses. When the county court authorized Gardner to build his second mill in the canyon it also included allowances for the canyon road, this time allowing him to charge twenty-five cents for each load of wood and poles hauled out of the canyon.[79] A month later, at the end of July 1852, the court granted the Gardner brothers control of the North Fork of Mill Creek Canyon—that is, the entire main canyon above Porter Fork—but took an additional step of requiring a bond to the amount of one thousand dollars for "the faithful making and keeping in repair the road."[80]

A year later, on 13 June 1853, the county court gave the road between the mouth of the canyon and Gardner's road in the North Fork to Daniel Russell, allowing him to erect a toll gate and charge twenty-five cents for each load hauled out of the canyon; however, it provided that after the expense of making and maintaining the road was paid the road was to be free to the public.[81] Thus the road belonged to the county court, but Russell was responsible for its maintenance at no expense to the county. However, canyon roads were difficult to maintain; they followed the streambed and often suffered considerable damage from floods due to snow melt or summer storms. In November 1853 the court received complaints about the road because it was continually breaking wagons that traveled over it. The court sent a representative to superintend the road and see that Russell complied with the conditions of the charter.[82]

To ensure that the grantee was honest in his dealings with the public and the county, the court required him to report receipts and expenditures semiannually. But reported expenditures usually exceeded receipts. In October 1855 Russell reported he had spent $561.22 for lumber, labor, and tools over the past six months, while he received only $445 at the toll gate.[83] Since frequent complaints had been received about the management of the road, the court requested that Russell relinquish his claim to the canyon. An interim custodian was appointed. The county court then made a major change in its handling of the Mill Creek Canyon road. Acting upon a petition for a county road from Mill Creek Canyon to the Jordan River, the court established a road running east from the state road (today's State Street) to the county road (today's Highland Drive), then by the best route to the mouth of Mill Creek Canyon and up the canyon to Gardner's upper mill. This step took considerable time, for the petition was presented in December 1855, but the resulting order was not made until September 1856.[84] The court also ordered that a person be employed to man the toll gate, he being paid a portion of the tolls collected, and that the remainder to be expended in keeping the canyon road in repair. This, in effect, placed the county court in the role of the grantee: it hired people to collect tolls and perform maintenance, paying for both from the toll proceeds.

It is not known who was maintaining the road, but it was not done to the satisfaction of the public. In March 1860, acting upon complaints about the state of the road during the winter months, the Salt Lake County Court ordered that the road be thrown open to the public during the months of January and February each year.[85] But the public would never enjoy the benefits of this order, for in October of that same year, following continual complaints about the condition of the road, the court ordered that the collection of tolls be abolished and the canyon be open to the public free of toll. It contracted with Archibald Gardner to make a good road and keep it in repair for two years for the sum of $400 per annum, to be paid out of the county treasury in quarterly installments.[86] This was the second big change in the management of the canyon road in only four years, but Mill Creek Canyon was very important to many people. Many of the county residents east of the Jordan River and north of Little Cottonwood Creek used the canyon for their source of wood. From one to two hundred cords of wood were said to be taken from the canyon daily.[87]

Gardner apparently did a good job, for the county selectmen expressed their satisfaction with the road.[88] But the expense to the county was more than anticipated. When a late summer storm caused extensive damage to the road and its bridges, estimated at $600, the county court made a special appropriation to Gardner for that amount.[89] It was perhaps the shock of this additional expense that caused the court to resort to collecting tolls again after Gardner's contract expired. Alva Alexander was given charge of the road, with authority to charge twenty cents for a single team and forty cents for a double team from March through November on the condition that he keep the road in good order. The following March Alexander turned his contract over to Benjamin Ashworth, who operated the road for several years under the same terms.[90] In August 1867 Ashworth was allowed to increase the toll to forty cents a load.[91] In October 1873 John Osguthorpe was given an appropriation of $200 to repair the road and make it good for travel by loaded teams from the canyon mouth to his mill, and to keep it in repair for three years. In exchange he was allowed to charge a twenty-five-cent toll on each load hauled out of the canyon, except that Archibald Gardner was to go free of toll with his own teams.[92] The variations in tolls over the years reflected the local economic situation.

It is likely that tolls were no longer collected after Osguthorpe's three years expired. By this time traffic in the canyon had decreased considerably; the railroad had arrived in Salt Lake City, most people were using coal for fuel, and lumber was being imported in ever-increasing quantities. The Mill Creek Canyon road received several small appropriations to repair bridges—$200 in October 1877 and $80 in June 1879—before Road District No. 21, East Mill Creek, was established, with John B. Fagg as district road supervisor.[93] With that step, the road became managed by the county and never again was placed under the responsibility of individual citizens.

*

About a mile south of Mill Creek Canyon a short drainage, known as Neffs Canyon, opens into the valley. It is flanked on the north by a ridge dividing its drainage from that of Mill Creek Canyon. On the south side it is shadowed by the imposing north face of Mount Olympus and the ridge running east from that peak, all reaching elevations well over 9,000 feet. Having a sheltering northern exposure, the slopes of the latter ridge provide a suitable environment for generous stands of timber, all within view of the people living in the newly settled Salt Lake City. Since most of the other Wasatch slopes that could be seen from the valley offered little more than sun-dried grasses, it was inevitable that the timber in Neffs Canyon would receive early attention. And so it did. At LDS General Conference in October 1848 Archibald Gardner, Brigham Young, and Amasa M. Lyman were appointed a committee to supervise the getting of timber from the canyon south of Mill Creek.[94] What this committee accomplished is not known, although it may well be that some timber did find its way from that canyon to Gardner's first mill on Mill Creek stream. By this time Gardner had built a road into Mill Creek Canyon and Samuel Thompson and his associates were building the first mill in that canyon. Perhaps it was the result of this committee's activities that encouraged Gardner to build his upper mill the following summer. In any case, no direct result of the committee's actions has been found, and Neffs Canyon remained unused, as far as is known, until December 1850, when the General Assembly of the State of Deseret approved an ordinance granting to James Rawlins the exclusive privilege of making a road and controlling the timber in the first canyon south of Mill Creek.[95]

Rawlins had arrived in the Salt Lake Valley in October 1848 with his family of fifteen, including a grandchild less than a year old. They settled at Mill Creek but by December 1851 had relocated to Big Cottonwood. When Brigham Young decided he wanted settlers for the south part of the valley, the family relocated to Willow Creek, known today as Draper. While they were at Mill Creek a grandchild was born who was to gain much more notoriety than did James Rawlins. The newborn, Joseph L. Rawlins, would go on to become a delegate to the United States Congress from the Territory of Utah in 1892 and one of the first senators elected from the new state of Utah in 1896.[96]

The grant James Rawlins received from the General Assembly allowed him to charge as much as twenty-five cents for each load of wood hauled out of the canyon, on the condition that he keep a road into the canyon in good order. The early concern about timber sources and reserves was still present, for the Assembly required Rawlins to see that "the timber and wood in said Kanyon are not wasted." How much Rawlins accomplished in the canyon is not known, but in November 1851 he transferred his claim and privileges to Orrin P. Rockwell and John Neff.[97]

John Neff had arrived in the Salt Lake Valley in October 1847 with his wife, seven of his children, and a son-in-law. At Winter Quarters he had built and operated a flour mill, and he left his son Franklin behind to operate it until the 1848 immigration season began. The following spring he constructed a flour mill on Mill Creek stream, near where 2700 East Street intersects the stream today. When completed, his mill was the best one operating; people brought their grain from great distances to have it ground there. As a result, John Neff was well known among the early settlers. Orrin Porter Rockwell was one of the first pioneers to enter the valley in July 1847. He became a close friend of John Neff and was acquainted with the entire Neff family. Some years later, on 3 May 1854, he married Neff's daughter Mary Ann in the parlor of Brigham Young's home.

That John Neff and Orrin P. Rockwell joined forces to receive Rawlins's grant might be questioned, for the following March, 1852, Rockwell and Franklin Neff, John Neff's son, petitioned the county court for official recognition of their rights to the first canyon south of Mill Creek, and also the right to build a mill on Mill Creek above Neff's mill.[98] In later entries in the county court's minutes the name Franklin Neff appears when the petition is presented, but John Neff is cited when the petition is acted upon, or the name John Neff is cited in a reference to a petition previously granted to Franklin Neff.[99] It is probable that John Neff's name was so well known that the scribe inadvertently used that name when the name of the son was intended. It is definitely known that it was Rockwell and Franklin Neff who petitioned the court and received the grant on 17 March 1852. The court, however, did reserve certain privileges in the canyon to Rawlins.

Apparently not much work got done on a road that year, for the next March the two men were back before the court asking, and receiving, control of the canyon for another year. The court allowed them "the privilege of erecting a toll gate as soon as they commence making the road and charging twenty five cents per load for each load of firewood and fence poles hauled out of said Kanyon and fifty cents per load for each load of sawing timber hauled of said Kanyon." This time they were required to build a good road to the acceptance of the court by 20 September. There were reasons for the time constraint other than the fact that Rockwell and Neff had not accomplished their goal the previous year—other people also wanted control of the canyon. The previous October one Horace Gibbs had petitioned the court for exclusive control, but his petition was denied because Rockwell and Neff already held such control. And in March 1853, at the same time Rockwell and Neff reapplied for control, W. W. Phelps requested the same thing. The court again granted the canyon to Rockwell and Neff, but this time with the time limitation.

That restriction worked, for on 21 September 1853 Franklin Neff reported that the road had been completed and was ready for inspection by the court. The court appointed an inspector, who came back in mid-November to report the road tolerably good and that the two men were still repairing it and were doing about as well as they could. Therefore he would report favorably and accept the road.[100]

Neff and Rockwell soon took steps to expand their control of timber in the canyon, for on 30 January 1854 they requested exclusive control of the right-hand fork of the first canyon south of Mill Creek. That the canyon was becoming popular as a source of wood for local residents was shown by stated opposition to the request by thirty-three citizens of the county. In spite of that, the court granted the petitioners their request, provided they make a slide to get down the timber by 1 December 1855, a deadline that was later extended to 1 March 1857.[101] In March 1856 E. H. Hiatt petitioned the court for control of timber in the "tributary of Neffs Kanyon lying on the South side and being the first Kanyon east of what is known as Neffs exclusive grant." This was the first use of the name Neffs for the canyon. Hiatt's request was granted, but apparently he never built a road into the tributary as he was required to do.[102]

In June 1860 Charles W. Thomas requested the right "to make a road in the second right hand kanyon, known as Hyatts Kanyon, being a tributary of Neffs Kanyon, for the purpose of obtaining the timber for lumber and shingles for the term of three years." In granting this request, the court withdrew Hiatt's grant since he had failed to comply with its conditions.[103] The Thomas grant has several interesting clauses. Even though thirteen years had passed since the first pioneers entered the valley, the Salt Lake County Court still had a concern for the conservation of timber. It required that Thomas bring the timber out of his tributary as fast as it was cut down and not let it remain to spoil with the bark off. The court further required that Thomas give Neff the privilege of sawing the timber, provided he did it as well and as reasonably as any others.

Although contemporary maps fail to indicate a name for this tributary of Neffs Canyon, it has been commonly known as Thomas Fork. Whether Thomas actually built a road into the fork is not known. It is a very steep drainage, and, while there is a faint trail leading up the fork today, there is little indication of a road. If Thomas took timber out of his fork, it is likely he used a slide to bring it down to the road in the main canyon just as Neff and Rockwell did at their exclusive grant, the tributary known today as Norths Fork.

While nothing appears in the county court records relating Norths Fork to anyone other than Neff and Rockwell, the name may have come from Levi North, who arrived in Salt Lake City in September 1852 and settled on a farm near today's Highland Drive and 3900 South, just west of Neffs Canyon. There he built a house and barn, for which he certainly needed timber. He also helped build roads and bridges in the area, for which he also needed timber, and he was known to be active in the canyons.[104] In June 1859, for instance, Edmund Ellsworth noted in his account book for Enoch Reese's sawmill that Levi North was paid twenty-seven dollars for nine days work.[105] On the other hand, the first use of the Norths Fork name that has been found was in 1898, by which time another North, Hyrum K., was actively filing and working claims in Parleys and Mill Creek Canyons and along the west face of the Wasatch.[106] Thus, although one may speculate on the origin of this name, the fact is that it is not known.

In June 1864 Samuel Rich requested and received the privilege of making a road into the upper end of Neffs Canyon, extending the road beyond that built by Thomas.[107]

While Neffs Canyon provided an important source for timber, it never had a sawmill of its own. Timber was taken down the road from Neffs Canyon to the Mill Creek stream a short distance below the mouth of Mill Creek Canyon, where several small mills were located. In December 1860 a mill located a half mile above Neff's mill was advertised for sale for $2,500, and the 1880 territorial census listed a sawmill on Mill Creek below the mouth of the canyon that was drawing timber from both Neffs and Mill Creek Canyons.[108]

CHAPTER 2

The Big Cottonwood Lumber Company

THE FIRST ATTEMPT TO USE the Big Cottonwood area for something other than a source of cottonwood logs came on 22 April 1848 when the High Council granted John Neff's petition for a mill site on Big Cottonwood Creek.[1] Neff had arrived in the valley the previous October and had chosen Big Cottonwood as the site for the flour mill he was ready to build. But that choice was short-lived, for three weeks later, on 13 May, he revised his request: "John Neff's petition to choose a mill site on Mill Creek was granted as the rattlesnakes were too thick at his site on Big Cottonwood."[2] Neff then moved to Mill Creek and built his mill above Gardner's sawmill, about where 2700 East Street crosses the stream today. It was not until October of that year that the Big Cottonwood area received special attention again. Included in the 1848 immigration was a group from Mississippi under the guidance of John Brown, who had been in the Mormon Battalion company that followed the pioneers into the valley in 1847, and who returned to Winter Quarters with Brigham Young to bring immigrants to the valley in the 1848 season. Brown wrote, "it was finally concluded for us to settle between the two Cottonwoods. I went to the canyon and got down some logs and built a cabin for the winter."[3] He did not further define the canyon where he got his logs, but this was likely the first time logs were taken from a canyon south of Mill Creek.

The period from mid-1848 until March 1852, when the county court took over the responsibilities of the county's resources, provides but little information about who was building mills or timber access roads. It is only through an oblique reference that it is known that James B. Porter had built a sawmill on Cottonwood Creek and on 24 February 1849 was granted the privilege of removing his sawmill two or three miles up the canyon.[4]

There is a bit of mystery surrounding James Buchannon Porter. He was born on 4 August 1805 in Buffalo, Pennsylvania, and in 1849 was forty-three years old. But his name does not appear on any of the extant emigrant rosters. Presumably he arrived sometime in the 1848 season with one of the several companies that did not maintain a roster or whose roster failed to survive. It is not certain that his mill was on Big Cottonwood Creek rather than on Little Cottonwood. However, at this early date people were just beginning to build up the Big Cottonwood settlement, and the name "Cottonwood" was generally used for Big Cottonwood. Porter probably discovered, as did Archibald and Robert Gardner on Mill Creek, that it was easy to get a mill up and running below the mouth of the canyon, but after hauling logs a great distance to the mill, it was found more convenient to have the mill where the timber stood and then haul the cut lumber out of the canyon.

It should be noted that today's perception of the location of the mouth of Big Cottonwood Canyon is not the same as that of the mid-nineteenth century. The Big Cottonwood Canal, which was constructed in the 1850 decade for the dual purpose of boating granite blocks to the city for the LDS Temple and to provide irrigation water, took its water from Big Cottonwood Creek at the mouth of the canyon. The headgate of this canal was just above Knudson Corner, or about where today's I-215 belt route crosses the creek, about a mile and a half downstream from the highway intersection that is today considered to be at the mouth of the canyon. The route into Big Cottonwood in the 1850s followed the streambed, which is well below the benches that carry today's highways. The traveler of that day was truly in the canyon long before he reached the western face of the mountains.

There is no other indication of activity in or near Big Cottonwood Canyon until after the county court was established in 1852. At the end of July in that year, Joseph Young, older brother of Brigham Young and senior president of the LDS Quorum of Seventies, petitioned the court for exclusive control of the wood, water, lumber, and stone of Big Cottonwood Canyon. On 12 August 1852 Young's request was granted.[5] A month later, on 21 September, in response to another request by Young, the court expanded and clarified the grant: "Be it ordered by this Court that the said Joseph Young, his heirs and assigns is hereby granted the exclusive control of the Kanyon known as Big Cottonwood Kanyon from its mouth to the summit of the Mountain on the East together with its tributaries... to have and to hold and to control the same." The court went on to declare that firewood and fence poles or any other timber that Young did not care to work up "in said mill or mills" should be free to the public, that Young had the right to determine what timber he wanted to use, and he could charge the public twenty-five cents per load for wood hauled out, the funds to be applied to the road running into the canyon.[6]

There are several interesting points in this grant. First, it covered a tremendous area, the entire Big Cottonwood watershed, although at this time it is probable that neither Young nor the court had any idea of its size, for Big Cottonwood Canyon had not yet been explored.[7] Second, the court reserved firewood and fence poles and such timber that Young did not care to use in his mill to be free to the public, although he could charge for use of his road into the canyon. This exception was removed the following year when, on 20 September 1853, the court granted Young's petition to amend the charter to give him control of fence poles and firewood as well as the sawing timber.[8] With this action, Joseph Young had full and complete control of Big Cottonwood Canyon.

The final point of interest is that the grant used the words "in said mill or mills," perhaps implying that a sawmill already existed and may have been mentioned in the request for the grant. This implication was confirmed a few weeks later during the general church conference when Joseph Young brought up the subject of the Seventies Hall of Science, a project he had been promoting but on which very little had been done for the past twelve months. In encouraging the brethren to support and help in the construction of the hall, he stated that he "has in his possession a Saw Mill which is ready to run, which will supply the lumber for the hall." He went on to say, "The mill that we have has an excellent natural dam, which was proved by the Freshet of last Spring—and the saw being circular, will run with great rapidity, for we can turn the whole of Cottonwood Creek upon the wheel if we wish."[9] While it is not known who built the mill of which he spoke, it may well have been the one moved into the canyon by James B. Porter two years earlier. Neither is the exact location of the mill known, but most likely it was in the lower two miles of the canyon. Appropriately, it became known as the Seventies Mill.

At that same general conference it was decided to start construction of the temple the following spring. While Brigham Young was in favor of building the Seventies Hall of Science, it became obvious that the hall would be competing with the temple for resources—both manpower and materials. In a discourse early in February 1853, Brigham Young put the Seventies on notice that "the Temple must be the first thing in our thoughts, and if I want all the funds that have been collected for the Seventies Hall, for the erection of the Temple, I calculate to use them."[10] The threat became fact in May when the church leader published a notice that he had "determined to suspend operations on the Seventies Hall for the present." Then he addressed the resources in Big Cottonwood Canyon: "As there is a large quantity of lumber in the Cottonwood Kanyon, where the mill is located, for the benefit of that building, I still wish to have that work go on, the road made, and perhaps another mill built, the lumber will be essential, and if it could be obtained now, would become advantageous to the Temple, if not to the Hall."[11]

In spite of Young's wishes, nothing much happened in Big Cottonwood Canyon until he took his own steps to make it happen. As soon as it was possible to travel after the next winter released its grip, the church president sent millwright Frederick Kesler on a mission to meet H. S. Eldridge in St. Louis and select and purchase mill and other machinery. Kesler returned with the equipment in August 1854. Meanwhile Brigham Young took steps to organize the Big Cottonwood Lumber Company, consisting of five prominent individuals besides himself: Daniel H. Wells, Abraham O. Smoot, Feramorz Little, Frederick Kesler, and Charles Decker. Later accounts by Kesler and Henry Culmer put John Sharp in the group in place of Decker, but extant records of the company counter that suggestion.[12] Culmer certainly knew Frederick Kesler and most likely got his information from the former millwright. Kesler was in his seventies when he told his story, and the events of which he spoke were almost forty years in the past. In 1865, some eleven years after the Big Cottonwood Lumber Company was formed, another group was organized to construct the Salt Lake and Jordan Canal, intended to bring Utah Lake water into Salt Lake City by way of a canal traversing the east side of the valley. The organizers were Brigham Young, Frederick Kesler, Abraham O. Smoot, and John Sharp. Most certainly Kesler got the personnel of the two organizations mixed up—a problem that underscores the dangers of accepting oral histories at face value.

The Big Cottonwood Lumber Company assumed Joseph Young's rights to the canyon for the consideration of $1,000.[13] It did not, however, go to the county court to effect a transfer of the rights. After the company was dissolved in the 1860s, the court was petitioned to transfer Joseph Young's rights to those who purchased the company's property. Throughout the existence of the Big Cottonwood Lumber Company, the canyon belonged to Joseph Young, at least in the eyes of the county court, although it was totally controlled by the company.

*

This is an appropriate point to step back and take a look at the men Brigham Young chose for this enterprise. The first was Daniel Hanmer Wells, a true leader of Utah Territory's settlement and development. Wells was born in Trenton, New York, on 27 October 1814. When he was twelve years old his father died, leaving him responsible for his mother and younger sister. After his father's estate was settled, he moved west, finally settling near Commerce, Illinois. There he cleared land for a farm and through his thrift and industry soon accumulated more land. His success as a farmer earned him the respect of his neighbors. As a young man he was elected constable, then justice of the peace, and became known as "Squire Wells," a title he carried throughout his life. When the Latter-day Saints fled from Missouri into Illinois, he welcomed them onto his land, platting part of his property into city lots for the refugees. He also provided the land for a temple. The new settlement became known as Nauvoo. A mutual respect and admiration grew between Wells and the settlers, and when the Saints came under attack again, he cast his lot with them, joined the Mormon Church, and became a leading spirit in the defense of the city. When Nauvoo was abandoned, he left his home and followed the exiles into Iowa. In 1848 he emigrated to Utah, serving as Brigham Young's aide-de-camp on this, the leader's second journey across the plains.

Once in Salt Lake City, Wells became active in nearly every major endeavor. He organized and for many years ran the Public Works Department on Temple Square. He helped organize the provisional State of Deseret, served on the first legislative council, was state attorney, and served as major-general and lieutenant-general of the state militia, the Nauvoo Legion. He intervened when troubles arose between settlers and Indians, serving as statesman or soldier, as circumstances demanded. He was in command during the Utah War, conducting a campaign that halted the U.S. Army without bloodshed on either side. In later years he served as mayor of Salt Lake City for ten years. He was involved to a greater or lesser degree in nearly all of Brigham Young's endeavors, and undertook many others on his own. While he was the president's advisor and associate, he did not serve in an official capacity until January 1857, when he was appointed second counselor to Brigham Young. That Daniel H. Wells should have been a part of the Big Cottonwood Lumber Company was almost to be expected, and throughout the early years of its existence he was the key man, the manager whose keen eye and strong hand kept the company under control and running smoothly. While he never took an active part in the daily manual labor in the canyon, he often visited the sites and was the ultimate authority in the operations. An example of this was given in December 1856, when Frederick Kesler promoted a retroactive increase in daily wages for many of the men working at the mills. The clerk entered the payments in the company books with the note "Amount allowed you making your wages at the rate of 3.00 per day," then added a parenthetical note, "(Proposed by F Kesler and agreed to by D H Wells.)"[14]

Feramorz Little was another versatile individual whom Brigham Young depended upon in many ventures. He was born on 14 June 1820 in Auriesville, New York. His mother, Susan Young Little, was an older sister of Brigham Young. When Little was four years old his father died, leaving his mother to support him and his two brothers. Several years later Feramorz went to live in another home and became separated from his family for some twelve years. As a result, he did not become a part of the budding Mormon Church when his mother and her brothers joined in 1832. He did not see his mother and uncle again until 1844 when he visited Nauvoo, Illinois. There he also met a young woman, Fannie M. Decker, who became his wife on 12 February 1846, with his uncle Brigham performing the ceremony. He took his bride to St. Louis, where they remained until 1850. Then he contracted with Livingston and Kinkead, Salt Lake merchants, to freight twenty-five tons of supplies from Fort Kearney to Utah. He had plans to continue on to California, but upon joining his and his wife's families in Salt Lake City he decided to stay. After settling on a farm west of the Jordan River, he took a contract to carry the mail between Salt Lake City and Laramie, Wyoming. With him in this endeavor were his two brothers-in-law, Charles F. Decker and Ephraim K. Hanks. The only settlement between Salt Lake and Laramie was Fort Bridger, and, except when in the safety of that shelter, the men were on their own, subject to the trials of Indians and Nature, heat and cold, rain and snow, risking injury at every turn. Between trips Little became involved in various ventures, gradually taking on more and more responsibilities. In 1853 he bought an interest in Brigham Young's flouring mill near the mouth of Big Kanyon (Parleys Canyon), and moved his family there. His brother-in-law Charles Decker built a residence nearby and joined him in that venture, as in many others that followed. In later years Little was a member of the Salt Lake City Council and in 1876 became mayor, holding that office for three terms.

Charles F. Decker was born on 12 June 1823 in Phelps, New York. After his parents separated in 1843 his mother, Harriet Page Wheeler Decker, married Lorenzo D. Young, one of Brigham Young's brothers, and accompanied her husband in the pioneering group that arrived in the Salt Lake Valley in July 1847. Charles Decker arrived in October of that same year, accompanied by his bride of eight months, Vilate, a daughter of Brigham Young. Two of his sisters, Clara Decker and Lucy Ann Decker, were wives of Brigham Young, the former also being among the July 1847 pioneers. Two other sisters, Fannie Maria and Harriet Amelia, were wives of Feramorz Little and Ephraim Hanks, respectively. Decker's part in carrying the mail across the mountains between Laramie and Salt Lake City, and his associations with Feramorz Little, have been mentioned. Charles Decker was an embodiment of the mountain men of earlier days, and he was a good choice to get involved in the nearby wilds of Big Cottonwood Canyon.

Frederick Kesler, born on 20 January 1816 at Meadville, Pennsylvania, was orphaned at the tender age of five years. At age fourteen he was in Ohio, apprenticed to a mill builder for five years before striking out on his own. After building mills on sites as widespread as Iowa and Mississippi, he found himself in Illinois, where he joined the growing Mormon Church. He was part of the pioneering company when it started west. When the first Mormon pioneers arrived at the Missouri River he built a large and substantial ferry boat, and then he was instructed to remain there in charge of transferring the Saints, their wagons, and their animals across the river. He built a flouring mill at Winter Quarters, and later built another one in Kansas for the Pottawattamie Indians.

When Frederick Kesler finally reached Salt Lake City in October 1851, Brigham Young immediately put him to work on mills in the territory, which included establishing the water power for the public works on the Temple Block, Brigham Young's new flouring mill on the Chase farm southeast of the city (the site of today's Liberty Park), Heber C. Kimball's mill north of Salt Lake City in Davis County, and the new sugar works on Big Kanyon Creek. On 29 March 1854 he left on a mission to the eastern states to purchase and bring back machinery needed in the various construction projects, including the proposed mills in Big Cottonwood Canyon. Frederick Kesler was heavily involved throughout the construction of the Big Cottonwood Lumber Company's mills, but once they were in operation, he left their management to others and went on to new projects.[15] He was ordained bishop of the Sixteenth Ward on 7 April 1856, about the middle of the Big Cottonwood construction period, and was forever after known as Bishop Kesler.

Abraham O. Smoot was thirty-nine years old in the late summer of 1854. He was a native of Owenton, Kentucky, having been born there on

17 February 1815. When he was thirteen his parents moved to Tennessee, where he was exposed to Mormonism and joined the church. He was at Winter Quarters when the pioneer company left for the Rocky Mountains, and he served as one of the captains and leaders for the second group that arrived in the Salt Lake Valley in September 1847. When the first nineteen wards were organized in Salt Lake City in February 1849, Smoot was ordained the Fifteenth Ward's first bishop. In 1852 he moved to South Cottonwood, then to Sugar House in 1854, serving as bishop at both places. The latter move was to allow him to manage Brigham Young's Forest Farm and to take charge of the construction of the sugar factory along Big Kanyon creek. The new building was popularly known as the Sugar House. When the ward was organized, some wanted it to be named the Kanyon Creek Ward, but, as the story goes, Smoot's wife, Margaret, suggested it be named after the area's most prominent feature, and so it became the Sugar House Ward. That name has been carried forward to this day. When the Big Cottonwood Lumber Company began operations, Abraham O. Smoot was heavily involved with the Sugar House construction. He never took an active part in the Big Cottonwood activities, and the company ledgers show only his marginal participation in providing equipment or materials.

*

In September 1854, shortly after Frederick Kesler returned from his eastern trip, Brigham Young gave his partners instructions to explore the unknown extent of Big Cottonwood Canyon to survey the timber resources and select locations for sawmills. This was a big order, for up to this time the canyon was considered to be "impregnable."[16] There are several accounts of this exploration. The first was a brief mention in the *Deseret News* in July 1856.[17] Then Henry L. A. Culmer described it in an article published in February 1892.[18] Culmer was not a part of the expedition; in fact, he was in England at that time and was only six months old. But in later years, after settling in Salt Lake City, he married Annette Wells, whose father, Daniel H. Wells, was a member of the exploring party, so his story may be accepted as a valid second-hand report from his father-in-law. Another account was given by Frederick Kesler, who wrote about it when the *Deseret News* was soliciting such stories from surviving pioneers in celebration of the fiftieth anniversary of the arrival of the first pioneers in the valley.[19]

The party started up Big Cottonwood Canyon but found it impossible to proceed beyond the foot of "The Stairs," a short distance above the existing Seventies Mill and its road. This may or may not have been the

Stairs that is known today at the foot of Storm Mountain. In those days there were both upper and lower Stairs, extending all the way from the Stairs Power Station to Mineral Fork. The upper part, from Mill B to Mineral Fork, may have been the most formidable section. This is difficult to imagine today, with the excellent road winding around the S-turn and up the north slope of the canyon, well above the stream most of the way. But in 1854, when the exploring party headed up the canyon, this section was narrow, barely wide enough at the bottom to contain a stream cascading over boulders and debris, and had steep, nearly vertical, slopes on both sides. At any rate, the men turned back and went up Mill Creek Canyon instead, where a road ran to Archibald Gardner's upper mill and to the timber farther up the canyon. From there, Kesler wrote, "we traveled on foot, carrying our blankets and provisions until we reached the top of the divide, where we made our first camp." Culmer said they came down into Big Cottonwood where Mill E was later built, just below Brighton. This is unlikely, however, since the south side of the ridge between the upper part of Mill Creek and Big Cottonwood drops into the Mill D drainage. It is more likely they came into the canyon where Mill D was later built. From there they explored downstream. Culmer wrote that it took them all day to descend scarcely a mile over or around the Stairs. Kesler remembered "climbing over the most rugged, rocky, and rough places I had ever traveled over." He also mentioned finding "some colonies of rattlesnakes, which were plentiful and some of them very large, but they did not harm us."

After their return Feramorz Little and Charles Decker were selected to superintend the building of a road up the canyon, while Kesler was to oversee the building of the mills. The mills were named by the letters of the alphabet, the first one being Mill A, the next Mill B, and so on. Since they were named as they were built or acquired, however, they were not in alphabetical order as one proceeds up the canyon. The site for Mill A was chosen at a point between eight and nine miles up the canyon. It may seem strange that they should have gone so far, necessitating the building of a road through so many miles of steep, narrow and rugged canyon, choked with vegetation and blocked by boulders and cascades. But that in itself may be the reason, for the site they chose was the first reasonably flat and open area above the Stairs that had a readily accessible source of timber. Little and Decker had a tough job ahead. The first step was to cut a pack animal trail so they could haul in equipment and supplies. Then the construction of the mill began, while a road was being opened in the canyon below. While the company hired men to work in the canyon, both Little and Decker also acted as independent contractors and had their own crews at work on the road. They didn't waste time; work on the trail and road began immediately, and the first mill was under construction by November 1854.

As soon as the road was passable a toll gate was set up near today's mouth of the canyon. The first tolls were collected by the company in early 1855, but they were not a substantial source of income and probably did more to control traffic in the canyon than to pay for the expenses of building and maintaining the road. Indeed, the tolls collected in the three years from 1855 through 1857 amounted to $520.11, while on a balance sheet of March 1859 the road was valued at $11,195.48.[20] The toll gate, described as "a swing pole," remained into the 1870s, even if it was no longer active.[21] A mining claim recorded in April 1874 referred to "the old tollgate or house," and the "old tollgate" was still there in June of the same year when the Brown and Sanford Ditch Company built its head gate on the stream nearby.[22] Only one later reference to it has been found: in a mining claim recorded in January 1876.[23]

One of the men hired to work on the first mill was Lorenzo Brown, who had arrived in Salt Lake City in 1848 with his parents, wife, and one child. In January 1855 he was thirty-one years of age and had a pregnant wife and three children to support. He was a farmer who seemed to do quite well with his onion crops, for he was always bartering onions for lumber, wheat, or other commodities. A year earlier he had worked as a carpenter on Brigham Young's mill on the Isaac Chase farm, and he later worked on Young's new home and stable. It was probably through these jobs that he was given the opportunity to work in Big Cottonwood Canyon. Brown wrote in his journal that early in the morning of 25 January 1855 he left with five or six others and "arrived in safety at the end of the wagon road 1 mile above Cooleys new mill about noon."[24] Hence, by this time the road into the canyon had extended well above today's Stairs, since John W. Cooley's mill was on the flat above the present Storm Mountain picnic ground. It is not known whether the road as far as Cooley's mill was a result of the road building efforts of Little and Decker or whether it had been built earlier, perhaps as a result of Brigham Young's discourse in February 1853 when, referring to the Seventies Mill, he expressed his wish to have the road made and perhaps another mill built.[25] Was Cooley's mill the other one that Brigham Young wanted? Perhaps so, for Joseph Young had full control of the canyon. There is no evidence that the county court gave Cooley permission to build a mill in Big Cottonwood Canyon, nor could it have done so without compromising or revoking Young's charter. The only way Cooley could have built a mill in the canyon was through the good graces and blessings of Joseph Young, yet Young would have given them if his brother Brigham insisted. While little information has surfaced about Cooley's mill, it may be that he dismantled the Seventies Mill and used its machinery. A year and a half later, in July 1856, when the Cooley mill was still standing, George D. Watt, a reporter for the *Deseret News,* wrote that Joseph Young's mill "has since been taken down."[26]

Whatever the origin of Cooley's mill, it would soon become the property of the Big Cottonwood Lumber Company. The ledgers for that company carry a credit entry for J. W. Cooley dated 1 May 1855: "By Your Mill with all the improvements, logs and right of Kanyon in Big Cottonwood Kanyon, sold us for $4000."[27] Although the Cooley mill entry in the company books carries a 1 May date, the actual transaction was made some time earlier. Despite the agreement, Cooley continued to work the mill. In a letter dated 23 April 1855 Brigham Young politely asked Cooley to deliver "the property I purchased of you" to Feramorz Little, who was working in the canyon and acting on behalf of the company.[28] On the same day Young also wrote to Little explaining he had learned "by Bro. Kesler that Bro. Cooley is taking quite active measures in Sawing, getting out lumber &c from his mill." He asked Little to go and take possession of all the property Cooley sold. Young said he "purchased from him [Cooley] all the interest which he possessed in the Kanyon, including mill saws, tools and all property belonging thereto or used there about, logs, lumber and every description of property," and he wanted it immediately. If it was not Cooley's understanding that the company was to have all the logs that were cut and all the lumber on hand, then he should come down and see Young so they could get an understanding before they went any further with the purchase.[29]

Since nothing else appeared relative to this matter, Cooley apparently gave up the mill and Young charged him for the logs he had cut and the lumber that had been removed. In a July 1855 balance sheet the amount expended for the purchase of the mill had been reduced by $153.88 "for articles afterwards purchased by J. W. Cooley, included in the Mill purchase."[30] Cooley then left the Wasatch scene. In March 1856 he was living in Tooele, and by December of the following year he had moved to Grantsville, where he engaged in farming.[31] By the time the Cooley mill purchase was consummated, Frederick Kesler had located the site for the company's second mill, Mill B, so the Cooley mill became the company's third mill and was designated Mill C. It is likely that the mill consisted of only the bare necessities of a sawmill, without the convenience of a roof or shelter. Considerable work was done on it after the company took control. George Laub wrote that in June 1855 he and Samuel Ensign were "sent to Big Cottonwood to repair a sawmill for President Brigham Young. This continued till sometime in September."[32] Since Mills A and B were just being built, Mill C was the only one that would have been in need of repair, and the company's account books substantiate this. Also during 1855 a mill house was built for Mill C. That mill provided lumber for the construction of both Mill A and Mill B. The company's books show payments for logs provided at the mill during the summer of 1857, and another payment for work on the flume in February 1859. There is little evidence the mill was being used at that late date, although in the company's balance sheet, dated 18 February 1862, the Cooley mill was still carried as an unencumbered asset, valued at $5,082.21.[33]

To go back to the afternoon of 25 January 1855, Lorenzo Brown and his comrades had to travel on foot from the end of the Big Cottonwood road. Their "provisions, bedding, tools, etc had of necessity be packed on mules a distance of 4 miles over rough & hilly road with considerable snow." Brown found it to be a "tedious walk," especially as he was without dinner. Fortunately one of his fellow workers and the "Boss of the crowd" shared some of their food, enabling him to reach his goal. That night he "took a bed on the floor and slept soundly." The next morning he was able to take a brighter view of his surroundings. It was a scene we can barely imagine today, with the canyon now so open, accessible, and changed by modern improvements. On 26 January 1855 Lorenzo Brown saw it this way: "This canyon from where we left our wagons is very narrow & was considered impassable until a mule path was worked, which must have cost near $100.[34] Of all places that I ever have been in for mountains, rocks, cataracts, precipices, hills & hollows I have never seen the 4 consecutive miles that could compare with this. The Kanyon is very narrow but generally filled with timber. The Companys claim begins about 2 miles below here where they expect to build another mill this season. Where we are at work the Kanyon is a little wider say 40 rods entirely.[35] The mountains on the S side which are very high & steep are densely covered with timber viz Fir White & red pine, spruce sufficient in sight of our house to keep the mill in operation 5 years."[36]

The forest he saw can scarcely be imagined today, for what we see now is what was left behind as useless by the loggers or what has grown since. The trees Brown was seeing had grown unmolested for centuries, save for the natural calamities of fire, pestilence, or scourge of insects. The size of the trees must have been impressive; some indications can be found in records that have survived. For instance, five years later, in April 1860, Captain Albert Tracy was detached from duty at Camp Floyd and was returning to Washington. En route he went up Parleys Canyon, crossed the summit, and spent a day at Snyder's Mill, near present-day Snyderville. There he had time to look over the mill and its yard and to talk with the lumbermen. One of them told Tracy "that some of the trees cut upon the heights measured—as I could see for myself by the logs—three feet in diameter. A large proportion were a full two feet."[37] Today it is possible to find trees two feet in diameter, but three feet is a rarity indeed.

Another example is given in the description of a mining claim. In 1888 the Lone Tree claim was filed, it being on the ridge between Red Butte and Emigration Canyons. The name of the claim was the result of a nearby "lone pine tree, the only one [of] like size, being 14½ ft in circumference, 3 feet from the ground."[38] A tree with that circumference is 4 feet 7 inches in diameter. But they grew even bigger than that: in 1898 the Great Ochre Spring mining claim was filed, it being in Mill B South Fork and located relative to "a large pine log about 6 feet in diameter and being the largest log in the canyon."[39]

Brown also described the canyon above Mill A: "About 2 miles above begins what is termed the basin. The kanyon there is very wide for 6 or 8 miles in length & very heavy timbered. This Kanyon & its various branches which all abound in choice timber extends up from this mill from 12 to 15 miles & is expected to employ some 7 or 8 mills."[40] Since he had not been up the canyon above the mill, he must have heard this information either directly or indirectly from those who had. One such source was Frederick Kesler, who had seen "the basin" during his exploratory trip the previous September, and who was frequently present in the canyon. For example, one day Brown wrote, "Yesterday Kesler the Mill Wright came up." The basin was the large open area at the mouth of Cardiff Fork. It is certain that at this time neither Kesler nor any of his associates had been above the basin; but from that vantage point it was easy to see that the canyon above was wide and heavily timbered.

Sunday was not a work day, so Brown amused himself with other small activities. The first weekend he wrote that he "strolled off by myself. Went up the mountain where some of the boys are chopping logs. They cut a tree some distance up the mountain & run it down, the whole length [of log] often 100 feet long or more. They run down with great velocity."[41] Here he was describing a logging slide, several of which were located on the steep south slopes of the canyon and still can be seen to this day. The logs that were brought down at this early date were used to frame the new mill.

Frederick Kesler was but one of the many pioneers who regularly wrote in their diaries. One of Kesler's diaries from 1857 survives and tells much about how he located a mill site. He was very careful to ensure that an adequate timber supply was readily accessible for the mill. In the case of Mill A, located at the mouth of Mill A Gulch, there could be little doubt that an adequate supply was at hand, for the entire south slope of the canyon provided an excellent environment for the dense forest that was found there then and is again found there today. When the woodsmen were taken up the slope to start cutting timber for the mill, Kesler was certainly there to direct where it should be cut and how it should be brought down. The slopes above Mill A are very steep, and one can imagine the workers complaining about the steepness of Kesler's mountain. And it did receive that name. When the mining era began, prospectors picked up the name and continued to use it; but, unlike the woodsmen, who could not see the summit from either the canyon bottom or within the wooded slopes, the prospectors called the mountain Kesler's Peak. Over the years the name gained another letter to become Kessler, and the peak, towering above Big Cottonwood Canyon on the ridge between Cardiff and Mineral Forks, is still known by that name today.

During January and February Brown was working at the mill. On his first day he wrote that they started by digging some two feet of snow from around the timbers and throwing in a quantity of boughs to stand upon. He later mentioned that while nights and mornings were very cold, the middle of the day was pleasant. But such weather would not continue long in midwinter. On 20 February he noted that snow was falling every day, and on the twenty-fifth he wrote: "This morning after consulting together for a time thought it time to be leaving as the snow was accumulating very fast & the prospect good for more."[42] Thus ended the work on the mill for that winter, but the men returned as soon as the snows melted in the spring. This working schedule was typical during the construction years that followed. In spite of early snows the men continued to work until the snow became too deep, then operations were suspended until spring. However, many of the workers had their own farms or gardens in the valley and their time in early spring was necessarily devoted to tilling their own soil and planting their crops. Only after completing that were they free to return to the mountains.

In the spring of 1855 construction began on the second mill, Mill B, located on a small flat in the narrow part of the canyon about a mile above Cooley's mill. The S-turn in today's highway is located on Mill B Flat. Some of the flat has been carved away for the lower half of the turn, while the other side of the flat has been built up for the upper half of the turn. The mill itself was located at the apex of the lower half of the turn. Mill B drew its timber from Broads Fork and Mill B North and South Forks. Remnants of logging slides or roads can be found in all three forks.

Lorenzo Brown returned in April 1855 to work on Mill A, which he described as being "large for this country being 30 by 50 feet with 6 bents above. There will be one upright & one circle saw above & shingle & lath machines below. Roof steep being square or half pitch."[43] He continued to work on the mill throughout the season, albeit with increasing discontent. In mid-June Brigham Young sent a letter to all men at work in Big Cottonwood Canyon, saying he was informed they were "in the habit of observing ten hour labor & making short time at that and do not perform faithful labor from morning till night," and threatened them with discharge if they did not faithfully perform their duties. Brown was sure some "evil designing persons" wished to curry favor with Young at the expense of others. He noted that Frederick Kesler required the men to work only ten hours and that he, Brown, instead of making short time had averaged eleven hours.[44] In September he complained that the millwright superintendent had been replaced and commented, "mean works & doings somewhere by somebody." Later that month he complained about the food: "I am perfectly disgusted with our fare & am sometimes quite sick & cannot eat all there are 25 or 30 men & but one boy to cook and every thing is nasty & greasy enough to sicken a Hottentot." That evening he walked down to Mill B and took supper there, "a good deal better than we have at Mill A."[45] He was likely reflecting the complaints of all the workers, for the working force diminished until by mid-October there were only two carpenters working at Mill A: Brown and Peter Sinclair.

Peter Sinclair was from Scotland, where he had been converted to Mormonism in 1851. Three years later he emigrated to Salt Lake City, arriving in October 1854. He was working for the Big Cottonwood Lumber Company as early as the spring of 1855 and continued through the end of 1857. He was twenty-three years old when he was working with Lorenzo Brown, who was ten years older. By this time Brown was blaming Kesler for the problems they faced in the canyon. When the head framer was sent on a mission to England, Brown thought it was the result of "an undue influence exercised by some person or persons supposed to be F

Kesler who finds fault with almost every man in the kanyon behind his back but has not sufficient moral courage to talk to a man face to face."[46] Perhaps it was only a private conflict between Brown and Kesler, for Sinclair made no mention of it in his diary. It came to a head early in December, however, when George Taggart, the new superintendent, told Brown that Kesler sent word for him to pack up his tools and leave the canyon. Brown wrote, "I felt to accommodate him as soon as possible & left next morning."[47] Thus ended Lorenzo Brown's affiliation with the Big Cottonwood mills. Sinclair remained to work on Mills A, D, and E over the next two years.

By midsummer the Mill D site had been established and construction begun there as well. At the end of September, in a letter to his son-in-law Edmund Ellsworth, who was in England on a mission, Brigham Young wrote that they had worked a road up Big Cottonwood Canyon that summer, had two fine sawmills running and two more that would be ready to run in a few weeks. He said they were preparing to build a few more and were "calculating on having from 8 to 10 good mills in the canyon."[48] He was referring to Mills A and C as the mills that were running, and Mills B and D as the two that would soon be ready. Mill B was completed that year, but Mill D would be delayed another year by circumstances that were only beginning to become apparent. The summer of 1855 yielded very poor harvests, mostly through a scourge of locusts, a hardship made worse when it was followed by an unusually severe winter. Early in 1856 the church's public works department had all but shut down since there were insufficient supplies in the tithing storerooms to provide for the employees. Workers "of every class were counselled to abandon their pursuits and go to raising grain. This we are literally compelled to do, out of necessity," wrote Heber C. Kimball to his son William on the last day of February 1856.[49] The situation affected everything and everyone, and construction work in the canyon would not resume until midsummer of that year.

Mill D was located in what Lorenzo Brown had described as the basin. It was at the mouth of Cardiff Fork, near the lower end of today's Spruces Campground. With the road built to this point, the most difficult part of the canyon was conquered. Above that point travel was relatively easy, but it was unknown territory, for it was above the place where the 1854 exploring party had entered the canyon. Nothing has been found to describe or date the survey of the upper canyon, but it must have happened in late 1855 or very early 1856. In view of the fact that the snowpack lasts much longer at those heights, the former period is most likely. It is also safe to assume that Frederick Kesler was involved in the exploration, for it was his duty to locate as well as build the mills. So far he had located the mills some distance apart, to allow each to have its own independent supply of timber. For the fifth mill, he went up toward the head of the canyon and found a suitable site less than a half mile below present-day Brighton. In looking over the timber resources he also found a flat with a beautiful lake, surrounded by mountain peaks, some with wooded slopes, others with bare granite surfaces. The lake was, of course, Silver Lake at Brighton, where thousands of people go every year to enjoy the same scenic wonders. It had no name on the day Kesler first saw it, but in 1857 he mentioned it in his journal and called it "Plesant lake," a name he probably did not tell to others, for no one else seems to have used it. The grandeur of the place could not have failed to impress him, as it has impressed so many since, and to cause him to carry word of this alpine wonderland back to the city.

It is not known when Brigham Young learned about the lake, but it is known that he and Feramorz Little went to visit the upper mill, Mill D, in April 1856.[50] Again, a month later, he left for a two-day trip to the Big Cottonwood mills, this time accompanied by Frederick Kesler, among others.[51] By this time he must have seen the lake and its surroundings and decided he had to show it to others. The first thing he did was to order the road extended from Mill D to the lake. The company's account books show an increase in road work during the month of June. The road was well under way, perhaps completed by 8 July 1856 when "President Brigham Young accompanied by Hiram B. Clawson [Young's secretary], Daniel H. Wells and Frederick Kesler started for Big Cottonwood Canyon. They took breakfast at Bro. Little's [on Parleys Creek below the mouth of the canyon] then called at the mills in the canyon and arrived at the lake at 5 pm."[52] On his return to the city Young had his office staff make out invitations, which were sent out on 18 July.[53]

PIC-NIC PARTY AT THE HEADWATERS OF
BIG COTTONWOOD.
[seal]

Pres. Brigham Young respectfully invites _____ and family to attend a Pic-Nic Party at the Lake in Big Cottonwood Kanyon on Thursday, 24th of July.

You will be required to start from the city very early on Wednesday morning, as no one will be permitted, after 2 o'clock, p.m. of the 23d, to pass the first mill, about four miles up the kanyon.

All persons are forbidden to make or kindle fires at any place in the kanyon, except on the campground.

G.S.L. City, July 18, 1856.

The restrictions placed upon the invited guests are both interesting and complementary. There was a considerable caution because of the need to avoid destruction of timber by fire, hence there were no fires allowed in the canyons. This concern was not without good reason: the summer thus far had been very hot and dry, and on 28 June a fire had broken out in Mill Creek Canyon.[54] Then on 10 July some men blasting a rock in Big Cottonwood Canyon, probably for the new road to the lake, had started a fire that burned for two days.[55] Also, the requirement that all pass the first mill, Mill C, before 2:00 p.m. ensured that they would reach the campground at the lake that evening and thereby avoid the need to camp in the canyon.

On 23 July there was a regular train of teams and carriages moving up the canyon. No one without an invitation was allowed past Mill B. By evening, seventy-one carriages and wagons, drawn by 201 horses or mules, had carried about 450 people to the lake, where they found a bowery, 24 by 40 feet in size, with a wooden floor for dancing, and two rafts on the lake. Many trees supported banners and flags, and a large pine next to the bowery carried the flag of the United States. The animals were turned loose to graze the grassy meadows, while a choir and three bands provided entertainment and music for dancing Wednesday evening and all day Thursday, the twenty-fourth.

On Thursday evening they had songs, prayers, and comments by Brigham Young. He closed with "I will now propose that we do not dissolve this meeting, if that is the feelings of those present, but adjourn it until the 23d day of July, 1857, to meet on this ground by 4 o'clock p.m., preparatory to celebrating the 24th; that virtually gives the people present an invitation, aside from those I shall invite hereafter." Then, in a reminder that even in those days overcrowding could become a problem, he admonished his audience: "The brethren and sisters who are now here and those whom I shall invite, I shall be happy to see here another season; but I give no liberty to any person to invite others to come here, without my permission. I have no question but that all have friends whom they would like to bring with them, but order and decorum must be maintained in this, as in all matters pertaining to the kingdom of God." He concluded with "Now you can dance as long as you please, but do not wear out the musicians." They danced into the wee hours of the morning.

At sunrise the revelers were awakened by cannon fire and band music. Soon the wagons were headed back to the valley, ending the first celebration at the head of Big Cottonwood Canyon. During the descent a bear was seen watching the wagons. When it approached the road, Charles F. Decker, one of the partners in the Big Cottonwood Lumber Company, dispatched it with his Sharp's rifle.[56]

Deseret News reporter George D. Watt described the facilities at several of the mills in the canyon. He wrote that there were three excellent sawmills in operation, which had already cut over 800,000 feet of lumber. This figure was a gross exaggeration, but the mills were producing. There was a blacksmith shop at the second mill, Mill B, and all mills had comfortable and commodious log buildings for the accommodation of the workmen and their families. The building at the third mill, Mill A, was a large log building, about 20 by 40 feet in size. Mill A sawmill had a large circular saw in addition to the upright saw.[57] The building at Mill A was destroyed several months later, on the night of 14 December, when a snow avalanche carried it and all the stabling a distance of 150 yards and pitched it into the millpond. Fortunately the occupants of the house had moved out a few days before. The building was further described as being one and a half stories high, with two heavy stone chimneys.[58] Several men, including Peter Sinclair, who had worked with Lorenzo Brown at Mill A, were working in Mill D at this time. They had run out of lumber, so they decided to start for home. They headed down canyon on Monday, 15 December, using snowshoes they had built for themselves. They arrived at Mill A about noon and found "a Tremendious Slide had ocoured that night—carying entierly off avery Strong Story and half house 52 feet by 22 also 70 feet of Stabels." They had considered staying at Mill A overnight, but now continued on to Mill B, where they arrived at 4 p.m., weary and with one man's feet "considerably frosted." After spending the night there, they continued into the city the next day.[59]

*

The year 1856 was the first year of substantial production by the mills. The lumber accounts in the company books show about 250,000 board feet of lumber flowing out of the canyon during the second half of the year. Logs used in construction were brought to the mills by men working for the company, but, when the mills went into operation, logging was done on contract. In August 1855 the company published a notice that it would let out the job of supplying two of their mills with logs.[60] The men who took this job were brothers Richard and Robert Maxfield. The contract, dated 13 October 1855, was to furnish saw logs to Mills A and B in sufficient quantities to keep the mills in constant operation for the next two years.

Richard Dunwell Maxfield and Robert Quorton Maxfield were born at Prince Edward Island, Canada, where their father, John Ellison Maxfield, had emigrated from England with his parents. At an early age they got practical experience in the lumber business at a sawmill their father owned. The entire family emigrated to Utah, arriving in Salt Lake City on 15 September 1851 and settling on the east side of the Jordan River, near where I-215 crosses the river today. While little is known about the early activities of brothers Richard and Robert, when they entered into contract with the Big Cottonwood Lumber Company they entered Big Cottonwood Canyon in a big way, creating a presence that continued throughout their lives. They were twenty-four and twenty-six years old, respectively, at this time.[61] The contract specified that they would be paid one-fourth each of cash, cattle, store pay, and lumber. They were to be held responsible for damage by fire caused by them or anyone working with them, and they were to "take the saw timber clear wherever they go in the Kanyon," thereby introducing clear cutting to the Wasatch Mountains. They were required to start at Mill B and take timber up the main canyon towards Mill A, unless they preferred to make a road into the first south fork above Mill B, in which case they were to observe the rule to take saw timber clear as they went. They were also required to haul the logs on wheels and not drag them on the road. The company had gone to great lengths to build the canyon road and didn't want it damaged by logs being dragged over it.[62]

Logs for Mill C were provided by Feramorz Little and Charles Decker, operating as Little and Decker. This partnership became the primary operating entity in the canyon except for the operation of the individual mills. It took care of the road repairs as well as the logging operations. Before the Maxfield brothers' contract had expired, they also were working through Little and Decker, and they continued to do so during the remaining years of operation of the Big Cottonwood Lumber Company. Over the years Little and Decker contracted with many other operators and companies for supplying logs, including E. Tuttle; Finley, Foster & Co.; George and Daniel Jacques and James Adams as separate contractors and as the joint Jacques & Adams; as well as R. Maxfield and Company. Employees of these companies built the roads and logging slides in the various tributaries of the canyon, and many of the names used today date from that period. Unfortunately, the lumber company books show only the name of the contractor and not the names of the people working for him, so it is difficult, if not impossible, to link the names with individuals at this late date.

Mill D construction had been pushed throughout the summer and fall of 1856, and it too was ready for operation about the time the winter snows ended the season. Actually, the snows stopped only the mills; much of the timber harvest went on throughout the winter. The logs were slid to the bottom of the slopes and snaked to the mills while the snows made those operations relatively easy. But snow choked the races or flumes so there was not enough water to run the mill.

In mid-April 1857 Frederick Kesler went up to Mill D on foot, and, even though the snow was still three feet deep at the mill, he got it into operation and sawed one log. "Strikes 200 per minute," he wrote in his journal. "Operates to a charm."[63] This comment implies that Mill D was fitted with a sash saw rather than a circular saw, and this seems to have been typical of early sawmills. Whether Kesler chose to build this type of mill or was instructed to do so is not known. A mill with a circular saw required some sort of gearing, for the saw had to turn much faster than the waterwheel. It is known that some years earlier Brigham Young had told Phineas Cook he was opposed to gearing, which had to be made of wood and soon would be worn out.[64] Belts and pulleys could be used instead of wooden gears, but belts were prone to stretching and had to be adjusted constantly. When Cook tried to convince Young otherwise, someone else was hired to supervise the job in Cook's place.

Circular saws had been used before—Joseph Young in speaking of the Seventies Mill in 1852 said its saw was circular, and they were being used in both Mill Creek and Little Cottonwood Canyons as well as in Mill A. But all four mills built by the company in Big Cottonwood Canyon had sash saws, including Mill A, which had both types of saws. This type of mill was relatively simple: for power they used a small-diameter, fast-turning flutter wheel with a crank mounted on one end of its shaft. A pitman arm (reminiscent of the function of the man in the pit under a pit saw) connected the crank with a gate carrying the saw or the saw blade itself on the floor above the waterwheel. In the first case, the gate slid in vertical tracks as it moved up and down. In the latter case, a long, stiff saw blade was used, guided at the ends but not stretched in a gate. This type was known as a Muley saw. The flutter wheel was designed to drive the saw at the desired speed—200 revolutions per minute in the case of Mill D, each revolution causing one stroke of the saw blade.

The waterwheels of the first three mills built by Kesler were very similar in design. When George Peacock, a judge in Sanpete County, requested information on the design of the mills, Daniel H. Wells wrote him a letter giving specifications.[65]

	Mills A and B	*Mill D*
Head	18 feet	23 feet
Penstock size	7 x 9 feet	5 x 9 feet
Waterwheel dia.	29 inches	32 inches
Bucket length	7 feet	9 feet
Number of buckets	8	
Crank throw	12 inches	

Mills A and B were nearly identical, while Mill D differed only slightly. The figures show a crank throw of twelve inches, meaning the saw blade moved up and down two feet with each stroke. The buckets were the waterwheel blades, eight of them extending radially outward from the shaft to the outer edge of the wheel. Each was seven feet long, making the waterwheel twenty-nine inches in diameter by seven feet wide. The saw blades were 6½ feet long and 9½ inches wide from the point of the tooth to the back of the blade. Kesler specified a two-inch pitch for the teeth, as the usual 2¼- to 2½-inch pitch was too coarse for the timber they were cutting in the Wasatch.[66]

*

Mill D was now operational. In early July Kesler noted that he had staked off the site for Mill E, leveled for the dam, and set the diggers to work on the mill pit. There is an existing photograph of Mill E, probably by C. R. Savage, that shows three peaks—Millicent, Tuscarora, and Wolverine—in the background. With the picture in hand it is easy to find the point where the peaks line up the same way and accurately locate where Mill E stood. There is little left to show what was once there except for a large granite boulder that is in the foreground in the picture and still rests where it did then. Above the mill site the race still exists, heavily overgrown with brush, as does the dam that impounded water for the mill. The dam has been breached but is easily recognized. The upper end of the millpond covered the area where the present-day road to Camp Tuttle crosses the stream. The mill was on the flat east of the creek and a few hundred feet downstream from the road crossing.

On 13 July, Kesler went up again. "Started up Big Cottonwood Kanyon travveled 15 miles in Kanyon on foot, arrived at Mill E at dark," he wrote. Then he added, "Tired out."[67] He "tarried" at the mill for one day, possibly recuperating from the previous day's efforts, then spent a day assisting Charles Decker in locating the second and third dance floors at the lake for the 24 July celebration, and put six men to work building the boweries. Feramorz Little had a sizeable group of men working on the road in preparation for the event.

As he had promised the year before, Brigham Young again arranged for a celebration at the head of the canyon on the twenty-fourth of July, this being the tenth anniversary of the entry of the first group of pioneers into the Salt Lake Valley. He again had invitations sent out, similar to those of the previous year. This year he listed more restrictions and requirements: fires were forbidden in the canyon, as before, but now also was the smoking of cigars or pipes. He requested that bishops accompany those invited from each ward and see that all were well equipped and prepared for the journey. Also, bishops were required to furnish a full and complete list of all persons accompanying them. It was reported that Young had sent thousands of invitations and that Lt. General D. H. Wells ordered a detachment of the Nauvoo Legion to patrol the canyon.[68] The invitations were much desired; people had heard about the previous year's celebration, and many who had not been there wanted to go. Recognizing this, Brigham Young gave remarks in the Bowery on Temple Square the Sunday before the twenty-fourth, cautioning those who did not receive an invitation to stay away. It is not, he said, because we do not desire your society but that it was inconsistent for everyone to go. So they gathered those who "ought to go" and could conveniently do so.[69] He might well have been concerned, for this year the "Pic-Nic Party" drew a great crowd. When it was over it was reported that 2,587 persons had attended, conveyed there in 464 carriages and wagons by 1,028 horses and mules and 332 oxen or cows.[70] They came from as far as Ogden to the north, and Provo and Springville to the south. On the evening of 22 July the mouth of the canyon held hundreds of wagons, their occupants camped there in anticipation of the next day's journey to the lake. Hosea Stout wrote, "The kanyon seemed full of men and teams." The next morning he started at daylight, without breakfast, and arrived at the lake about one p.m. A large number of people were already there, and teams continued coming in all evening.[71]

One of the guests was Lorenzo Brown, the same man who had worked on Mill A in 1855 and who wrote of the basin farther up the canyon. But all his work with the company was on Mills A and B, so this was the first time he had seen the upper regions. "It is astonishing to see the quantity of timber," he wrote in his journal, "The kanyon being wide and filled with white & red pine balsam, Large quaking asp etc., all of large size and immense quantities & enough to last for years."[72] His reaction was shared by many others. Hosea Stout wrote, "The kanyon is well set with heavy forests of timber from the base to the summit of the mountains."[73] President James N. Jones of the Utah Stake was surprised to see the amount of timber that was there; and Wilford Woodruff noted, "I never saw finer timber in my life."[74]

The festivities followed the pattern established the previous year, except this time there were three dance floors and many more dancers. This was the celebration that was interrupted by the arrival of Judson Stoddard, A. O. Smoot, and O. P. Rockwell from the east. Most accounts written in later years say these men brought news of U. S. Army troops marching towards Utah. Accounts written at that time, as found in diaries or journals, either don't mention the event at all or say it was a report that the postmaster at Independence had orders not to deliver the mail to Stoddard or Rockwell, who were carrying the mail between Independence and Salt Lake City. In fact, the coming of the army was known before this time and accounts of the events had been published in the major east coast newspapers. While it may have been of great concern, it was not new information. But the refusal to allow the mail to pass through to Utah was news and was an indication that the situation might be worse than it at first appeared.

For most people at the celebration, the news did little to spoil their day. After evening prayers the bands began to play and the people danced well into the night. The next morning the first teams headed down canyon at daybreak, not too long after the dancing ended. Frederick Kesler, who was present as millwright for the company and bishop of the Sixteenth Ward, wrote in his journal: "Dancing until 4 ocl. A.M. At 1/2 past 4 ocl. I moved down the kanyon." By midday all was quiet once again at the lake near the head of Big Cottonwood Canyon. Only the crushed grass, the many cooling firepits and the three boweries with their rough plank floors indicated that anything unusual had happened. And a road account entry in the company's books charged: "Repairs on road after 24th July, viz. 18 days work @ 2 36.00."

With the celebration behind them, the men of the Big Cottonwood Lumber Company could get back to routine work. Construction on Mill E continued rapidly until it was completed at the end of the year. On 3 August Frederick Kesler went up to Mill E again, where he spent three days making a thorough assessment of the timber reserves before locating a site for the next mill. The records of his explorations are interesting because they show the kind of preparations he made before choosing a mill site, and they provide one of the first records of hiking in the Brighton area. On 4 August he wrote: "Started at 8 ocl. to explore a North East Kanyon past over a High Ridge and past Down to Find little Streams of water past up Kanyon to Summit where I Could overlook a large Section of Country & Could See Provo & Weber Rivers Traveled North on High Divide to North fork of Kanyon past Down to forks & then to main Kanyon there is a great quantity of good pine in the abov Kanyon & is easy to get at arrivd in Camp at 4 ocl. P.M."[75] In this narrative he described going over the hill where today's Guardsman Pass road climbs out of Brighton and dropped down into Mill F East Fork, where he found the little streams of water. He went up that fork to the divide, where he had a grand view to the east. He followed the divide north, over one high point and down to Scotts Pass, then down the drainage that he called the "north fork of kanyon," or the north fork of Mill F East Fork. He followed it down to the main canyon, which is Mill F Flat today, then went up the canyon to Mill E.

The next day, 5 August, he wrote: "Left Camp at 8 ocl. went Down main Kanyon 2 miles went up a South Dry Kanyon found plenty of Good timber past Down a High mountain to left to find water found a Good Spring & Lake found 2 Horses past up a High mountain to our Left Had a fair view of Plesnt Lake followed Ridge South & west & deceded Down to 2 Lakes full of mountain Aligators past Down Kanyon to plesant Lake Rested a little on Old Camp Ground many reflections past through my mind. arvd at Camp at 1/2 past 3 ocl."[76] This time he went down to where the Solitude ski area is located today and up the bowl in Queen Bess Hill, the bowl in the center of the ski area. While it doesn't look much like a canyon today, it probably did then when the slopes were heavily timbered. He turned left and dropped down into the upper end of Mill F South Fork, which has several heavy springs, then followed the fork up to the lake at the end: Lake Solitude.

Kesler climbed Mount Evergreen, which may not be so high but is very steep on the side he was climbing. When he reached the ridge he looked to his left and had "a fair view of Plesnt [Plesant] lake," or Silver Lake. He followed the ridge to the southwest, then dropped down to the Twin Lakes, which in those days were separate and distinct. His mention of "mountain aligators" is interesting. During the celebration a few weeks earlier, young Winslow Farr wrote in his diary: "I with my Cousin O. Badger went to some Lakes about 2 miles from camp I caught a fish with 4 legs I brought to camp alive in my hand."[77] He probably went to the Twin Lakes too. What both he and Kesler saw were salamanders, animals that once populated many of the Wasatch lakes. They were also known as Dog Fish and are the source of the name for both Dog Lakes. From the

Twin Lakes Kesler went down to his Plesant Lake, where he rested and reflected on the events of two weeks before when more than two thousand people were camped there and enjoying themselves.

On both trips Kesler found good stands of timber. There can be little surprise that he decided upon a site for the new mill and placed it the next day. On the sixth he wrote: "Left at 7 ocl. traveld Down Main Kanyon 3½ miles Zigzagd about up & Down mining [minding?] the Best Location for a mill Settled on a sight for Mill F about 1¾ miles Below mill E. . . ."[78] This placed Mill F at the confluence of Mill F East Fork and Big Cottonwood streams, where the large condominium buildings at Solitude ski resort are found today.[79]

Selecting the site for the mill was as far as that project went in 1857; Kesler still had work to do on Mill E and on several flouring mills in the valley that year. Meanwhile tensions were rising, leading to the Utah War. When the Nauvoo Legion was mobilized, Kesler was leader of his battalion and spent time drilling and instructing them before they went to spend the month of November in Echo Canyon. Early the following year the residents of northern Utah were advised to abandon their homes and move south before the army arrived. A few men were left behind to watch the city and be in a position to burn it to the ground if the army chose to take possession or plunder. As events unfolded, matters were settled peaceably; the army entered the city on 26 June but marched through and out the west side, continuing south and west until it stopped and established Camp Floyd in Cedar Valley. By the time the northern Utah residents returned to their homes, their society and economy had been badly disrupted. While the five sawmills in the canyon continued to run and produce lumber, the company chose to defer the building of the sixth mill.

While the populace was displaced in Utah County and points south, the Big Cottonwood Lumber Company mills served the public by providing lumber for the construction of shelters. Lumber from the mills had been taken into town to the tithing storehouse and yard in the past, but now it was taken directly to Utah Valley. The need for lumber was acute, so much so that Daniel H. Wells reported they had taken down the Bowery as well as the fences around it and the tabernacle on Temple Square to salvage the lumber.[80]

Following the period of negotiations that settled the Utah War, and after the army troops had passed through the city and gone to Cedar Valley, the tensions of some were momentarily relieved by a retreat in the mountains near the lumber company's mills. A total of fifty-two people took part, including Brigham Young, his counselors Heber C. Kimball and

Daniel H. Wells, seven of the Twelve Apostles, and other prominent per-
sonalities, including all of the principals of the lumber company except
A. O. Smoot. On 26 July they drove a short distance into the canyon,
where they spent the night. The next morning they took breakfast at Mill
C, then went to the lake at the head of the canyon. The following day they
went down to Mill D, then up the South Fork (Cardiff Fork) about a
mile, where they camped. On the fourth day they headed back to the city,
visiting the mills on the way.[81]

When the federal troops came west the new territorial governor, Alfred Cumming, and his wife came with them. After Cumming was settled
in Salt Lake City, Brigham Young invited him and his wife to a Pic-Nic
excursion in Big Cottonwood Canyon. On 26 August 1858 a fairly large
party rendezvoused at a shady spot near Mill D, where the lumber company had erected a bowery with accommodations, presumably a wooden
floor, for those who wished to dance. The party included women and children, most of whom traveled in carriages and wagons. Mrs. Cumming,
however, chose to ride her spirited pony and was thereby able to enjoy
more of the scenery than could those who preferred the comfort of the
carriages. She also achieved a measure of notoriety when her chosen mode
of travel was reported upon in the newspaper. The previous winter, when
Mormon resistance and the coming of winter had stalled the army's westward progress and it had to winter near burned-out Fort Bridger, Cumming and his wife suffered the winter hardships there with the soldiers.
Mrs. Cumming wrote letters describing their sufferings from the bitter
cold and short supplies. One of her letters survives to make interesting
reading. But now, the following summer, she and the governor enjoyed
the music, dancing, and song that enlivened their visit to the mountains.
The group returned to the city after the outing of two nights.[82]

After the army troops settled at Camp Floyd, there was need of much
in the way of building materials for the camp. This was a new market for
the Big Cottonwood Lumber Company, whose lumber had mostly been
delivered to the tithing yard in the past. Now, however, many loads of
lumber headed west from Big Cottonwood Canyon, destined for the new
military camp. The company ledgers show 461,000 board feet of lumber,
with a value of $23,062.65, charged to the U. S. Army Quartermaster ac-
count in October and November of 1858.[83] While this may have been
good business for the company, it also gave grounds for criticism. When
Horace Greeley traveled from New York to San Francisco in 1859 and
wrote a book about his travels, he wrote about the lumber business with
Camp Floyd:

[At Camp Floyd] the boards for roofs, finishing off, etc., supplied by Brigham Young and his son-in-law [Feramorz Little], from the only canyon opening into Salt Lake Valley which abounds in timber (yellow pine, I believe) fit for sawing. The territorial legislature (which is another name for the church) granted this canyon to Brigham, who runs three saw mills therein, at a clear profit of one hundred dollars or so per day. His profit on the lumber supplied to the camp was probably over fifty thousand dollars. The price was seventy dollars per thousand feet, delivered. President Young assured me, with evident self-complacency, that he did not need and would not accept a dollar of salary from the church—he considered himself able to make all the money he needed by business, as he had made the two hundred and fifty thousand dollars worth of property he already possesses. With a legislature ever ready to grant him such perquisites as this lumber canyon, I should think he might. The total cost of this post to the government was about two hundred thousand dollars.[84]

Although Greeley's account is full of errors, the average reader of the day would not have known that and would have accepted what had been written. Big Cottonwood was not the only canyon opening into the valley abounding in timber. Lumber was flowing from mills in City Creek, Mill Creek and Little Cottonwood Canyons, with the latter also delivering large quantities to Camp Floyd. The charge for lumber delivered to the camp was five cents per board foot, or fifty dollars per thousand feet.

*

In 1860 Brigham Young and the Big Cottonwood Lumber Company again hosted a celebration at the Big Cottonwood lake on the twenty-fourth of July. This was similar to the two earlier celebrations but did not have as many participants as the last one, in 1857. Where that event drew more than 2,500 people, this one had about 1,100. Access to the canyon was not as rigidly controlled as before. To avoid crowding in the canyon, those invited were given permission to travel as best suited them; therefore, many people went up on 22 July. The rest, including the presidency of the church, arrived the next day. There were again three dance floors and boweries, twenty-four by forty feet in size, and at least four bands to entertain the visitors. Unfortunately, Mother Nature was not in an entertaining mood and the late afternoon activities were interrupted by "the sudden bursting of a thunder storm over the mountains," scattering the people to their tents and wagons. Late in the day the skies cleared and the celebration resumed as planned. The next day dawned with a three-gun salute under heavy clouds. At nine o'clock a thirteen-gun salute, one for each year since 1847, was given, the bands taking turns playing selections between each firing. The rest of the day and much of the night was spent in dancing, singing, and the enjoyment of various other amusements. The next morning the camp was awakened by a martial band, and by nine o'clock the last of the revelers was on the way down the canyon.[85] One can only imagine how the meadow around the lake looked after this great number of people and their animals departed. Indeed, two months later Sir Richard Burton, who was on a visit to Salt Lake City, was brought to Mill E and the Cottonwood lake by Feramorz Little. He mentioned "the signs of last season's revelry—heaps of charcoal and charred trunks, rough tables of two planks supported by trestles, chairs or rail-like settles, and the brushy remnants of three boweries."[86]

Before the celebration took place a *New York Herald* correspondent in Salt Lake City reported the forthcoming event and said, "I hear that some few Gentiles are to be invited; if so, you will be furnished a report of the great day."[87] Apparently the invitation was given, for the celebration was reported at great length in the New York newspaper. The reporter obviously was not in tune with the Mormon beliefs, but his enjoyment of the event showed through his thinly veiled sarcasm. Mother Nature got his attention, however, and he reported:

> "...a most terrific thunder storm burst over the mountains and scattered everybody to shelter. It was a terrible shower while it lasted, forcing the rain through double thick wove covers and tents. The ex-Governor and family were at table at the time under a temporary bowery, and were forced to get up and run for shelter. I passed shortly after the storm, and saw the loaded table completely drenched. So unaccustomed to rain, there was poor provision made for shelter, and probably very few escaped a complete drenching. When the weather cleared up every family seemed to have clothes to dry, and the camp was for an hour or two like one great bleaching field.[88]

He also reported occasional showers the next day, "enough to dampen feelings as well as clothes," but the merry dance continued.

Wilford Woodruff, an ardent fisherman, was among those who went up the canyon a day early. He spent the next morning fishing in the lake, where he caught eighteen trout. Many others also were fishing and numerous trout were caught. Woodruff noted the hard rain in the afternoon. "Most of the people in camp were wet as many were not prepared for a rain storm. There was much thunder and lightening."[89] The celebration was marred by one other event. During the descent down the canyon one of President Young's carriages carrying part of his family turned over and landed in the creek. Fortunately, the only damage to the occupants was a few slight bruises and some measure of inconvenience.

*

In 1861, when the first transcontinental telegraph line was being constructed, Salt Lake City was planned to be the junction between the western and eastern divisions of the line. In December 1860 Edward Creighton, superintendent of the Pacific Telegraph Company, which was building the eastern portion of the line, visited Salt Lake City to negotiate with Brigham Young to furnish telegraph poles for 250 miles east and the same distance west of the city.[90] Eventually Young received the contract to build the line west to the Ruby Valley in Nevada, while Little and Decker took the contract to build the line east to South Pass in Wyoming. The poles were cut by the Big Cottonwood Lumber Company and hauled to their sites by teamsters. The first pole was set on East Temple Street (Main Street) in front of Livingston, Bell & Company's store on 10 July 1861, and the line rapidly moved in both directions.[91] By the end of August there were about thirty ox and mule teams on the western line, under the charge of Brigham's son John W. and Arza Hinckley.[92] The eastern connection was completed to Salt Lake City and the first message sent on 18 October 1861; the western connection was made on 24 October 1861.[93]

In 1861 Brigham Young took steps to disband the Big Cottonwood Lumber Company. On 5 March Frederick Kesler noted in his journal that he had attended a meeting of the Big Cottonwood Lumber Company in the church president's office where they discussed dividing the canyon, each member having a share and each share controlled by its owner.[94] Why Young chose to do this at this time is not known; perhaps he had bigger things to occupy his time and felt the sawmills could continue operating under private ownership. Although nothing more happened that year, early the following year a balance sheet was prepared for the company, showing the profit over its seven years of operation to have been $36,431.72, and that sum was distributed to the six partners pro-rated according to the capital they had invested. Charles Decker was no longer listed as an individual partner, his name having been replaced by the firm of Little and Decker. The company's property was valued at $40,728.39, including the five mills, the canyon road, houses, and other buildings. This sum, too, was distributed among the partners.[95]

At a meeting of the company on the first day of March Brigham Young again proposed dividing the canyon among the six partners. Frederick Kesler wanted to sell his interest, which was calculated as $2,607.06. He was a millwright who wanted to build mills, not operate them. On the other hand, Brigham Young wanted mills operating and wouldn't agree to Kesler taking his share in cash. The Big Cottonwood Lumber Company was dissolved and the canyon was distributed among the other partners, with Wells taking Mill E, Little and Decker taking Mill D, and Brigham Young taking Mills A, B, and C. Neither Kesler nor Smoot took part in the distributions within the canyon, although Smoot took at least some of his share in lumber, which he used on his farm.[96]

As for Kesler, Brigham Young had another plan. At the church conference in October 1861, at a time when the Gardner brothers were still running their flour mill on the Jordan River at Big Cottonwood, Robert Gardner was called as one of 309 men to relocate and establish a settlement in southern Utah in the Saint George area.[97] Gardner concluded to sell his half of the mill. Brigham Young offered to buy it in partnership with Frederick Kesler, provided he could buy all of it. Archibald Gardner was reluctant to sell his half, but after a period of negotiation, Young and Kesler bought the mill for the sum of $13,000 and took possession on 5 November.[98] Kesler moved one of his families to the mill, renamed it the Excelsior Mill, and ran it for the next five years. In November 1867 he sold his half of the mill to Brigham Young for $8,000 and moved to his ranch in the northwest part of the Salt Lake Valley, close to both the Great Salt Lake and the Oquirrh Mountains, where he remained until his death on 12 June 1899.[99]

CHAPTER 3

The Lumber Industry in Big Cottonwood Canyon

AFTER THE BIG COTTONWOOD LUMBER COMPANY WAS DISSOLVED the sawmills continued to operate as before, with most of the same people involved. But the organization was different, for now the mills were independent of one another, although the canyon road was maintained jointly. For a few years, until the mining boom began, there was little activity other than the lumbering operations, and precious little news was published about that.

Mill A continued to run as before, probably by the same men who had run it for the lumber company; but sometime during the next few years, before 1867, it was sold to brothers Richard and Robert Maxfield. It is not known if they had a lumberyard in the valley to dispose of their product, but they did do some business with the county in furnishing bridge timbers and piles.[1] Then in June 1872 they sold the mill to one William Howard, a resident of the Cottonwood section of the Salt Lake Valley, for the sum of $8,000.[2] At the time the Big Cottonwood Lumber Company was dissolved, the mill was valued at $9,595.84, but that represented all its costs, the total amount invested in the mill over the years of its existence. It is not known what the Maxfield brothers paid for it. At the time of their sale the mill property also included a blacksmith shop and a dwelling house. Howard held the mill for less than two years, during which time several additional houses were built, before he sold it to Philander Butler.[3]

Butler was the eldest of six brothers—Philander, Leander, Alma, Alva, Neri, and Eri—and one sister, Miranda, who arrived in Utah Territory with their father, Samuel, in September 1857.[4] They settled in South Cottonwood, but within a year Samuel and Miranda moved on to California, leaving the boys behind. Four of them appeared in the 1870 census, Philander as a blacksmith, Eri and Leander as teamsters, and Alma as a laborer, all living at Big Cottonwood.[5] In June 1871 the county court granted Philander a license to sell liquor at the mouth of Big Cottonwood Canyon, where the brothers had settled, and one year later it approved another license for him to make and sell beer.[6] It was reported that P. Butler had a fine brewery and was doing a lively business, and a later newspaper article revealed that he also had a hotel there.[7] By mid-1874 the area was known as Butlerville, and by the following year a post office had been established there.[8] Among the other settlers in the new community was the McGhie family, consisting of four brothers and a number of sisters. There is a tale that when the election was held to choose a name for the settlement there were more Butlers than McGhies, so the name became Butlerville.[9]

The Butlers had operated Mill A only about one year when, at 5:00 a.m. on 18 March 1875, a gigantic snow avalanche came rushing down the denuded slopes, crushing the mill and all the buildings surrounding it. Daniel Caine, with five members of his family and a boarder, were sleeping in one of the houses. Caine managed to get out of the wreckage and went to the Dolly Varden Mine for help. Several miners responded and all victims were soon extricated, shaken and bruised, but alive.[10] However, the sawmill was destroyed. Only its name survived, carried on to this day in Mill A Gulch on the north side of Big Cottonwood Canyon and Mill A Basin at the head of the gulch, cradled under the east slopes of Mount Raymond.

Daniel Caine was in the news again a year later when he left his home at Argenta, not far from the wreckage of Mill A, to go to Silver Springs farther up the canyon. That afternoon a man working on the canyon road saw a body in the stream, which was running very high and swift, but was unable to recover it. The body was found the next day; it was Caine, who had fallen into the stream and drowned. He had only one arm, the other having been lost when a car, presumably a railroad car, had run over it some years earlier. He had come to Utah from Wisconsin about three years before his demise and was engaged in mining in Big Cottonwood. He left a wife and four children.[11]

*

There is no evidence that Mill C was operating or even existed when the Big Cottonwood Lumber Company ceased operations, even though it appeared on the company's closing balance sheet, where it was valued at $5,082.21. Presumably this amount included the $4,000 paid to John Cooley for the mill and his rights to the canyon. But the mill left its name behind: the flat above the Storm Mountain picnic area is known as Mill C Flat.

Mill B, however, was an operating property. Brigham Young placed it in the hands of his eldest son, Joseph A., who hired William S. Covert as sawyer and caretaker, he being paid one-half of all lumber and slabs he made.[12] The fifty-four-year-old Covert had come to Utah in 1848 with his wife and five children. Covert apparently ran the mill for many years; as late as 1871 Mill B was also known as Covert's mill.[13] After the mining boom began Covert and his family took part in it by filing a number of claims in the canyon, many of them in the immediate vicinity of Mill B.[14] Covert's offspring went on to file many more claims in the mining district in the thirty years that followed.

When Joseph A. Young went on a trip to New York in 1864, Brigham Young had George W. Thatcher run the mill, which, as he wrote, "frees my mind of that much care in business matters, and gives br. George an excellent opportunity for increasing his means and his experience in conducting affairs."[15] Twenty-four-year-old "br. George" was Brigham's son-in-law, he having married Luna, Joseph A.'s sister, three years earlier. He would marry another of Brigham's daughters, Fanny, two years later. How much experience George Thatcher derived from his association with Mill B is not known, but he went on to a distinguished career in banking and railroading. When Joseph A. Young returned from his travels, he took over the mill again and became a recognized lumber merchant in the city. He later added to his holdings a sawmill at Lambs Canyon and another in North Mill Creek Canyon in Davis County. He had a lumberyard on South Temple Street, directly south of Temple Square, where he did business until the coming of the railroad changed the face of the Utah lumber business. However, in 1869 Mill B was no longer listed as one of his sawmills.[16] Apparently the mill was idle for a number of years. In 1874 it was reported as in the process of being repaired, with materials for that purpose being on the ground.[17] The next season the mill was in operation again.

Travelers in the canyon were soon complaining about the operator sliding logs down the mountain near Mill B. The timber came down with tremendous velocity, according to the complaint, driving rocks and earth before it and obstructing the passage of teams. The bridge across the creek at the mill was being torn to pieces by logs being dragged over it, and the travelers wanted the activity stopped.[18] The operator at the time was most certainly Richard Maxfield. He had sold Mill A in June 1872 and must have negotiated to buy Mill B immediately thereafter. He ran it for several years before selling it to Archibald Gardner for $3,000 in March 1877. Gardner immediately sold half of it to David Brinton of Big Cottonwood for $1,500.[19] Gardner had recently closed his mills in Mill Creek Canyon, but it is not likely he spent much time at Mill B, for he had several other sawmills to attend to, including those in Little Cottonwood Canyon he had recently acquired or built. He held his interest in Mill B only until January 1879, at which time he sold it to Alva Butler, brother of Philander Butler, who owned Mill A at the time of its demise.[20] How long Butler and Brinton operated the mill is not known, but they held the property until 1894 when, through a convoluted series of transfers, the mill and its rights fell into the hands of the Big Cottonwood Power Company. With that action Mill B was permanently out of use for the lumber business, although it probably had not operated for some years previous. The mill itself apparently remained as a reminder of the old lumbering days, however: a mining claim in January 1904 was located relative to Mill B, implying the structure still stood at that time.[21]

*

The only result of the change of ownership of Mill D, at least to the outside observer, was the establishment of a new lumberyard in downtown Salt Lake City. It was opened by Feramorz Little on the corner of Second East and First South Streets, adjacent to his home. In the canyon it was business as usual. It is not known if Charles Decker was active in the canyon after this time, although he continued to be involved in a number of enterprises in Salt Lake City. In 1864 he brought the first steam sawmill into the territory and set it up in Lambs Canyon.[22] The steam sawmill featured a portable circular saw and a steam engine. The first consisted of a wood or cast-iron frame with a suitable mechanism to hold and control the saw, and a track on which rode a carriage for the log. The steam boiler and engine were connected to the saw with a large belt that rode on pulley wheels on each of the major components. Steam sawmills could be very efficient; they worked well and could be relocated easily and quickly when the need arose. They had been used for many years in the eastern states but were relatively unknown in Utah, primarily due to the difficulty of transporting the bulky and heavy equipment. However, by 1864 the transcontinental railroad was pushing westward and the distance between the railhead and the Salt Lake Valley was gradually growing shorter. An experienced hand at moving heavy loads with wagons and teams, Decker undertook to bring the first steam sawmill into the valley. His was quickly followed by others, with Joseph A. Young bringing two to supplement and eventually replace his Mill B operations.

Charles Decker went into business with one Zenus Evans and sold lumber through Latimer and Taylor, a Salt Lake City partnership making sashes and doors. That arrangement continued at least through 1867.[23] He also operated a saw and planing mill at or near his residence in the Eighth Ward, a venture that came to an abrupt end on 23 June 1868 when the shop caught fire and everything except two chests of tools was destroyed.[24] The following year he was listed as a lumber dealer, working at the corner of Sixth South and Fifth East Streets. In September 1873 he was the proprietor of the Sandy Railroad Hotel, a restaurant and hotel at Sandy Station. That venture lasted several years until it too was a victim of fire. On 14 April 1880 a fire started in the restaurant of the Decker House. Fanned by strong winds, it not only consumed the hotel but spread to the railroad station, which also burned to the ground.[25] Decker later settled in Parleys Canyon, but there is no evidence that he had any further association with Mill D or the Big Cottonwood lumber industry.

In the spring of 1862 a young man by the name of Francis Armstrong approached Feramorz Little for a job. He had emigrated from England to Canada with his parents in 1851. Eight years later, at age nineteen, he left his family and came to the United States. He found his way to Richmond, Missouri, where he found a job in a sawmill. He worked there until the spring of 1861, when he left for Utah, arriving in Salt Lake City in September. After working at a number of odd jobs during the winter, he tried to get back into the lumber business by seeking a job at Mill D. A story is told that when Armstrong approached Little for a job, Little told him that all his men paid tithing. Armstrong answered, "I'm not a Mormon, but I'll pay tithing."[26] It may be amusing, but is only a story, for Armstrong had joined the LDS Church while in Missouri, which was why he left there and moved to Utah.

While Armstrong continued to work at Mill D, Feramorz Little became more and more involved in the business community in the city. In March 1864 he became proprietor of the Salt Lake House on the east side of East Temple Street between First and Second South, and he ran that establishment throughout the decade of the 1860s. By then he had determined to leave the lumber business, so in 1871 he sold Mill D to Francis Armstrong for the sum of $21,000.[27] Armstrong immediately formed a partnership with Charles S. Bagley, and Mill D continued to operate under the firm of Armstrong and Bagley.

Charles Steward Bagley was a native of New Brunswick, Canada, where as a young man he was engaged in the lumber business. In 1855 his parents and family, who had joined the Mormon Church some years before, moved to Salt Lake City. Almost immediately Bagley joined the Big Cottonwood Lumber Company, doing work on the canyon road and at Mills A, D, and E. He later worked for Feramorz Little at Mill D and now joined Armstrong as a joint owner. The two men took over Little's lumberyard in the city, which they used as an outlet for their lumber until 1875, when they moved to a larger yard west of the downtown area on First South Street.[28]

*

After Daniel H. Wells assumed ownership of Mill E, he opened a lumberyard adjacent to his residence at the corner of South Temple and East Temple Streets, where he marketed his lumber as long as he ran mills in the canyon. He also undertook the construction of Mill F, which Frederick Kesler had located and designed in 1857, before the Utah War upset plans for its addition to the Big Cottonwood Lumber Company's properties. In 1863 Wells sent a bill of materials prepared by Kesler for the new mill to St. Louis, and in the spring of 1864 construction began in earnest. However, at the end of April 1864 Wells left for Liverpool, England, to replace George Q. Cannon, president of the LDS British Mission, who had become ill and was returning home. Supervision of the mill construction was left in the hands of Wells's clerk, James Jack, and his foreman, Arza Hinckley. Hinckley immediately approached Feramorz Little and negotiated a trade of timberland within the canyon. When the canyon was divided between the mill owners, Little was given everything on the north and east sides of the stream in the upper canyon, while Wells was given the south and west sides. However, when Kesler had selected the location for Mill F, it was planned on the basis of timber reserves found on both sides of the canyon. In the trade, Little was given all the timber in the main canyon below Mill F in exchange for the side canyon to the east of the mill, the one that became known as Mill F East Fork.[29] While this gave Wells a large amount of timber convenient to the mill, it also became important several years later in 1873 when Armstrong and Bagley brought a steam sawmill into the canyon and set it up at Silver Springs, drawing on timber in Silver Fork and on the surrounding south slopes of the main canyon.[30]

By midsummer 1864 there were ten carpenters and millwrights at work on Mill F, and a number of woodsmen were hauling logs to the site, for both its construction and its operation. The machinery was set in motion and the first log sawn on 13 September. By the end of the month the mill was in full operation, even though the machinery was not completely enclosed. The operators expected to saw about one million feet of lumber the next season. The expense of the mill to that time was $13,620.[31]

Arza Hinckley had hired one Henry H. Davenport, an emigrant sawyer from Iowa, to run the mill for the rest of the season. In January Hinckley and Davenport entered into an agreement whereby the latter would lease and operate Mills E and F during the following season. Davenport did serve as sawyer in 1865, but the snowfall the previous winter was so heavy that it had collapsed the roof of the new mill.[32] Due to damage to the mill he cut less than half of what had been expected. Although the mill could run without a roof, Arza Hinckley arranged for Nelson Wheeler Whipple to go up and make other necessary repairs.

Whipple, a native of New York state, came to Salt Lake City in 1850 when he was thirty-two years old. He was a member of one of the first parties to enter the valley by way of Big Kanyon (Parleys Canyon). For the next ten years he worked at various mills in the territory, including brief periods at Mill D. He built and operated two different mills and shops at Mountain Dell in Parleys Canyon, then took a job with Hinckley to make repairs to the "Esquire mill," as he called Mill F, referring to Wells's title from earlier days in Illinois. Whipple arrived at Mill F with his family on 10 July 1865. He had no formal education and yet he was a prolific writer. While his spelling was poor, he managed to get his thoughts across very well. The next morning he wrote, "we found ourselves in high cool country surrounded with a most butifall Ceinery [*sic*] of thickly timbered hills on every side of every shade of green from pale to the deep evergreen of the mountain pine." He built a shanty for his family, then set to work repairing the mill and changing the gearing. In its previous operation it had been found to have been geared too high. After the repairs were completed he went down to Mill D to do some work for Feramorz Little. However, he wrote, he felt inclined to put a new roof on the "Esqr. Mill," so he agreed with Hinckley to take the job for $350, to be paid mostly in lumber. He described the mill, saying it was "as good a piece of mechinry as could be made out of the timber in this country, the mill was one hundred feet long about 33 ft wide and 3 stories high the roof off and it lying entirely exposed to the weather." He and two of his sons cleared away the wreckage of the old roof, cut and raised thirty logs, twenty-five feet long and a foot in diameter, for rafters, then covered the roof with 7,000 feet of lumber. They did the job in two weeks and one day.[33]

In the previous year, 1864, construction was started on the new tabernacle on Temple Square in Salt Lake City. The first tabernacle had been built in 1851, but a rapidly expanding population had outgrown the building. The new tabernacle, 250 feet long and 150 feet wide, would seat nearly 9,000 persons. The roof was to be self-supporting, with no interior pillars. To do this it was constructed with many semicircular arches, giving the finished building the appearance of an oblong hemisphere. The structure required an enormous amount of lumber, placing a heavy burden upon the local sawmills. The roof, which had an area approximately 20 percent greater than a regulation football field, was to be shingled. To prepare for this job, Brigham Young asked Whipple to go to Mill F and saw 300,000 shingles. Accordingly, he went up Big Cottonwood Canyon early in July 1866, but when he got to Mill F he wrote that "the man who was runing that mill (who was a Gentile) much oposed to my puting in the shingle Saw thier." The man was none other than Henry Davenport, who had leased the mill for the new season. Whipple's parenthetical comment about Davenport being a gentile (non-Mormon) was indicative of his firm devotion to his religion and his distrust of those who did not share his faith. Years later, when Daniel Caine fell into Big Cottonwood Creek and drowned in 1876, Whipple noted in his journal that Caine was not a member of the church and was "a bitter enemy of the Mormons, often cussing and abusing the position of Brigham Young, etc., etc.." He then added, "Pity a good many more of that stripe had not fallen in at the same time."[34]

After his confrontation with Davenport at Mill F, Whipple went up the canyon to Mill E, which was standing idle at the time, and installed his shingle machine there. He had to build a new waterwheel and a frame for his machine, so he was not able to start sawing shingles until the end of July. Bishop Alexander McRae of the Eleventh Ward came up with a team to haul logs to the mill and shingles to the city. Brigham Young thought the shingles should be delivered at twelve dollars per thousand, so Whipple and McRae agreed to split the proceeds. Between 26 July and 20 October 1866, when heavy snows drove them to lower elevations, Whipple sawed 270,000 shingles, which his sons packed and McRae delivered. The following season, after finishing his spring planting and making repairs to his home, Whipple returned to Mill E to complete the job, sawing 192,000 shingles for the church and another 50,000 for himself.[35]

During the summer of 1867 Whipple also provided board for four of D. H. Wells's men who were rebuilding Mill E. They probably installed a circular saw at this time; as early as October 1864 Arza Hinckley had recommended that the mill be given a new wheel and circular saw, as there was timber above the mill to last for several years.[36] Wells was in England at that time and didn't return for another year, on 7 October 1865. Now, in 1867 he was personally taking control of his mills again and preparing to operate them in a serious manner. In June of the previous year Wells had petitioned the county court for a grant to Big Cottonwood Canyon in an attempt to formalize his rights and privileges there.[37] The request placed the court in a difficult position, for it had given the entire canyon to Joseph Young in 1852. Yet a request from Daniel H. Wells was not one to be ignored. Since January 1857, when he was appointed second councilor to President Brigham Young on the death of Jedediah M. Grant, he had been a member of the First Presidency of the LDS Church. And shortly after his return from England he was elected mayor of Salt Lake City, a position he would hold for four terms. No action was taken on the request for many months, although there were surely discussions behind the scenes. Finally, on 5 March 1867, Brigham Young wrote an ambiguous note stating, "This certifies that I have sold and disposed of all my right and title to Big Cottonwood Kanyon above the mouth thereof to parties now occupying the same."[38] While he signed the note, it was not addressed to anyone. Yet, the same day the court addressed the request again. "In consideration of the large amount of money and means already expended and to be expended in future in making roads & bridges and in the erection of mills & machinery & keeping roads in said kanyon, &c, &c," the court minutes read, "The exclusive right & control of said kanyon is granted to Danl. H. Wells, F. Little, Seymour & LeGrand Young & Richard Maxfield . . ."[39] The grant also allowed them to charge a toll of seventy-five cents for single teams and $1.50 for double teams, except during the months of January, February, and March. Wells, Little, and Maxfield were the operators of Mills E and F, D, and A, respectively, at this time. Seymour and LeGrand Young were sons of Joseph Young, who held the original grant to the canyon. At the end of 1862 Brigham Young had placed all his mills in the hands of his son Joseph A., who hired his twenty-three-year-old cousin LeGrand Young to look after the City Creek mill.[40] In 1867 Seymour and LeGrand Young were listed as lumber dealers in the Seventeenth Ward, on the east side of Second West between North and South Temple Streets.[41] How serious this business was is not known; it is possible the brothers were operating Mill B at that time, but their serious involvement could not have lasted long, because shortly after this time Seymour went on to study surgery and became a prominent physician and surgeon, while LeGrand studied law and became equally prominent in his chosen field.

By the end of 1867 Mill E had joined Mill F in producing lumber again.

They were reported to have provided several hundred thousand feet of finishing lumber for the new tabernacle.[42] In 1871 Mill F was threatened by an enormous forest fire that consumed much timber in the lower part of Silver Fork and upper Big Cottonwood Canyon, extending as far up as Mill E. It was said that it began when some loggers had tried to destroy a yellow jackets' nest by setting fire to the brush around it. A visitor to Alta and Silver Fork commented about the mountaintops being shrouded in smoke. When he traveled across the divide and down Silver Fork he described the view bounded by great swaying sheets of flame. Miners were felling timber and starting backfires to save their mine buildings, the small community of Silver Springs, and Mill F. It was reported that several times the stack of lumber at the mill caught fire but each time was saved. A change in wind direction slowed the march of the fire, and a subsequent rainstorm quenched the remaining flames. Years later the damage from this fire was still apparent.[43]

Both Mills E and F continued to produce for the next ten years, but great changes were taking place in the lumber industry. The coming of the railroad in 1869 provided relatively cheap and easy transportation, allowing lumber to be brought in from distant sawmills. New lumberyards appeared in the city to challenge those offering lumber from local sources. The railroad also brought another, less obvious change that challenged the right of local sawmill operators to cut timber in the vicinity of their mills.

Typically, the establishment of a territory was followed within a short time by the creation of a local office of Surveyor-General of Public Lands and a U.S. Land Office to make public lands available to settlers. However, that did not take place in the case of the Territory of Utah. While Congress provided for a Utah Surveyor-General in 1855, some five years after the territory was established, it did nothing to provide for a Land Office.[44] In spite of requests over the years by local and federal officials, both Mormon and gentile, a Utah Land Office was not created until 1868.[45] The Pacific Railroad Act of 1862 provided for a land grant to the railroad company of ten sections, 6,400 acres, for each mile of railroad built. In 1864 this was increased to twenty sections as an incentive to the builders. Alternate, odd-numbered sections were given to the railroad company, while even-numbered sections were kept by the government. This is why many maps, such as those showing national forests, exhibit a checkerboard pattern along the path of the first transcontinental railroad that extends many miles to either side of the railroad right-of-way. As the railroad approached Utah, it was imperative that the territorial Land Office be established to assign the railroad lands. But, once created, it also faced the tremendous task of taking charge of all public lands as well as making them available to settlers. Part of taking charge included the protection of government timberlands. And the laws made no provision to allow government timber to be cut and used for profit or resale. Whereas previously the county court had authority to grant great tracts of timber to individuals, that authority now was usurped by the United States government, and it did not recognize earlier grants by the county courts.

At first the local Land Office was too busy or had insufficient staff to pay attention to area timber depredations. And when it finally added a timber agent to its staff, he could do little more than publish notices of what was and was not allowed relative to cutting timber. After all, he had the entire territory under his jurisdiction and little or no resources to go into the field to enforce the laws. What finally forced the issue was antagonism between the gentiles and Mormons. It started in the pages of the *Salt Lake Tribune*. By 1874, the newspaper's fourth year of existence, the *Tribune* had developed into a rabidly anti-Mormon publication. Its managers had an intense dislike for Daniel H. Wells, who held great power and influence in the city due to his holding the office of mayor of Salt Lake City and his position in the First Presidency of the Mormon Church. The paper seized upon his operations in Big Cottonwood Canyon to attack him but could do little other than publish accusations and innuendos. When it wrote, "It can be clearly proven that Wells, and others in Utah, have been using the Government timber for speculative purposes...," it was correct, but its editorial finger was pointed straight at Wells.[46] A month later the paper read, "It is high time ... to bring the pirates of the canyons to justice, and among them the big rascals of the hierarchy."[47] When a heavy snowfall closed the canyon for the winter, the editor printed, "Government timber is safe from the depredations of the old Church robbers for four months."[48] In January and February 1875 Mayor Wells sold $464 worth of lumber to the city for use at the bath house and city cemetery, a fact that only added fuel to the *Tribune*'s fire.

With a change of personnel at the Land Office the *Tribune* gained a new ally. The Land Office was typically run under the management of two men, called the register and the receiver, both being political appointees. The register, who was nominally in charge, maintained the records of lands surveyed and sold, while the receiver was responsible for the financial transactions. The men occupying these positions never stayed long, either leaving after using the job as a stepping-stone to better opportunities, or being replaced due to political changes in Washington. The new Land Office register was one Col. Oliver A. Patton, a distant relative of U.S. President Ulysses S. Grant. The *Salt Lake Herald,* noting the register's long black hair, called him a "long-haired swell cousin of the government."[49] Patton was a vain and boisterous man who arrived with a strong premeditated dislike for Mormons, and he immediately became the hero to the *Tribune* and almost the devil reincarnate to the pro-Mormon *Herald* and *Deseret News.* Perhaps encouraged and inspired by the *Tribune,* Patton rallied behind the newspaper's attacks on Wells and charged valiantly ahead. On 19 June 1875 Patton and U.S. Marshal George Maxwell went up Big Cottonwood Canyon and seized all logs and lumber at Wells's mills "as forfeit to the U. S. for stumpage due on timber feloniously cut upon the public domain."[50] The confiscated property was to be sold at public auction.

In a letter of instruction to the local Land Office in 1870 the timber laws were reviewed and the methods to be followed in enforcing them outlined. It authorized officials to compromise with the guilty parties, allowing them to pay expenses and a reasonable stumpage fee, not to be less than $1.50 per thousand feet. Only when such compromise could not be made should the timber be seized and sold at public auction. The instructions created a curious situation in which the Land Office could not permit the cutting of timber on public lands under any circumstances, with or without payment of fees, but if someone did illegally cut timber, the officials could then sell that which was cut to the guilty party, and in fact they were instructed to do so. Only if the guilty party refused to buy the timber he already held could it be seized and sold at public sale. In time the payments were considered to be stumpage fees, while in fact they were fines for breaking the law. Despite the instructions, the going rate for lumber was one dollar per thousand board feet. The instructions creating this confusing situation were reaffirmed in July 1874.[51]

Whether there had been an attempt at compromise before the seizure is not clear, but, after notice of the impending sale of the timber was published, an attempt by Wells to compromise the situation resulted in Patton declaring he expected to collect at the rate of $2.50 per thousand feet not only on the lumber that was seized but on all that had been taken in the past, presumably since the formation of the Big Cottonwood Lumber Company.[52] Wells chose to fight the matter in the courts. After the lawyers for the two parties got together, Wells posted a bond in the amount of $8,000, putting a halt to the sale of the timber. The matter languished in the courts until 13 July of the following year when the charge against Wells was quietly dismissed.[53]

As a result of the events of the preceding year, the *Tribune* started using a new name for Daniel H. Wells. It had often mentioned the "pirates" who were stealing government timber, but now it focused upon a minor physical impairment that afflicted Wells. He had a slight cast to his left eye, which may be seen in some of his portraits. It may have caused sufficient annoyance to cause him to wear an eye patch on occasion, although no portraits showing him wearing one have come to light. The *Tribune,* however, seized on this strabismic impairment when it branded him the "One-Eyed Pirate of the Wasatch," a title it continued to use for nearly two years.[54]

Colonel Patton took steps to avenge his defeat in court. In August 1875 he had appointed a deputy timber agent for Big Cottonwood Canyon alone to keep the pressure on the sawmill operators there. This agent assessed stumpage tax and agent's fees every month. On some occasions, it was stated, the fee amounted to more than the tax.[55] In August 1876, however, the agent failed to make an appearance. On 9 September Patton again seized the lumber at Wells's mills, charging nonpayment of stumpage fees.[56] He scheduled another public sale for 25 September. Wells immediately filed a suit of replevin to halt the sale, posting a $9,000 bond to secure it.[57] The register was given five days to file a counter bond but did not do so. The *Herald* reported Patton was unable to find anyone to sign for his bond. "Men of means," it said, "are not becoming liable for money to suit the whims of Oliver A. Patton, even though he may be the last of the president's eternal line."[58]

With this latest turn of events Wells assumed that the entire matter was closed and ordered his men to resume hauling lumber into the city. Timber Agent Richard Greenway reported this activity to Patton, who became furious and swore revenge. He went to Camp Douglas and asked the commanding officer, General John E. Smith, to provide a force of men to hold the lumber until it could be sold. The general refused the request, explaining that civil authority must be exhausted before the military could act. Patton then telegraphed the Land Department in Washington, claiming the territorial courts had failed to enforce the law. He also swore out a warrant for the arrest of Wells, charging him with stealing timber from public lands. The warrant was served and Wells was released upon posting a $1,000 bond.

Patton, having lost all reason in the matter, organized an armed posse and headed into Big Cottonwood Canyon to conduct the advertised sale of lumber. Meanwhile Wells went back to court, where Chief Justice Michael Schaeffer granted a writ enjoining Patton from going ahead with the sale.

While deputies started toward the canyon to serve the newly issued order, Patton and his men arrived at Wells's lower mill. They already had one man in custody, a teamster who was going down the canyon with a load of lumber. They forced him to unload his wagon and go back up the canyon with them. At the lower mill they found four more teamsters sleeping in a room. They were awakened and handcuffed, even though three of them were boys aged fourteen to eighteen. The posse then went to the upper mill where they accosted three more men, who were shackled and forced to march to the lower mill. There they met the deputies who served the restraining order forbidding the sale. The *Herald* and *Deseret News* reported that Patton swore and said the court had no right to grant the writ and that he would teach the court its business. The *Tribune,* still trying to support him, wrote that Col. Patton declined to respect the order of the court. At 10 o'clock on 28 September Patton sold the lumber to one John Nichols, a miner from Cottonwood. Having accomplished his goal, he placed his handcuffed prisoners, all eight of them, into a wagon and took them to town, where the court immediately released them on $500 bail each.[59]

From this point the battle moved into the courtroom, but it became increasingly confusing. The court asked Patton to show cause why he should not be punished for contempt in disregarding the order enjoining him from selling the lumber he had seized. His attorneys argued that he and Greenway, the timber agent, were acting as United States government officers, and in that capacity the court had no jurisdiction over them. Attorneys for Wells presented several arguments, claiming the territorial courts were instruments of Congress and of the government, that the Fifth Amendment to the Constitution prohibits taking of any individual's property without due process, and repeatedly stressing that if the arguments of the Land Office agents were sustained they would be beyond the reach of the courts and become a law unto themselves. Patton's attorneys responded that appeals of land officers' actions should not be made through the courts but through the officers' superiors. When the court interfered with the officers' actions, it was creating a conflict between the government and the territorial judiciary.

Chief Justice Schaeffer, finding the simple contempt proceeding rapidly escalating, ruled that his court did have jurisdiction and that the register and timber agent had placed themselves in contempt, something they could easily have avoided by postponing the sale and presenting their arguments to dissolve the injunction prohibiting the sale. But to avoid the appearance of conflict with any other government department, the judge imposed the lightest penalty possible, requiring the defendants to simply pay the costs of the contempt proceeding.[60] Patton then asked that the injunction restraining him from selling the lumber be dissolved, using the same arguments presented in the contempt hearing. While this was a rather belated request, the court allowed his request and quashed the order. Wells's attorneys immediately asked the amount of a bond on appeal. However, since the decision being appealed was based upon the lack of jurisdiction of the court, Patton's attorneys questioned whether the decision could be appealed. It was a thorny question, to be sure. The judge took the matter under advisement.[61]

While the judge was pondering his position, John Nichols, the man who had purchased the lumber at the sale, filed an action in replevin against Wells to protect his investment. Wells immediately posted a counter bond to the amount of $1,800 to indemnify Nichols against loss, thereby maintaining control of the lumber he still considered to be his.[62]

In the midst of all these proceedings the *Herald* printed an account of a man from Evanston, Henry Job, who went to the Land Office to pick up his patent on a tract of government land. He had paid all fees involved for the land but now was required to pay an additional fee of one dollar. While he questioned the propriety of the additional fee, he chose to pay rather than contest it. However, he was told the paper currency with which he tried to pay was not acceptable—only a silver dollar would do. From a description provided by Job, the *Herald* reporter concluded that the Land Office representative was none other than Oliver Patton, the receiver it had often charged as being above the law. It predicted Patton was headed for a fall. Unless there is a serious misunderstanding, the *Herald* reporter wrote, "he of the raven locks and elegant manners [Patton] will soon have to rustle for hair oil with the common herd."[63] Patton responded with a letter to the *Salt Lake Tribune* denying the accusation and branded the writer of the article a "liar, knave and scoundrel." That same day Byron Groo, the reporter who wrote the article, met Patton on East Temple Street and the two got into a scuffle. When friends of Patton restrained Groo, Patton attempted to draw a derringer from his pocket before he was restrained. The two men were arrested and fined ten dollars each for their ungentlemanly behavior.[64]

While this incident may appear to have been inconsequential, it proved to be the proverbial tip of the iceberg. About the same time Patton and Groo were paying fines for their Main Street indiscretions, there arrived in the city a special commissioner from the Interior Department, Maurice M. Kaighn. Acting upon numerous complaints of irregularities in the operations of the local Land Office, the department sent Kaighn to investigate and report on the situation in Salt Lake City. His report, forwarded to the assistant secretary within ten days of his arrival, was of such significance that an immediate order was issued suspending the register and receiver. On 26 October Patton received a dispatch directing him to turn over the books, papers, and money of his office to Utah Territorial Secretary General Moses M. Bane, who in turn was directed to take over temporary operation of the Land Office.[65] Patton fought the suspension, going so far as to telegraph the president to ask for the immediate recall of Kaighn, charging him with conniving with Mormons to secure the removal order. While the *Salt Lake Herald* gloated over the turn of events, the *Tribune* suffered mixed emotions, having discovered its hero had feet of clay. While the paper was Patton's strongest supporter and may well have been the inspiring force behind the register's actions, it now backed away from the disgraced official. It wrote, "We have heartily endorsed Register Patton's vigorous dealing with the boss timber thief, Squire Wells, and only regret that after so meritorious a proceeding we cannot now stand by him in his trouble."[66] And in an editorial the next day it recited the complaints "of arbitrary and tyrannical conduct of the Register, of irregular proceedings, and a studied neglect of the public interest." It went on to brand Patton's actions as "a reproduction of Grantism on a small scale." It even went so far as to suggest that Patton held the belief that he was master and not a public servant, allegations it had ridiculed when they were made by the *Herald* and *Deseret News* in the past.[67]

The dismissal of Register Patton came as a relief to many people, certainly including Chief Justice Schaeffer. The court case between Wells and Patton appeared to disappear with the disgraced official, and well it might, for its purpose was to prevent the sale of the lumber, something it failed to do, and it had since been replaced with another action between Nichols and Wells. But that case quietly languished in the backlog of court cases until late the following spring when a jury awarded John Nichols $1,800— the amount of the bond that had been posted.[68] Certainly Daniel H. Wells was severely injured by the proceedings of the previous year, but so too was the entire lumber industry in the Wasatch. While dealings between the operators and the timber agent became more businesslike, gone were the days when the operator was in total control of the timber in his part of the canyon. The *Tribune* couldn't resist one last swipe at Wells's operations when it wrote, "The One-Eyed Pirate is viewing the wreck of timber along the Big Cottonwood Canyon."[69] Its sarcasm was simultaneously the poignant truth.

Wells was delivered another blow on 29 August 1877 when his long-time friend, confidant, business associate, and spiritual leader Brigham Young died. In spite of all the hardships, he continued to run the mills for a few more years, but their days were numbered. In 1879 Wells was still listed as a lumber dealer, but his operation closed some time after that.[70] In August 1882 L. L. Despain wrote to Wells saying he understood Mill F was not running and asked to lease it for the rest of the season and the next year.[71] It is not known if Despain did in fact run the mill, but it was likely sporadically run by lessees for a few years. By the end of the decade it was referred to as "old Mill F" by one prospector.[72] Its presence remained for another decade, when the site was referred to as "what was known as SawMill F."[73]

*

Although Daniel H. Wells was personally attacked by the Land Office Receiver in 1875 and 1876, Armstrong and Bagley were also affected by the changing situation. Their problems, however, began with the canyon road. When the Big Cottonwood Lumber Company was operating, the company took care of the road. However, after it ceased operations and the canyon was divided among the new mill owners, it became their joint responsibility to maintain the road. The only people using the road above Mill A were the operators of Mill D and Mills E and F, the former being first Feramorz Little and then Armstrong and Bagley, and the latter being Daniel H. Wells. At first this was not of much concern, but after the mining boom began the road received much more traffic and required more maintenance. In the spring of 1872, for instance, the recorder for the Big Cottonwood Mining District, H. C. Hullinger, reported that the canyon road was in terrible condition: high water had torn down embankments and floated away bridges, but, at at the present rate of progress by Messrs. Armstrong, Bagley and Maxfield, the latter still in possession of Mill A, the road would be open for teams in another week. In the interim the Maxfield brothers used a pack train to haul supplies to Silver Springs.[74] The following year Hullinger reported that Armstrong and Bagley deserved great credit for their energy and perseverance in making the roads safe at such an early date.[75]

There were other pressures on both the road and timber reserves due to the Camp Douglas appetite for firewood and construction timber. In September 1874 it was reported that the contractor supplying the camp had seventeen teams hauling wood out of the canyon and that if things continued at that rate much longer it would be almost impossible for teams to get up or down the canyon.[76] From all indications it appeared that Armstrong and Bagley were doing more than their fair share of work to keep the road open. Apparently they felt so too, because in 1875 they let it be known that they were moving their sawmills out of the canyon.[77] Why they decided to abandon their mills is not known, but it may well have been a result of the land register's attempts to enforce the laws prohibiting the cutting of timber from public lands. Indeed, in July 1875, shortly after the lumber and timber at D. H. Wells's mills was seized by Col. Patton, Armstrong and Bagley had paid the U. S. Land Office forty dollars for 40,000 feet of timber.[78] Since the company had kept up the roads and bridges, principally at its own expense, the owners planned to remove some twenty bridges they had constructed over the stream and salvage the lumber that they contained. This provided cause for concern, especially within the mining community. Patton ordered his deputy timber agent to seize the bridges; since the timber used to build them was cut from public lands, he contended that the bridges were government property. The threat and the resulting action may have been concocted by the news media, since nothing much happened on either side except that a toll was again charged for all teams using the canyon road. And while Armstrong and Bagley did not abandon the canyon immediately, the company did consolidate its operations by moving its steam sawmill to Mill D from Silver Springs.[79] The move may have been necessitated by a lack of timber at the old site. This was, after all, the area that had been ravaged by the big forest fire of 1871. Although the mill had been operating for only two years, it had all but stripped the adjacent slopes of timber. More than two and a half decades later, in July 1902, Albert F. Potter was making a survey of forests on government lands in Utah. He visited both Big and Little Cottonwood Canyons and reported, "At Silver Springs a sawmill has been operating which cut everything off the point south. There is a fair reproduction of young trees coming in."[80]

Armstrong and Bagley continued to operate for a few more years. In 1879 Francis Armstrong still had his lumberyard in the city, but by this time he had bought out one of the partners, Wm. H. Folsom, in Latimer, Taylor & Co., a company formed in 1864 with Charles F. Decker, one of the original operators of Mill D. Decker was no longer with the company, having been replaced by Folsom. In October 1881, upon the death of Latimer, the remaining partners purchased his interest and changed the name of the firm to Taylor, Romney and Armstrong.[81] By this time Armstrong had closed his mills in Big Cottonwood Canyon and his lumberyard in the city, the latter being taken over and operated by Col. Elijah Sells. The disposition of the steam sawmill is not known, but several years earlier there were reports of plans to ship it to the Black Hills when that area was opened to settlers and miners.[82] Mill D's remains were used as a reference point for prospectors as late as 1901.[83] Both it and Mill E were shown on the U. S. Geological Survey's Cottonwood Special Map, dated 1907. The map was based upon surveys made in 1903 and 1905, so some of the mills' detritus must have been visible at that late date.

*

In 1867, while Nelson Wheeler Whipple was sawing shingles at Mill E, he learned through Francis Armstrong that Feramorz Little might be amenable to selling a piece of timberland across the basin from Mill D. Whipple did come to terms with Little, and after he finished sawing shingles he spent a month at his new acquisition building a cabin and cutting timber for a mill.[84] The next spring he and his sons began building the mill. The biggest job was digging the pit and races. He gave the dimensions in his journal, writing that the headrace was 43 rods (over 700 feet) long, two and a half feet deep, six feet wide at the bottom and eight at the top. The pit for the waterwheel was fourteen feet long, eight feet wide, and ten feet deep. The tailrace, carrying the water away from the mill after it passed over the wheel, was 10 rods (165 feet) long and from one to seven feet deep. The mill itself was thirty feet long and fourteen feet wide. Building it took much of the summer. The workers had the shingle machine running by mid-August and the lumber saws on 1 October. Still, they were able to saw 100,000 shingles and 115,000 feet of lumber before the season closed. In keeping with the spirit of mill names in Big Cottonwood Canyon, Whipple named this one Mill G. He ran it for several years before selling the sawmill and shingle machine to Philander Butler for $3,000 on 3 December 1872.[85] The location of this mill has not been definitely established, but it is believed to have been in the southwest corner of the basin, on Cardiff Fork stream. There is a faint road on the northeast flank of Kessler Peak, running between the Kessler Peak North Trail and the north face slide area, that the U. S. Land Office surveyor in 1902 identified as an "old wood road."[86] It may well be that this road was used to bring out logs for Whipple's Mill G.

Whipple ran the mill for Butler in 1873, during which time he modified it and installed a twenty-three-inch turbine wheel, and for part of 1874, the year Butler bought Mill A from William Howard. In March of that year Butler mortgaged most of his properties, including Mill G, in favor of Howard to finance the purchase of Mill A.[87] The mortgage was canceled in August of the same year. Meanwhile Whipple thought there was enough timber in a fork farther down Big Cottonwood Canyon to justify building a small mill there. Accordingly, he quit Butler in September 1874 and moved down to his new site. It is not known how long Mill G lasted, or if it even ran after Whipple left. The mill left little lasting impression, although on some older maps the flat at the mouth of Cardiff Fork is listed as Mill G Flat.

Whipple's new site was located about midway between Mill C Flat and Mill B, at a place he called Gravely Hill. Immediately upon moving there he and his sons built a cabin, ten by fourteen feet—small, but good and warm, he wrote—then constructed a dam, race, penstock, and flume, and began the mill itself. It was sixteen feet wide and forty-four feet long, and it was nearly ready to begin operating by the time they quit for the winter, on 1 January 1875. The next spring they finished the mill and built a road up the canyon to the north, the one known today as Whipple Fork, and began cutting and hauling logs. This was the year that Land Office Register Patton seized Daniel H. Wells's timber and lumber the first time, but Whipple seemed totally unaffected by it. At least he never mentioned it in his journal. Two years later he did make an entry saying that Timber Agent Greenway had called for stumpage. "I agreed to pay him $15 soon," he wrote.[88]

The winter of 1876–77 was quite mild, so Whipple stayed at the mill and continued working, selling most of his lumber through David B. Brinton's Big Cottonwood Cooperative store in Cottonwood. When the weather was too cold to saw logs, he worked in his shop, where he had installed several small circular saws and a turning lathe. Not having found the winter experience too unpleasant, he stayed the following winter as well. He kept busy making a table, cupboard, flour box, rocking chair, and shingle machine, among other things, and attending to his twenty chickens. Sometimes the stream was frozen and the mill could not run. Once the stream overflowed due to freezing and flooded his hen house and stable. Another time he worked half the day to get the mill started, then sawed only two small logs before quitting because it was too cold. Situated at the side of the road as he was, he had frequent visitors. Molly Wall stopped here to warm her feet, he once wrote. She was going to the Bates house at Cardiff Fork. On another cold night two miners came in and Whipple gave them some supper and lodging.[89] He continued wintering at the mill for two more years, but then he wrote, "This is the fourth winter I have been here alone on Christmas and New Years, but I have determined not to be here another winter, for I do not enjoy it much." He also complained frequently about severe stomach pains, with symptoms of lead poisoning.

In 1880 he and his boys installed and worked a shingle machine at Bates Fork (Cardiff Fork) and sawed shingles through the autumn and early winter. But by then Whipple was sixty-two years old, and as his health deteriorated he was able to do less and less. By 1884 he had rented his mill to Alvin W. Green, but he and his sons continued to run it for the season. After that his lumber business deteriorated as his health failed and his sons left for lives of their own. Whipple died on 5 July 1887 of consumption, which he had said was from living so long at high altitudes.[90] Whipple's mill didn't outlast him by many years. In 1895 a mining claim description referred to the "Old Whipple Mill Site," implying the site was all that remained.[91] What did remain, however, was the name Whipple Fork for the canyon north of the mill where Whipple got most of his logs.

*

Sometime in 1876 or 1877 Philander Butler and David B. Brinton acquired a steam sawmill from Francis J. Pascoe, who ran a lime quarry and kiln north of Salt Lake City, and moved it into a north fork of Big Cottonwood Canyon, midway between Mills A and D. Nelson Wheeler Whipple visited the mill in June 1877 to pick up some lumber Butler owed him. He wrote, "The mill is situated in the midst of a butiful ceinery about 1½ miles from the main kanyon."[92] Butler kept the mill there only one season before moving it to the head of the canyon, above William S. Brighton's hotel, leaving behind his name for the fork that has ever since been known as Butler Fork.

The new location for the mill was at the base of today's Mount Millicent. By this time the area around Brighton's hotel and Silver Lake was much used and enjoyed by summer visitors, but the mill operated without disturbing them too much. Only one complaint has been found, and that when Butler's employees became a bit too rowdy while partying. One of the vacationers wrote, "The night was made hideous by the drinking orgies of the mill men. Their shrieks and oaths vividly suggesting Dante's Inferno rendered sleep impossible."[93]

Mill workers made the news on the Fourth of July 1883 when three young loggers, or wood choppers, went boating on Silver Lake. While they were having a "scuffling frolic," the boat overturned. One of the men could not swim, and in the confusion that followed two men drowned. Curiously, the survivor was the non-swimmer, who was rescued after he managed to clamber onto the overturned boat.[94]

The operation of the Butler mill came to an abrupt end on 1 October 1884 when the steam boiler exploded, sending scalding water, steam, and metal fragments in all directions. In an instant one man was killed and

four others, including two of the Butler brothers, Neri and Alva, were injured or burned. Several of them were thrown as far as twenty feet by the
blast, and Neri's eyes were seriously injured. And then, as if to hide the
disaster, the mill shed collapsed with a crashing of timbers. Justice Edward Sims came over from Alta to hold an inquest before the body of the
dead man, John Smith, was taken down to the valley.[95]

It transpired that Daniel H. Wells had considered buying the mill's
steam engine in 1880, before he got completely out of the lumber business, and had Thomas Pierpont go up to inspect it. Pierpont had been
master mechanic of the Utah Central Railroad when it began operating
in 1870 and later organized the Salt Lake Foundry and Machine Company in Salt Lake City. He was one of the foremost experts in steam engines in the territory and a logical choice to inspect the engine in Big
Cottonwood Canyon. He found the equipment deficient and could not
give a favorable report. He later said the boiler was unsafe and should
have been condemned. Pierpont gave his opinion that many other unsafe
boilers were operating in similar out-of-the-way places.[96]

In his history of pioneer sawmills, Asa R. Bowthorpe wrote that it was
thought that Neri Butler might lose his eyesight, but it returned and he
repaired the engine and went right on sawing lumber.[97] What he wrote
may be fact, but nothing else has been found to verify that the mill was
run again. In 1888, four years after the explosion, the Butlers and David
Brinton were in court, the defendants in a case brought by the United States,
alleging that they had been cutting and removing timber from public
lands in the vicinity of their steam mill in the years 1883 through 1885.
Since 1885 was after the demise of the steam mill, the suggestion could be
made that this shows that the mill was rebuilt. But the timber might have
been cut for another mill; the Butlers did have a water-powered sawmill
at the mouth of Big Cottonwood Canyon that they had been operating
for some time and continued to operate after 1885. The brothers Philander, Leander, and Neri Butler had filed a claim on Big Cottonwood Creek
for a mill site in 1872.[98] In 1880 a mining claim used the Butler flume as a
reference point, suggesting the mill was running at that time.[99] It continued to run until near the turn of the century, when its water rights became important first to the newly established electric companies and
then to Salt Lake City for culinary water. As for the court case, the Butlers, after numerous persons testified in their behalf, were acquitted.[100]

*

Another water-powered sawmill that operated briefly around the turn
of the century belonged to Alvin Green. It was located between the stream

and the road near the upper end of today's Spruces Campground. It was used as a reference in a mining claim filed in September 1900.[101] In July 1902 Albert F. Potter reported, "Two miles above Argenta there is a large valley known as Mill D Flat. There has been a good body of timber here on the south side of the canyon but it has been pretty closely trimmed out by the sawmills. At the upper end of this valley is located the sawmill of Mrs. Green. She saws a few lots for settlers once in a while, but does not run steady. Has a few cattle and is making her home here."[102] Mrs. Green would have been the wife of Alvin Green, who, according to Bowthorpe, built the mill and later sold it to Julius Kuck. There were three generations of Alvin Greens who were active in the canyon at one time or another. Alvin G. Green was one of the pioneers of 1847, arriving in the valley in September 1847 with his parents and younger brother and sister. He married and had a family of eight children, one of them being Alvin W. Green. He, in turn, had a son Alvin R. Green. Alvin G. Green was filing mining claims in Big Cottonwood Canyon from the earliest days of the mining boom, 1870, through 1895.[103] His son Alvin W. was prospecting with his father as early as 1871, when he was but seventeen years old.[104] It was undoubtedly Alvin W. and his wife Alice who ran the sawmill. At the time Potter passed through, they would have been a little over fifty years old. Green's mill is shown on the U. S. Geological Survey's Cottonwood Special Map, dated 1907, and his name survives today in Greens Basin, where he got most of his logs.

Julius Kuck was a German immigrant who came to Utah intending to continue his work as an attorney, only to find he was not sufficiently versed in American law. He thought he could remedy the situation by spending a short time studying at an American law school. To cover the expenses of schooling and supporting his family, he moved his wife and daughters to the Wasatch Mountains with the idea that he could make a quick windfall in lumber and minerals. Indeed, in 1895 he filed three mineral claims, all in Willow Patch Fork, where he built a cabin to house his family.[105] Willow Patch Fork is today known as Willow Creek or Little Willow. It is located on the north side of the canyon a short distance above the Silver Fork community. While it is not known if Kuck did any lumbering, he did work his claims; but his quick reward eluded him. His wife found the winters long and cold and the isolation unbearable, so she took the daughters and left, never to return. She later divorced him; in recorded documents dated 1909 and later he was referred to as "Julius Kuck, divorced."[106]

With his family gone, Kuck lost all aspiration to return to his former professional life and gave himself to his mountain home. After that he never left the canyon except to replenish his supplies. He filed three mineral claims in 1899 and seven more in 1901 and 1902, all in the vicinity of Willow Patch Fork. Comfortable in his surroundings, he made a homestead application and in 1904 received a patent on a quarter section, 160 acres, covering much of Willow Patch Fork.[107] He mortgaged his land in 1909, probably when he bought the sawmill from the Greens. Kuck then used the mill to saw aspen logs into lumber to make furniture that he sold to canyon residents and those who passed by. He put a big sign on the front of his mill proclaiming it to be the Wasatch Aspen Works. But he coveted his solitude, and he soon became known as the Hermit of the Wasatch. When asked why he preferred the hermit's life, he said he liked the scenery in Cottonwood. And when hikers passed by and wanted to look into his cabin, he complained that "Hermiting ain't what it used to be."

According to Bowthorpe, Kuck's mill was later destroyed by fire. Just when this happened is not certain, but it may have been around 1920. In that year Kuck deeded his land to Frank E. Bagley, a son of the Charles S. Bagley who had been one of the partners in Mill D many years before. Frank Bagley had held the mortgages Kuck placed on his property eleven years earlier, and he now became owner of the land.[108] But Kuck continued to live in his little cabin in the woods until 1934, when it caught fire and burned down. He was taken to the city for treatment of burns, but soon returned to the mountains to rebuild his cabin. Forest Service personnel wanted him to move closer to the mouth of the canyon in the winter so they wouldn't have to go so far to check on his welfare. He was, after all, seventy-four years old. Then in April 1937 his cabin burned again. He escaped, but while trying to salvage some of his possessions he was badly burned about the head and shoulders. Carrying his few remaining possessions on his back, he trudged nearly two miles through deep snow to the ranger station near Days Fork.[109] He was taken to the city and placed in the county hospital, where he remained until he died on 28 September 1938. After his death, county social workers located his widow and a daughter in California and burial arrangements were placed in their hands, although it was reported that he had accumulated enough money to cover the costs of his funeral. He was buried in Mount Olivet Cemetery.[110]

Figure 1. The Mormon pioneers' first means of providing lumber was the basic and labor-intensive pit saw. In some parts of the world the pit saw is still in use today, as seen in this picture, taken in Nepal in 1986. These men propped the log up at an angle with a post to allow the "pit-man" to work below while the sawyer works above. (Author's photo)

Figure 2. Organizers of the Big Cottonwood Lumber Company. Brigham Young, top left, was instrumental in starting the company and opening Big Cotton-wood Canyon to exploit its timber resources. Daniel H. Wells, top center, continued to run mills E and F as his own after the company ceased operations. Feramorz Little, top right, supervised the building of the first road up the canyon and construction of the first mills. After the company ceased operations he continued to run Mill D. Frederick Kesler, lower left, was responsible for the construction of mills A, B, D, and E. In this portrait he is wearing a uniform as battalion leader in the Nauvoo Legion. Abraham O. Smoot, lower right, was one of the organizers of the company but was busy building the Sugar House at that time and contributed very little to the lumber company. His search for timber in Mill Creek Canyon led to Church Fork receiving its name. (LDS Church Historical Archives, hereafter cited as CHO)

Figure 3. Big Cottonwood Lumber Company Mill B, which was located at the apex of the lower half of the S-turn on the present canyon road. (*Salt Lake Mining Review*, 31 July 1899, hereafter cited as SLMR)

Figure 4. Big Cottonwood Lumber Company Mill E. The mill was located just below Brighton. Mount Tuscarora, Mount Millicent, and Wolverine Peak are visible, left to right, in the background. (Utah State Historical Society, hereafter cited as UHS)

Figure 5. Mill E site, as it appears today. The boulder in the left foreground is the same one the man is leaning on in the preceding photograph of Mill E. (Author's photo)

Figure 6. Wasatch Aspen Works, formerly the Green Sawmill, where Julius Kuck made aspen furniture. It was located at the upper end of Spruces Campground. (UHS)

Figure 7. Slopes in Mill D North Fork in the early twentieth century showed the results of logging in earlier years. Trees were cut in the winter when the snow was deep, hence the high stumps. (UHS)

Figure 8. The Woolley brothers operated sawmills in Little Cottonwood Canyon. John Mills Woolley, left, ran the mills from 1854 to 1864. His brother Samuel Amos Woolley, right, hauled the lumber into the city. After John Woolley's death in 1864 Samuel took charge of the mills until 1876, when he sold them and became a farmer. (Courtesy of Preston W. Parkinson)

Figure 9. Site of Woolley's lower mill. Little Cottonwood Creek flows from right to left, and the Hogum Fork stream flows in from the bottom. Notice the tiny town of Hog-Um, source of the name of Hogum Fork. Today the road along the north side of the stream is a popular hiking and biking trail. (From map of 31 July 1874, 472-B, BLM, Salt Lake City office.)

Figure 10. Woolley's mill in Little Cottonwood Canyon. C. W. Carter photograph, circa 1870. This is believed to be the turbine mill at Hawley Flat. (CHO)

Figure 11. Hauling logs from Albion Basin. One end of the log was mounted on the two-wheeled carriage while the other end dragged on the ground. Note the size of the logs. Devils Castle looms in the background. (CHO)

Figure 12. A logging camp in Albion Basin. The presence of several women in the foreground indicates that some loggers had moved their families into the logging camp. (CHO)

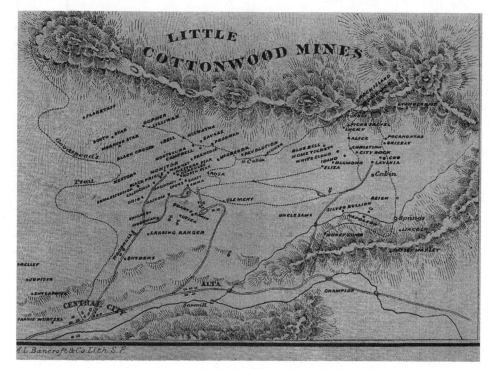

Figure 13. Bancroft map of Alta, circa spring 1871. Goodspeed's Trail, climbing to the ridge from Central City, went to the Reed and Benson Mine and crossed the ridge well above and to the east of Cardiff Pass, where the trail of today crosses. Also note the sawmill at the confluence of Little Cottonwood and Grizzly Gulch streams. This was Woolley's steam sawmill, which was sold to the Emma mining company at about this time.

Figure 14. Froiseth Alta City plat of 1873. It was extended to include Central City. The Alta City plat filed with the *Salt Lake County Recorder* in July 1873 included only Alta City from Mill Street on the south (called Cottonwood Street on that map) to Charlotta Street on the north and from the eastern boundary to Fifth West Street. (B. A. M. Froiseth, "Froiseth's Map of Little Cottonwood Mining District and Vicinity, Utah …," 1873)

Figure 15. Sketch of Emma Mine, made in the early 1870s. (From T. B. H. Stenhouse, *The Rocky Mountain Saints*, 1874)

Figure 16. Alta in the early 1870s; looking west on Walker Street. The tall trees were a prominent feature during Alta's early years, but were destroyed by fire in August 1878. (CHO)

Figure 17. Alta in the mid-1870s. View looking southeast, with Albion Basin in the background. Notice that many trees had been cut from the slopes above town. (CHO)

Figure 18. Clasbey and Read's store on Walker Street in Alta in the period 1872–1878. (*Salt Lake Tribune*, hereafter cited as SLT)

Figure 19. Alta in the mid-1870s. View looking west on Walker Street. Compare this with the earlier photograph taken from nearly the same spot. The tall sign on the right is for the Grand Hotel, with the California Brewery on the uphill side, between the hotel and the drug store. Sam Gee's Washing & Ironing sign is the first one on the left, with the Miner's Restaurant sign above and behind it. (SLT)

Figure 20. Alta after the 1885 avalanche. View from the tramway station. Notice how the slopes had been denuded, leaving only stumps. C. R. Savage photograph, 3 July 1885. (SLT)

Figure 21. An artist's conception of the ride down the tramway from Alta. He must been impressed by the steepness of the canyon walls. Note the snowshed. (UHS)

Figure 22. Heading up the canyon on the tramway. Two-mule teams, as seen here, were typically used. (SLT)

Figure 23. Tramway ore cars at Alta. Ore was sacked before shipping to allow easier handling. (CHO)

Figure 24. Coming up the Alta tramway. On this trip three mules were harnessed in tandem to provide the power. (SLMR, 30 October 1904)

Figure 25. Two tram carloads of passengers ready to leave Wasatch to go up the canyon. Notice the narrow gauge coach behind the man in the foreground. (UHS)

Figure 26. Railroad loading terminal at Wasatch, near the mouth of Little Cottonwood Canyon. This was the terminus of the railroad from Sandy. Since the tracks in the picture are standard gauge, the picture was taken either in the 1890s or after 1914. (UHS)

Figure 27. A Shay locomotive of the Little Cottonwood Transportation Company. This railroad, running from Wasatch to Alta, operated from 1918 through 1921. (UHS)

Figure 28. Jitney on the Salt Lake and Alta Scenic Railway at Wasatch. The driver is Elbert Despain, with his wife at his side. Wells Culmer, son of Henry L. A. Culmer, is in the right rear seat, while Robert Marvin is in the left rear seat. This vehicle, and others like it, carried passengers and mail on the narrow gauge tracks from Wasatch to Alta for about ten years between 1918 and 1928. (UHS)

Figure 29. A Shay locomotive of the Little Cottonwood Transportation Company heading up the canyon from Wasatch. Note the radiator cap of a jitney in the foreground. (SLT)

CHAPTER 4

The Lumber Industry in
Little Cottonwood Canyon

As SETTLERS MOVED FARTHER SOUTH in the Salt Lake Valley Little Cottonwood Canyon became more important as a source for timber. Except for its distance from the city, it was much more accessible than its neighbor to the north. The two canyons shared a common name, but they were wildly different. While Big Cottonwood Canyon was a deep, narrow gorge, in many places barely wide enough to contain its stream, Little Cottonwood was broad and open, offering plenty of room to move up and down the canyon without having to fight the flowing water. In spite of the easier access, however, it took several years before any attempts were made to develop the canyon.

On 15 January 1850 the General Assembly granted the petition of William Crosby and three others for the control of a canyon south of Big Cottonwood.[1] It is not known which of the canyons south of Big Cottonwood Crosby had in mind, but nothing came of the grant. Then, almost a year later, on 7 January 1851, another grant was made, this time in the form of an ordinance granting Little Cottonwood Canyon to Benjamin L. Clapp and Charles Drown for mill purposes. As was usual, they were required to make a road into the canyon and keep it in repair, and they were authorized to charge a toll for wood and poles removed over their road. They were to have exclusive control of the timber to supply a sawmill that they might construct on the creek.[2] At the end of 1851 there was a road into the canyon, as evidenced by a diary note made by Warren Foote, who had arrived in the valley in September 1850 and settled in Big Cottonwood. In spite of his location, he went to Little Cottonwood to get his winter supply of wood, and while there, he wrote, he had been "working some on the road in that kanion."[3]

There is little evidence to show whether Clapp and Drown carried out the full intent of their grant. But the canyon was theirs for two years, until 21 January 1853, when the territorial Legislative Assembly passed an act repealing the ordinance that granted the canyon to them.[4] Looking back over a century and a half, this appears to have been an unusual action, but the purposes behind it probably had little if anything to do with Little Cottonwood Canyon. Benjamin Lynn Clapp was a very early member of the Mormon Church and had long held high positions within the Quorum of the Seventies. He was the chaplain of his company when he crossed the plains in 1850 and immediately upon settling in Salt Lake City he assumed important secular and spiritual roles. When the General Assembly passed the ordinance incorporating Great Salt Lake City on 9 January 1851, Clapp was named as one of the first councilmen even though he had only been in town for a few months. And at the annual spring and fall general conferences he was regularly sustained as one of the Quorum of Seventies. It may well have been his high standing that enabled him to get the grant of Little Cottonwood Canyon. But then he began to question his faith, to express feelings that were, in the eyes of his peers, opposed to the truth. At the April 1852 conference his name was not among those presented to be sustained in their positions; his standing in the Seventies was laid over until he would make a "humble confession of his folly."[5] His name reappeared at the October conference, but as an associate to Joseph Young, president of the Seventies.[6] It seemed he was on his way down and out. In January 1853 the act was passed repealing the ordinance granting him Little Cottonwood Canyon. He was then sent to the San Pete County settlements, where his sagging beliefs got him into difficulties with the local bishop. He was dropped from the Seventies and finally, at spring conference in 1859, was excommunicated from the church.[7] He left his family and went to California, where he died in 1860.

Not much is known about Charles Drown, except that he came west in 1849 shortly before his thirty-fourth birthday. It is not likely he knew Benjamin Clapp before he came to Utah, since he came from New York and Clapp was from Alabama. But after the two men met in Utah they became very close friends. In 1857, after Drown had gone on a church mission to California, Clapp's wife gave birth to twin boys. Clapp revealed his esteem for Charles Drown by naming his sons Elija Charles and Elisha Drown. It may well be that Clapp followed Drown to California, but Drown returned while Clapp did not. Drown died in Salt Lake City in 1859 or 1860.[8]

There may have been another reason for the legislature to repeal the Clapp and Drown ordinance. A year after the ordinance was passed the legislature created the county courts, giving them responsibility for the resources within their respective counties. It may have been recognized that Little Cottonwood Canyon had been placed beyond the reach of the county court by legislative action. In that case, the act repealing the ordinance freed the canyon from this legislative conflict and allowed it to go into the hands of the county court where it belonged.

Two months after the act was passed Winslow Farr and Jeter Clinton presented the court with a petition for exclusive control of the timber and water of "Little Cottonwood Kanyon South."[9] The request was deferred until the next meeting of the court. At that time a number of residents of Little Cottonwood presented a remonstrance against the petition, so the petition was tabled indefinitely.[10] The court minutes do not give the reasons for the residents' objections, but in November the petition was presented again, this time in Clinton's name alone. With only a one-day delay, the court granted the petition in the following words:

> On the above petition the Court granted to Jeter Clinton the right of the water privileges in said kanyon sufficient to carry on the sawmilling & other machinery business to any extent he may occupy, also the control of the sawing timber to that extent that the same may not be monopolized by individuals disposed to go ahead cut & slash timber and leave it in the way &c to the detriment and inconvenience of others wishing to operate in said kanyon. Said Clinton is required to make and keep in repair the road in said kanyon as the wants of the community may demand and shall have the right to charge (25 cts) per load for wood, poles, timber & lumber hauled out of said kanyon provided that the rights heretofore accrued by the citizens of South Cottonwood and Little Cottonwood Wards are not interfered with.[11]

With this action the court tied up the Little Cottonwood Canyon timber resources for the next twenty-five years, much as it had done in Big Cottonwood Canyon with the Joseph Young grant the year before.

Jeter Clinton was the same person mentioned earlier in connection with the bucket-making machinery in Mill Creek Canyon. Although that venture would take place several years later, Clinton was a rather prominent man about town. He had returned from a two-year mission to England in 1850, bringing back with him 150 English emigrants. In January

1851, when the General Assembly passed an ordinance incorporating Great Salt Lake City, Clinton became one of the city councilors; he was re-elected two years later. For several years he, as Dr. Clinton, was in charge of the quarantine hospital near the mouth of Emigration Canyon, where immigrants afflicted with infectious diseases were held while they recovered. There is no evidence that he had ever worked as a millwright, or ever would in later years, but he went to work immediately in Little Cottonwood Canyon and had one sawmill running in time for the 1854 season, with another under construction. He was not to enjoy the fruits of his labor, however, for in June he was appointed to a mission to the eastern states. Within two weeks he had completed an agreement to sell his Little Cottonwood operations to John M. Woolley, Nathan Davis, and one B. Snow.[12]

John Mills Woolley was thirty-one years old when he entered the lumber business in Little Cottonwood Canyon. A member of a large Quaker family that had converted to Mormonism, he had come to Utah in September 1847 with his wife and their baby. When the city was laid out and building lots were assigned, he took one in the Ninth Ward, in the block between Fourth and Fifth South and Third and Fourth East Streets. He was able to get adjacent lots for two of his brothers, Edwin D. and Samuel A., who arrived in 1848, so their families could live close to one another. Edwin later relocated, but John and Samuel remained at their initial Utah home site for the rest of their lives.

Nathan Davis had become a part of the Woolley family when he married an older sister of John Woolley's some eighteen years earlier. When he and his wife, Sarah, came to Utah in October 1851 Brigham Young appointed him to take charge of the machine shops in the public works center on Temple Square, a job he held at the time he and John Woolley bought the sawmills. For that reason he did not play a strong role in the daily operations in the canyon, a job that was ably carried out by John.

B. Snow was undoubtedly Bernard Snow, who arrived in Utah with his wife in 1851. He was variously a millwright and a carpenter, and he also established a considerable reputation as musician, poet, and actor. He wrote music and text for several hymns and once played the lead in a local presentation of *Damon and Pythias*. His fame as an actor was such that the Deseret Dramatic Association put on a benefit for him in February 1856.[13] Although it is not known how he became associated with Woolley and Davis, he took an active part in the operation of the mill until he left on a mission to Europe in September 1856. By that time he had withdrawn from the partnership, which continued with the two remaining members.

On the day after he made the agreement with Clinton, John M. Woolley sent men and teams into the canyon to haul materials to the unfinished mill. For the next four or five weeks they hauled logs down to the lower mill to be cut and took the lumber up to finish the new mill. By mid-August they no longer needed all their lumber for the new mill and could start taking some to the mouth of the canyon and into the city. They installed a shingle machine and set the new mill into operation. The lower mill was abandoned in the years that followed.

In his diaries, John M. Woolley repeatedly mentioned the upper and lower mills during 1854, but after January 1855 he wrote only about "the mill." The location of the lower mill is not known, but the new mill was at a large flat on the south side of Little Cottonwood Creek at Hogum Fork. When a Land Office surveyor was surveying the bottom of the canyon in April 1874, he mentioned in his field notes that "In a garden near Wooley's saw mill they raise onions, turnips and radishes," and that the place "is known as White's Flat, named after Robert White, the first settler."[14] He sketched a map that showed the canyon road running on the north side of the stream. A bridge gave access to the south side where the road curved back onto the flat. Another branch of the road continued up canyon on the south side of the stream. This branch crossed to the north side again farther upstream, beyond the extent of the surveyor's sketch. The parallel road on the north side, across from the hogback, was not there in the early days of the mill; it was added some years before the survey was made.

Remnants of the roads can still be seen. The flat has unusual heaps of granite boulders where they were stacked to give clear space to pile lumber and logs. The detritus of many years of use can still be found in the form of crockery shards, wrought nails, and bits of metal parts from wagons and machinery. In the tracing from the surveyor's map (Figure 9), the Hogum Fork stream flows in from the bottom, while Little Cottonwood Creek flows from right to left. At the time of the survey Little Cottonwood Creek divided for a short distance at its confluence with the Hogum Fork stream. At that point the creek flows in a fairly broad floodplain, and over the years it has often changed its bed due to occasional heavy flooding. The hogback, noted on the east side of Woolley's mill, is a significant feature in this part of the canyon and is mentioned in several mining claims that used Woolley's mill as a reference point. The mill presumably took its water from the Hogum Fork stream. Faint traces of a flume still can be seen on the hillside below the hogback.

Once the mill was producing, life went on quietly. Loggers cut the timber high on the south slopes and sent the logs down slides to the canyon bottom, where they were hauled to the mill. Finished lumber, slabs, and shingles were hauled to the mouth of the canyon, where they were stacked and held until other teamsters could haul them into the city. Woolley made use of all the help that was available, including Davis's oldest sons. When Nathan Davis or John's brother Edwin came out to the mill they usually hauled a load of lumber back to town with them, as did John himself whenever he went home. In the city the lumber was stacked next to his house until it could be delivered to the customer.

There were a few moments of excitement; one happened on 3 September 1855. While one of the mill workers, David C. Williams, was carrying planks away from the circular saw, he accidentally let the end of a board drop on top of the whirling saw blade. The plank was drawn violently past the saw and Williams, who gripped it too tightly, was thrown upon the saw and then projected another eighteen feet. He was severely cut and mangled and survived only a few minutes. Woolley took the body into town the next day for burial. His partner, B. Snow, later said he had stopped the saw a few hours before the accident and cautioned Williams about his careless actions, stressing that the possible consequences included severe injury or death. Sorrowfully, he soon witnessed a horrible demonstration of the accuracy of his statement.[15]

In the spring of 1858, at the time Salt Lake City was evacuated during the Utah War troubles, Woolley moved to American Fork and commuted between there and Little Cottonwood Canyon, hauling a considerable amount of lumber to Utah Valley to build shelters for those who had fled there. Later that summer, after he brought his families back to Salt Lake City, he did a considerable business with the army, with many loads of his lumber going south and west to Camp Floyd.

After Snow left on his mission in 1856, his work was assumed by Rees Williams, who became ever more important in the daily operations, if not a partner in the business. As early as January 1857 the concern was advertised as Davis, Woolley, Williams & Co. When they advertised for men to stock the mill with logs, they asked responsible parties to contact N. Davis or J. M. Woolley in the city, or R. J. Williams at the mill. Later that year, when they were advertising for "a person acquainted with running circular saws" to run the mill, the ad also was for Davis, Woolley, Williams & Co.[16]

Rees Jones Williams had come to Utah from his native south Wales in 1851 with his wife and infant son. He had worked at the mill since 1854, ever since Woolley himself was there. He had the ability to do any task at hand and to do it well. He was a capable machinist, a good sawyer, and a strong hand when logs had to be brought down off the mountain. But on Thursday, 31 May 1860, tragedy struck. He was working as the sawyer that day and had just cut a log that was too big for the circular saw. This was a common occurrence, in which the saw could not quite cut through the log, and the plank, or slab in this case, would still be attached by a thin strip of wood above the cut. The procedure was to take an ax and strike the log above the cut, splitting the residual wood and allowing the plank or slab to fall away. On this day, as Williams swung the ax over his head it struck an overhead beam with such a violence that it threw him backwards onto the spinning four-foot-diameter saw blade, which nearly severed his right arm and shoulder. When John M. Woolley rushed to his aid Williams said, "O, brother Woolley, I am done now." A doctor was sent for, but it was a long way to travel. Rees Jones Williams died about eight hours after the accident, a half hour after the doctor arrived.[17]

With Williams gone, John M. Woolley had a heavier burden at the mill, one that was relieved slightly when his brother Samuel returned from Parowan after closing the bucket factory there. Samuel hauled lumber from the canyon to the city through the winter, but in the spring he was sent east on a mission. When he returned in the fall of 1861 he went back to hauling lumber, making three or four trips a week, often with two teams and wagons. It took all day to go from the center of the city to Little Cottonwood Canyon, load the wagons with lumber or shingles, return to the city, and unload the wagons. It was a trip that was tiring for man and beast. One day he noted that he made the trip in 8¾ hours. Some days he would go up to the mill and make several trips from there to the mouth of the canyon, replenishing the supply at that staging point before taking a load into the city.

To break the boredom of the trips, Samuel Woolley sometimes took his young son, Amos Mervin, for company. Thursday, 16 June 1864, was one of those times. After he loaded the wagon with lumber they started home. It was a hot day with a very strong southwest wind, the kind that seems to suck the moisture out of everything in its path. When they got to a steep grade, Samuel and his son got off to walk. As Amos walked on the lee side of the wagon, in the midst of all the wind-blown dust, his father told him to come around to the other side, out of the dust. No sooner had the boy done so than a strong gust of wind blew him off balance, throwing him onto the road, where the rear wheel of the wagon ran over him, passing between his legs, over his body and off the right side below his arm, severely smashing the lower part of his body. Grief-stricken,

Samuel unloaded the lumber from the wagon and got a woman who was walking along the road to sit in the back and hold the child on her lap while he drove his team as fast as he could, making the fourteen miles to the city in an hour and a half. It was an endless trip for the father, not knowing if the boy would live long enough to reach the doctor. He did, however, and when Dr. Anderson examined him there seemed to be no broken bones; but there was no way of knowing what his internal injuries were. Samuel wrote that his sorrow was as great as he wanted to endure. He blamed himself for advising the boy to walk on the side out of the dust, which proved to be almost his destruction. He guessed that 1,500 pounds of weight had passed over the child. Amos Mervin was only four years and four months old.[18]

Two days later Samuel went back out to retrieve the lumber he had left by the side of the road. When he got there, he wrote, "it brought a very painful & heart sorrowing feeling for me to see the spot where I had [seen] the wagon wheel pass over the body of my son & to handle the lumber that born down the wheel that crushed his little form to earth."[19] However, in spite of injuries that seemed more than a tiny body could bear, Amos Mervin survived. Nine days after the accident he stood up for a few moments, and after that he continued to improve. He improved so well, in fact, that he went on to live to an age of eighty-nine years. When he died he was the last Woolley descendent still living on the block where John M. Woolley settled in the first years of Salt Lake City.[20]

Samuel Woolley went back to his routine of hauling lumber. Then, almost before the shock of his son's injuries had worn off, he received word from the canyon that his brother John had met with an accident. On Tuesday, 16 August 1864, John Woolley had taken several of his men above the mill to inspect a bridge at the base of one of the logging slides. While they were there a large log came down the slide at a frightening speed and struck a pile of loose rocks at the bottom, scattering them in all directions. One struck Woolley on the side of the head, knocking him senseless. His companions immediately carried him to his house at the mill and administered to him while word was sent to the city. Dr. Anderson left for the canyon immediately. The next morning Edwin and Samuel Woolley went to the Little Cottonwood mill to find John still unconscious. At one o'clock the following morning Samuel returned to the city, arriving there at daybreak. After gathering some supplies, he started for the canyon again in company with his wife Catharine and his sister Sarah, Nathan Davis's wife. When they were about halfway there they met a messenger who told them John M. Woolley had died. Under a new cloud of grief they continued to the mill. There a wooden coffin was made and John's workers carried him down the canyon and placed him on Davis's buckboard for his final trip into the city, where he was buried late Friday afternoon. Some forty carriages and over 200 people followed his remains to the cemetery. As Samuel noted in his diary, it was a loss much lamented by the people in general.[21] John M. Woolley left three widows, one with six children and another with three.

It was a loss that might have been lamented by Samuel most of all, for he not only lost a brother but also assumed the burden of his brother's business. It was immediately decided that Samuel was the most logical choice to carry on the business, at least temporarily.[22] However, it was soon discovered that John M. Woolley had died intestate, and his estate went into probate. The court appointed Edwin D. Woolley and Isaac Groo as administrators of the estate. Edwin was a brother of the deceased, while Isaac Groo had been the foreman at the mill since February 1864 and as such probably knew more about the daily operations than anyone else. These two men recommended to the court that the interests of the estate and heirs would be best served by a continuation of the partnership known as Woolley & Davis. While the two administrators represented the John M. Woolley estate, Samuel Woolley was the man who actually ran the business. He went at it with zeal, possibly handling it better than his brother ever did, expanding it and showing himself to be an excellent businessman. Even though this originally was to be a temporary responsibility, he continued in the job for nearly ten years.

He started by leaning heavily upon the men who had run the mill with his brother: Isaac Groo, the businessman; Thomas Gerrard, who ran the sawmill; and Solomon J. Despain, who ran the shingle and lath mill. Groo had come to Salt Lake City in September 1854 and settled in the Ninth Ward, where John M. Woolley was bishop. He taught school during the winters and labored in the summers until 1859, when he became supervisor of streets and general watermaster in Salt Lake City. In February 1864 he resigned and went to work for Woolley at the mill in Little Cottonwood Canyon. At the time Samuel Woolley took over, Groo was also one of the administrators of the John M. Woolley estate. He stayed and worked with Samuel until December 1865, when he returned to the employ of Salt Lake City.[23]

Little is known about Thomas Gerrard, except that he was with John M. Woolley as early as October 1858 and stayed at least until the spring of 1871. He was the sawyer, and since mills often were given the name of their sawyer rather than their owner, the mill was sometimes known as

Gerrard's mill. Indeed, a Land Office map dated 5 August 1871 shows the mill below Hogum Fork as being "Girards Saw Mill."[24] Thomas Gerrard lived at the mill with his wife and children. In March 1868 his wife died during childbirth at their cabin in the canyon. Her body was taken to the city for burial.[25] Gerrard later remarried, and when he left Woolley's employ, probably in the spring of 1871, he moved into the city and took a job with Zion's Co-operative Mercantile Institution. After forty years at that job, he retired to his home on Third Avenue, where he died on 28 March 1912.[26] Coincidentally, one of his daughters, Florence Belle, who was born in Little Cottonwood Canyon only three months after John M. Woolley was killed, went on to become the bride of Taylor Harrar Woolley, one of John's sons.[27]

Solomon J. Despain brought his family to Utah from Alabama, arriving in Salt Lake City in August 1861. While camped in the city he was asked what he and his sons could do; he said they could chop wood, so they were sent to work for Bishop John M. Woolley, who was looking for woodchoppers. They located at the mouth of Little Cottonwood Canyon, where they began an association with both Big and Little Cottonwood Canyons that extended well into the twentieth century. After John Woolley died, Solomon Despain ran the shingle mill for several years before he quit and built his own small shingle mill.[28] His sons continued logging for Samuel Woolley for another decade.

In 1865, the year after John Woolley's death, the construction of a new mill was started, but it had barely begun before the two administrators stopped it because of lack of means. Undaunted, Samuel Woolley started another new mill several miles farther up the canyon in 1866 but did not have it running before winter halted his work. The following spring he went up to check the new mill but found there was such a large snow slide covering it that he could see nothing. He estimated the snow was forty feet deep where the mill was. Two months later he went back to find the mill had been destroyed in the slide.[29]

Samuel Woolley also took steps to bring the canyon grant up-to-date. The original grant was to Jeter Clinton, and, while the purchase of the mills and canyon privileges gave the rights to Woolley and Davis, the county court had never formally changed the recipient of the grant. So Samuel, as agent for Woolley & Davis, petitioned the court and got control granted to Nathan Davis and the administrators of the estate of John M. Woolley for all the saw timber from their mill up to the summit.[30] This was in April 1866, a time when changes were beginning to take place in the canyon. Prospectors were appearing in ever-increasing numbers, scouring the hills and digging their little prospect holes. Woolley never wrote complaints about these intruders, probably because at first there were not that many of them. Also, by the time their numbers increased, they had become an excellent source of business. As early as September 1866 he noted he and Isaac Groo had gone up to "the Gentile mining camp," and had done some business.[31] Then, as traffic over the canyon road maintained by Woolley and Davis increased, a notice was published in every issue of the *Deseret News Weekly* during the summer months of 1867: "Notice to All Whom it may Concern. From this date, a Toll of Fifty Cents will be charged for each Wagon to all persons traveling in Little Cottonwood Kanyon. Woolley & Davis, June 1, 1867."[32]

In May 1869 Samuel Woolley located and built a section of road on the north side of the creek from the original mill to the first bridge above it, allowing through traffic to avoid the mill yard and its activity.[33] Eventually this small section would become the primary road, while the original road on the south side of the creek would be abandoned. Today, the original road still can be followed, but a short distance upstream, where the hogback presses close to the creek, it has been completely washed out. That was a problem in Woolley's day as well. On the last day of July 1869 a heavy storm caused the stream to flow over its banks, washing out much of the road and destroying all but one of the bridges. Woolley put a crew of men to work repairing the damage, but it took nearly two weeks to put the road into a condition where they could haul logs to the mill again. Then, on 19 August, Nature did it again and washed out the road worse than before, including the section around the hogback. Woolley had his men rebuild that portion of the road as well as others, a task that kept them busy well into September.[34]

To prepare to supply the ever-increasing demands of the miners in the canyon, Samuel Woolley ordered a steam sawmill. When it arrived in the spring of 1868 he took it to the head of the canyon and set it up at the confluence of the streams where the town of Alta soon would be located. Once installed, this mill became the property of Nathan Davis. Two years earlier, in February 1866, Davis had requested that the value of the canyon properties be ascertained in order that the assets of the partnership could be divided. He agreed to abide by the amount set by Edwin D. and Samuel A. Woolley; however, when they returned with a figure, he refused to accept it.[35] The matter ended in probate court. The steam sawmill had been ordered before the case came before the judge. There the administrators of the John M. Woolley estate claimed Davis owned only one-third of the company. This implied that back in 1856, when Bernard Snow left on his mission, Woolley had assumed Snow's share of the partnership. Davis maintained that he was entitled to one-half of the assets. The court agreed and appointed three arbitrators to go to the canyon, evaluate the properties, and, if possible, divide them. One of the arbitrators, incidentally, was Feramorz Little, who at this time still owned Mill D in Big Cottonwood Canyon. After examining the canyon, the three men decided that the timber in the upper part, above the miners' cabins, was worth as much as the timber in the lower part. They gave Davis the upper part, the J. M. Woolley estate the lower part, and decreed that the personal property be divided and both parties share in the expense of maintaining the road up to the dividing line once Davis began to haul lumber.[36] Now that the steam mill was installed, it was in Davis's part of the canyon and became his property, but Samuel A. Woolley took the responsibility of running it, just as he had for the partnership's mills during the past few years.

The turn of events over the steam mill was probably not to Samuel's liking; for several years he had been trying to build a second mill, and, now that he had done so, it no longer benefited his brother's estate. Accordingly, the next season, 1869, he picked out another site for a new mill for the "JM Estate," as he called it. The site he chose was on a large flat about where the lower end of Snowbird Resort is today. The mill built there drew its timber from Gad Valley on the south side of the canyon. Woolley used the name Gad Valley as early as June 1869 when he was looking for a site for the new mill.[37] The name was a mining term, a gad being one of a miner's tools, a pointed steel bar used to break up or loosen ore. By this time many of the older mills in the territory had been or were being converted to use turbines rather than the older style waterwheels. Hence, Woolley's new mill was likely built with a turbine to drive it; as early as June 1871 it was called the Turbine Sawmill, although Woolley didn't refer to the turbine mill until 25 June 1872.[38] It also became known as Hawley's mill, after C. B. Hawley, who ran it for Woolley for several years. The flat where the mill was located became known as Hawley Flat, a name that was used well into the twentieth century.[39]

Samuel Woolley tried several times without success to sell the mills and canyon grant so that he could get out of the lumber business. After the steam sawmill had been running and supplying lumber to the mines for a year and a half, he had an opportunity to sell part of the operation. Joseph R. Walker, one of four brothers who were prominent merchants and bankers in Salt Lake City and who had recently become involved in mining ventures, approached Woolley to purchase the steam mill and the upper canyon rights for the Emma Silver Mining Company. By this time

Nathan Davis had sold half of his rights to William Folsom and George Romney, who had a planing mill and sash and door factory in the city. As a result, Woolley spent quite a few days running about town to get all parties in agreement and all papers in order. On 27 January 1871 the sale was completed. The rights to the timber in the upper part of the canyon and the mill went to Walker, who paid $3,600—half going to Davis and half to Folsom and Romney.[40]

But Samuel Woolley couldn't just walk away from the steam mill. In April he wrote a proposition to the Emma company to supply their mill with logs for the year. Before a contract was signed, he also had agreed to run the mill and furnish 500,000 feet or more of lumber to the mine.[41] By the end of the month he had teams of loggers on the job cutting timber, most of it coming from the slopes south of the mill. He did such a good job that by the middle of September the Emma superintendent asked him to stop, as they thought they had enough timbers for the season. However, by mid-October they had the mill running again and worked it for another month before closing the contract. Woolley proposed running the Emma company's mill again for the year 1872 but failed to get the contract. He tried again in 1873, but by that time the Emma was past its zenith; it had been sold to a British consortium and was now in the hands of promoters and speculators. It was being run by people on a continent an ocean away who cared much more about what was happening in British financial circles than they did about the details of an underground mine in the Wasatch Mountains of Utah. There is no evidence that the Emma sawmill ran in 1872, but the Davenport Mine, located a short distance above Alta in Grizzly Gulch, did have a steam sawmill running that year.[42] It was installed by John H. Ely, owner of the Davenport, and was "running day and night" by mid-August.[43] This may have been the Emma's mill being run by the other company, because the Davenport Mine did not run a sawmill again. In 1873 the Emma was again sawing timber, this time under the direction of H. P. Mason, the superintendent of the mine, and making lumber available to others in Alta.[44]

Meanwhile, Samuel Woolley was busy with his turbine mill farther down the canyon, hauling much of his product up to the mining community. In October 1872 a correspondent for the *Salt Lake Herald* observed that the timbering capacity of the canyon had been strained to the utmost and noted that a large amount of California lumber was being brought in for building purposes, as the sawmills in the canyon were unable to supply the demand. Woolley must have recognized this too, because in April 1873 he ordered a new steam sawmill for the JM Estate through Nathan Davis, who now was an agent for Griffith & Wedge vertical steam engines and sawmills.[45] The new equipment arrived in June, only two months after it was ordered. Back in the days when John M. Woolley started working in Little Cottonwood Canyon it would have taken the best part of a year to get equipment from the States. Such was the march of progress.

By the end of August the steam mill had been installed, and the following month it was producing lumber. Woolley was very much pleased with its operation. At the end of September he wrote, "This steam mill is a splendid one, would with a full set of hands saw 10,000 [feet] per day on one shift only. We have sawed this year with the old mill 215,541 ft & with the steam mill 81,401 ft. . . ."[46] That was remarkable, for the steam mill had been working only one month, while the turbine mill had been running all season. It worked so well that Woolley started to dismantle the turbine mill and move its stable and house down to the steam mill. The exact location of the steam mill is not known, but it is believed to have been at the Gad Valley stream, just a short distance below the turbine mill. It drew most of its logs from Gad Valley.

The loggers would usually go up on the mountain early in May, when the snow was still deep. They would cut the trees, then snake the logs to the slides and slide them down to the canyon bottom, where they would remain until they could be taken to the mill. The logging was done on contract for a certain quantity at a given price; for instance, 150,000 feet at six dollars per 1,000, as was the case with John Vance's contract in May 1872.[47] The logs were scaled to determine how many board feet of lumber were in each one. When the snow melted sufficiently that the logs could be moved to the mill, two-wheeled horse-drawn carts were used, where one end of the log was placed over the cart axle, while the other end dragged on the ground.

A logger's life was a lonely one, for loggers would go onto the mountain, living in small shanties or cabins, and stay there until they finished their contract or were driven out by adversity. This was the case with Vance and W. T. Fletcher at the end of May 1875, when they came home in the night because their shanty caught fire and burned all their provisions, bedding, Vance's watch, and about ten dollars in cash that they had left in their pockets.[48]

While the sawyers and other workers at the mills had their families in the canyon while they were working there, the loggers worked alone until they came down off the mountain. Occasionally their families would stay with them in the canyon for short periods. Over the years, Woolley contracted with a number of loggers, among them John C. Vance, Hyrum J. Smith, W. T. Fletcher, and Henry Despain. John Camille Vance was one of the 1847 pioneers, arriving in the Salt Lake Valley that September when he was eleven years old. He married Samuel A. Woolley's daughter Rachel Anna on 4 October 1869. She died at age twenty on 5 December 1872, two days after giving birth to a stillborn child.[49] Vance stayed on and worked with his father-in-law as long as Woolley remained in the canyon. Hyrum Smith came into the Salt Lake Valley by a different route. His parents had taken the ship *Brooklyn* and traveled to California when Hyrum was only one year old. His father died in San Francisco, and his mother later married Isaac Harrison, one of the Mormon Battalion members who was making his way up the California coast on his way to rejoin the pioneers in Utah. At age eighteen Hyrum went to work for John M. Woolley as a logger; he stayed on to log for Samuel. Henry Despain was one of the sons of Solomon Despain, who ran the shingle and lath mills for John M. Woolley years earlier.

The slides that brought the logs off the mountain probably had been given names since they were first used, but John M. Woolley never mentioned them in his diaries. Once Samuel took over, however, he began to use the names he invented or inherited. The earliest and most-used slides were Red Pine and White Pine, named after the two types of trees that were being cut. Red Pine was the common name for Douglas fir, while White Pine was the name used for Englemann spruce, in both cases the name indicating the color of the wood in the tree. The heights above the slides became known as Red Pine and White Pine Mountains, respectively, and the names are still used today for the drainages closest to the slides: Red Pine Fork and White Pine Fork.

*

In 1875 Archibald Gardner made an appearance in Little Cottonwood Canyon. He had purchased a large steam sawmill in 1874 and set it up temporarily in Harkers Canyon in the Oquirrh Mountains to supply lumber to make flumes and gates for the Galena Canal, which he was building under contract along the Jordan River.[50] His mills in Mill Creek Canyon were still running, but the timber supply there was rapidly becoming depleted. This was a problem facing mill operators in many parts of the Wasatch Mountains. Several years before, in August 1872, the superintendent of the Emma Mine asked Samuel Woolley to provide 500 cords of wood. Woolley went up the North Fork (Grizzly Gulch) to look for timber, but concluded, "not much chance up there." And in December of the same year he noted that there were fourteen men cutting logs on the Emma grant without permission.[51] Timber was getting scarce, and men would cut it any place they could find some.

Gardner was also looking for a new timber source. In March 1875 he had approached the county court asking for and receiving the privilege of making a road and collecting tolls in Little Willow Canyon, between Big and Little Cottonwood Canyons.[52] He was likely trying to exploit the timber in that drainage and, like others before him, found that the difficulties were outweighing the potential benefits. Moving farther south, he found what he was seeking in the White Pine tributary of Little Cottonwood. He purchased another steam sawmill from Thomas Taylor in Tooele County and prepared to move it into the Wasatch. Curiously, Samuel Woolley and three of his men had traveled to E. T. City in Tooele County at the end of April to look at a sawmill and timber. He didn't mention what kind of mill it was, but he said it belonged to Taylor and Cutter. It was very possibly the same mill that Gardner was now bringing into Woolley's territory.[53] When Samuel Woolley learned of Gardner's activities he contacted him about selling or leasing the JM Estate's rights on White Pine Mountain. On his way home from the canyon on 25 June 1875 he stopped to see Gardner, but they could come to no agreement. Then on Tuesday, 6 July, he learned that Gardner was preparing to put a mill on that very mountain.

Woolley still believed that the lower part of the canyon belonged to the estate of his brother by virtue of the county court grant in April 1866. He may have known of the legal proceedings against Daniel H. Wells in Big Cottonwood Canyon over Wells's use of timber from public lands, but he never mentioned it in his diary. Still, he must have been aware of the U. S. Land Office taking control of the timber on public lands, since authorities of that office had been in the canyon demanding stumpage fees for the timber being cut.[54] In spite of these things, Woolley felt strongly that Gardner was trespassing. The very next day he went to see his brother Edwin, and the two of them went to G. B. Wallace, president of the LDS Church's Salt Lake Stake High Council, and charged Gardner with unchristianlike conduct in cutting timber and placing sawmills in Little Cottonwood Canyon without permission of the J. M. Woolley estate. Since Gardner and the Woolley brothers were all of the same faith, and in fact each one of them was the bishop of their respective wards, they chose not to pursue the matter in the judicial courts, opting instead to use the ecclesiastic court of the High Council. Curiously, however, the trial was held in the city hall. Samuel Woolley was the main witness; he contended that the county court had never withdrawn the canyon grant and that he had tried to negotiate with Gardner to sell the JM Estate mill and timber. Gardner admitted that was so, but said that he was offered the mill only if he also bought the timber, which he refused to do. He said he had gone everywhere in the Wasatch to find enough timber to set up his mills, and now that he found the timber he had as much right to it as did Woolley. The president of the High Council opined that the grant of the county court was void because the federal government had since taken over control of public lands; there was enough timber for both parties for several seasons, so they both ought to take out as much as they could. The High Council upheld this opinion. Samuel Woolley was not very happy with the result; it was, he thought, the most one-sided decision he ever heard.[55]

Although his loss at the trial seemed to drain Woolley's enthusiasm for the lumber business, he did return to run the mill for the rest of the year. Big changes were taking place in the canyon, quite aside from Gardner moving in with his mills. In recent years a railroad had been completed to the mouth of the canyon, and it was being used to haul mining equipment and ore, as well as stone from the granite quarries. As early as May 1873 Woolley had taken "the cars" from the city to Sandy, and then from Sandy to Granite City at the mouth of Little Cottonwood Canyon.[56] Now, in 1875, a track was being laid up the canyon to Alta, opening new business opportunities for the sawmill. The first contract was for ties, but then a decision was made to cover the track with snowsheds for nearly the entire length of the canyon. Woolley soon noted that he had taken a contract to furnish the railroad company a great amount of lumber, most of it four-by-six planks to cover the tracks. The following spring he sent a team of loggers onto the mountain to cut and slide logs; he then began to negotiate selling the sawmill to the superintendent of the railroad. On 8 June he noted he had sold the estate steam mill for $3,250.[57] He made an offer to run the mill for the summer but did not get that job. Except for some unfinished business with the loggers, his work in the canyon was done. Late in the summer he and two of his sons went up to the old turbine mill to see what they could salvage, but they found little of value. He loaded what lumber was left and rode out of the canyon, closing a big chapter in his life. Samuel A. Woolley spent the rest of his life as a farmer, working a large plot of land he acquired in the Sugar House Ward.

After the High Council trial Archibald Gardner finished moving the small steam sawmill into White Pine Fork and began operations there. He also moved the large mill from Harker Canyon into Little Cottonwood and set it up at Tanners Flat. With two mills running he was in a position to supply all the lumber required to shed the railroad tracks as well as supply the mining industry. As with his operations in Mill Creek Canyon, he employed many of his family members at the mills. His daughter Polly and an adopted Indian girl, Fanny, were doing the cooking, while his son Neil was the sawyer at the Tannersville mill and another son, Archibald (Archie), hauled logs to the mill.[58] Sometime during the summer of 1876 Archie hauled a sawmill his father was said to have bought from Woolley out of the canyon, to be set up near Knudsen's mill on Big Cottonwood Creek below the mouth of that canyon.[59] Woolley never noted having sold anything to Gardner, however; if this had once been his steam sawmill, Gardner must have bought it from the railroad company.

On the morning of 4 December 1876 Archie and Heber Clark, a young man from Pleasant Grove who was employed at the mills, went up White Pine Fork to bring down some logs. It was cold up there, and when they got to the White Pine mill they went into the boiler house to warm their feet. While they were there the boiler exploded, killing young Gardner and severely injuring Clark. The engineer was scalded and thrown out of the mill but was not otherwise seriously injured. The first reports stated that the safety valve had been closed, but later reports put the blame on the elder Gardner, saying the boiler was old and defective. A doctor who was called from the city met the injured men at Sandy Station, where he treated them. Clark was taken to the Catholic Church's hospital, where he remained for several months. His leg was so badly mutilated that it was amputated at the thigh a week after the accident.[60] After he got out of the hospital, Clark sued Gardner for $20,000, using the reports of the boiler being defective as the basis for his suit. He later compromised, settling the case for $1,500.[61]

The following season the remains of the damaged mill were moved from White Pine to the Little Cottonwood stream below, where it was set up and a waterwheel built to power it. An Alta correspondent for the *Salt Lake Tribune* noted in August 1877 that a "company is putting up a water power sawmill a few miles below here and it will be running in a few days."[62] The same correspondent also reported a steam sawmill at Alta, which started running late in July. Two mining claims were filed using this "steam sawmill in Alta" as a reference.[63] It is not likely that this was the old Emma company mill, which had not run for a number of years and was most likely taken away before this time. No further reference to this Alta steam mill has been found, except for a suggestion that it had been destroyed by a snowslide.[64]

Archibald Gardner's water-powered mill operated for several years before being destroyed by a snow avalanche. His big steam mill continued to operate in the canyon for a few more years. During that time the mining operations were slowing down and Gardner was left to do most of the maintenance on the canyon road himself, which also was often the case in Mill Creek Canyon. And, as he did there, he enlisted the aid of the county to help. In 1878 the county road commissioner reported that he had examined the road, found a few repairs unavoidable, and had engaged A. Gardner to make those repairs.[65] The following year, the commissioner reported that Gardner was repairing the road and was of the opinion that the county should contribute at least one hundred dollars.[66] Late in 1881 Gardner gave up and moved the big steam sawmill to the foothills east of Draper, thus ending the lumber industry in Little Cottonwood Canyon.[67] By this time there were precious few trees left. In 1902, when Albert F. Potter was making his survey of forests on public lands, he wrote, "Tannerville is an old sawmill camp and most of the timbered area south of it has been cut over, all of the good sawlogs being cleaned out."[68]

CHAPTER 5

Mining Comes to the Wasatch

A HISTORY OF MINING IN THE WASATCH could occupy a whole volume itself. There are many publications treating local mining operations and their history, and others are in preparation.[1] In this narrative the intent is to address only a few of the more prominent mining operations for their impact upon the local society and economy and for the artifacts or some remembrances they may have left behind. When the subject of mining is raised, many usually think in terms of precious metals. However, mining encompasses much more than the extraction of silver and gold. It includes the removal from the earth of any sort of material that might be useful to commerce or daily survival. In this context, over the years the Wasatch has provided limestone, red and white sandstone, granite, lithographic stone, fire clay, lead, onyx, gypsum, tripoli, copper, molybdenite, sand, gravel, and other materials. However, most of the surviving evidence of mines and mining in the Wasatch Mountains is the result of the search for precious metals. And that mining era began in the early 1860s.

After the Civil War began the troops encamped at Camp Floyd, by that time known as Camp Crittenden, were called east to take part in the great conflict. They left Utah in July 1861. There was, however, concern about the protection of the overland mail route and new transcontinental telegraph line, so the secretary of war called upon the governor of California to provide troops to guard the route from the Carson Valley to Salt Lake City and on to Fort Laramie. The Third Infantry California Volunteers, under Col. Patrick Edward Connor, was formed and moved into Utah. The troops passed through Salt Lake City on 20 October 1862 and camped on the east bench overlooking the city. Two days later Camp Douglas was established at the mouth of Red Butte Canyon. Connor

brought with him a firm dislike for Mormons; he considered them trai-
tors and therefore expanded his assignment to include keeping an eye on
them, as well as to solve, if possible, the "Mormon question."

Within Connor's command were many veterans of the California and
Nevada gold fields, miners and prospectors who were eager to continue
their previous trade. In the summer following their arrival in Utah, ar-
gentiferous galena, a silver ore, was found in the West (Oquirrh) Mountains, at Bingham Canyon. Soon after the discovery, at a meeting at the
Jordan wardhouse on 17 September 1863, the West Mountain Quartz Mining District was formed. The district was very large, extending from the
Jordan River westward to the 114th degree of longitude, just a few miles
shy of today's Utah-Nevada border. It extended south to the 40th degree
of latitude, which runs below the southern extremity of Utah Lake, and
north to the south shore of the Great Salt Lake and the 41st degree of latitude, or the northern boundary of Tooele County today.[2] Its bylaws,
written by General Connor based upon California mining laws, became a
model for other mining districts that followed.[3] Curiously, in spite of
Connor's dislike for Mormons, LDS Bishop Archibald Gardner was elected
recorder of the newly formed district.

There are two versions of this discovery of silver ore. One gives credit
to the wife of a surgeon at Camp Douglas. The lady reportedly had gained
some knowledge of minerals in California, and while strolling about the
mountainside during a picnic outing she picked up a piece of ore. Her
companions immediately prospected the area and made their location
of the ore. One might question why this lady from Camp Douglas would
be on a picnic in Bingham Canyon, but there is an explanation. At the beginning of the year 1863 the garrison at Camp Douglas included four
companies of infantry and four companies of cavalry. There was not near
enough forage for that many animals near the camp, so other pasture-
lands had to be used. Many of the animals and a large contingent of men
went to Rush Valley south of Tooele to occupy rangelands that had been
used in earlier years by the troops at Camp Floyd. A smaller herd was
sent to the vicinity of Bingham Canyon, together with a company of men
to guard them. When the picnic party was organized, Bingham Canyon
was chosen as the location since some troops were already there.[4]

The other version has the discovery being made by a man named
George Ogilvie, who was logging in the canyon. He sent a piece of the ore
to Connor, who had it assayed and, finding it to be of some value, went to
the site with a party of officers and ladies from the camp and made the
location.[5] There is some support for each story, for on 17 September

1863, the same day the West Mountain Mining District was formed, Bishop Gardner recorded three location notices. One showed George B. Ogilvie as the discoverer, another showed Mrs. Robert K. Reid as discoverer, while the third, with the greatest number of names, also named Ogilvie as discoverer. Mrs. Reid was the wife of Robert K. Reid, a surgeon from Stockton, California, and one of the California Volunteers.[6]

Conner probably suspected all along that precious metals could be found in the mountains of Utah Territory. Now that a discovery was made he saw a possible solution to his perceived Mormon question. He formulated a plan where, by publicizing and promoting mining in the territory, he could "invite thither a large Gentile and loyal population, sufficient by peaceful means and through the ballot-box to overwhelm the Mormons by mere force of numbers, and thus wrest from the church—disloyal and traitorous to the core—the absolute and tyrannical control of temporal and civic affairs."[7] He further instructed the commanders of his various posts and detachments to give their men the greatest freedom to prospect the country in the vicinity of their posts as long as it would not interfere with their military duties.[8] The result was the formation of mining districts throughout the territory.

On 18 November 1863, shortly after the West Mountain Mining District was organized, a meeting of miners was held at Great Salt Lake City to form the Wasatch Mountain Mining District. It included all of the Wasatch Mountains from the Weber River down to the fortieth degree of north latitude. Its western boundary was the eastern shores of the Great Salt Lake, Utah Lake, and the Jordan River, while the eastern boundary was the Weber River and a line running due south from its head.[9] This area proved to be much too large to be practical. On 20 July 1864 a group of miners met to create a new district within the limits of the Wasatch Mountain Mining District. It is interesting, but not too surprising, that this meeting was held at Camp Douglas, an indication that most of the miners of that day were members of the military establishment. The new district, called the Mountain Lake Mining District, extended from the south bank of Parleys Creek to the head of Utah Lake, and from the Jordan River to the eastern extremity of the Wasatch Mountain district.[10] Since the name was taken from the many lakes found along the crest of the Wasatch, it was inevitable that in popular usage the name would assume a plural form, the Mountain Lakes Mining District.

The Mountain Lakes Mining District lasted quite a few years before it too was broken into smaller, more manageable districts. At a miners' meeting held in Little Cottonwood Canyon on 30 August 1869 the district bylaws were amended to considerably shrink its area, deleting all the former area east of the crest of the Wasatch, which became the Uintah Mining District. But the district now took in some of the mountains north of Parleys Canyon, probably at the request of miners operating there. It ran from the Hot Springs, just north of Salt Lake City, south to the southern boundary of Salt Lake County, and from the county line along the crest of the Wasatch on the east to the Jordan River on the west.[11] But even before this happened a group of miners had broken away and created the Cottonwood Lakes Mining District, formed in August 1868 within the bounds of the Mountain Lakes Mining District. Only a year later, on 20 September 1869, the Cottonwood Lakes Mining District became the Little Cottonwood Mining District. It included the entire Little Cottonwood watershed to the base of the mountains on the west, and it has remained essentially unchanged to the present day.

Wasatch Mountain Mining District—18 Nov. 1863–??

    Mountain Lake Mining District—20 July 1864–(ca. 31 Oct. 1871)

        Cottonwood Lakes Mining District—Aug. 1868–Sept. 1869

            Little Cottonwood Mining District—20 Sept. 1869–(present)

        Big Cottonwood Mining District—17 Mar. 1870–(present)

            Argenta Mining District—22 Aug. 1872–10 July 1876

        New Eldorado Mining District—31 Jan. 1871–(1880)

            Mill Park Mining District—1880–??

    Great Salt Lake Mining District—27 Aug. 1866–??

        Hot Springs Mining District—9 Dec. 1870–(present)

            Adams Mining District—3 July 1873–1880

Mining districts were formed by the miners to serve themselves. They were an interesting form of self government that sometimes worked and sometimes didn't. In the long term they provided a basis for proprietary rights to properties, especially when sufficient capital was available to engage the legal community. In the short term, however, they suffered all the failings and frailties of the individuals involved. When new miners with new ideas appeared, or a group objected to the controlling group, the organization of a district could easily be upset, usually by forming a new district. That was probably the case with the Cottonwood Lakes Mining District. While it is not clear where its boundaries were, its recorder was James F. Woodman, the same man who became the first recorder of the Little Cottonwood Mining District when it was formed a year later.[12] The recorder of the Mountain Lake Mining District at that time and for many years before was James Wall, one of the soldiers who came into Utah with Connor.[13] That made him one of the true old-timers and a representative of the military contingent. Woodman, on the other hand, was a relative newcomer, one of the "large Gentile and loyal population" Connor had hoped to draw into the territory.

The following spring, on 17 March 1870, the miners of Big Cottonwood met at Slate Springs to form their mining district. On 10 May they met again at the mouth of the canyon to pass bylaws forming the Big Cottonwood Mining District.[14] Its boundaries took in the entire Big Cottonwood and Neffs Canyon watersheds. On 22 August 1872 the Big Cottonwood district was fragmented by the formation of the Argenta Mining District, which took in that part of the parent district lying west of the Reed and Benson Ridge on the east side of the South (Cardiff) Fork.[15] The new district lasted only about four years before being absorbed back into the Big Cottonwood Mining District.[16] After that time the district remained essentially unchanged, although the northern boundary was somewhat vague, if not in the bylaws, certainly in the minds of the miners.

The next district to the north was the Great Salt Lake Mining District, taking in the entire northern half of Salt Lake County. It was formed on 27 August 1866, at a time when a report of gold discovered in City Creek Canyon set off a flurry of mining excitement in that canyon.[17] But, as the excitement died, so did the mining district. It was replaced four years later by the Hot Springs Mining District, formed on 9 December 1870 and taking in the area between Emigration Canyon and the northern boundary of Salt Lake County.[18] Not to be outdone, the miners in the Mill Creek Canyon area held a meeting and on 31 January 1871 adopted bylaws for the New Eldorado Mining District. It took in the area between Emigration Canyon and the divide between Mill Creek and Big Cottonwood Canyons, including Neffs Canyon, even though the latter had been included in the original boundaries of the Big Cottonwood Mining District.[19]

When mining districts failed to yield good ore deposits, or interest in the area waned, the organization that had been created to serve the miners generally faded and was either forgotten or its records scattered, misplaced, or destroyed. In those cases, when activity resumed, it seemed easier to simply start a new district. Such was the case with the New Eldorado Mining District, which by 1880 was displaced by the Mill Park Mining District; but it lasted less time than its predecessor.[20] Miners in Mill Creek or Parleys Canyon who were inclined to register their claims went either to the Big Cottonwood or Hot Springs districts, as suited their fancy.

Another splinter group formed the Adams Mining District within the territory encompassed by the Hot Springs District on 3 July 1873. The district was very small, taking in forty-nine square miles that covered most of City Creek Canyon.[21] It operated about seven years before being quietly absorbed back into the Hot Springs District.

*

There are several variations about the first discovery of ore in the Wasatch. Some accounts say General Connor found the first silver-bearing lead ore in Little Cottonwood and in Mountain Lake (Big Cottonwood) in the summer of 1864.[22] While the date is probably correct, it is unlikely that Connor himself made any such discovery; he was the commanding officer of troops spread across Nevada, Utah, Idaho, and Wyoming and most likely had precious little time to tramp about the mountains in search of minerals. His name did not appear on any of the claims recorded in 1864, and it appeared on only three the following year. Most of Connor's mining activity was in the area around Stockton, on the west side of the Oquirrh Mountains.

Another account has Lt. D. R. Firman of the First Nevada Cavalry and other soldiers locating a number of mines on Emma Hill on the north slope above Alta.[23] Firman may have been active before the Mountain Lake Mining District was formed, but if there ever were records for the Wasatch Mountain Mining District they failed to survive and we have no record of such activity. The Mountain Lake records begin the day the district was formed, 20 July 1864, but Firman's name does not appear until 1871, and then on only one claim. As shall be seen, it is unlikely anyone prospected in the upper extremities of Little Cottonwood Canyon before 1865.

A more plausible account was given by James Wall. He said that the first party of soldiers, encouraged by General Connor, who provided pack animals, provisions, and arms for protection, arrived in Little Cottonwood on 8 July 1864, only two weeks before the Mountain Lake Mining District was organized.[24] Even with Connor's encouragement and support, however, the first year's prospecting did not leave a very impressive record. The first locations recorded in the new district's books were at the head of Big Cottonwood Canyon and in the vicinity of Red Pine and

White Pine Forks in Little Cottonwood Canyon. In both cases the claims were as far up the canyons as roads extended at the time. The year 1864 ended with only eleven claims recorded in Little Cottonwood Canyon and eighteen in Big Cottonwood.

While 1864 may have been significant in bringing prospectors into the Wasatch Mountains, it was also important to the California Volunteers, who made up the prospecting force. Most of the volunteers had signed up for three years in 1861, so the year 1864 saw the expiration of many enlistments. While some signed on for another year, and a few took their discharge and remained in the area, most returned to their homes in California. Out of about 125 individual names recorded in the Mountain Lakes Mining District in 1864, only one-fifth of them appeared the following year, and only a dozen the year after that. This is not to suggest that all of the original 125 or so individuals were in the Wasatch Mountains prospecting for silver. The laws of the mining districts allowed claims to extend 500 feet on each side of the lode but only 200 feet along the lode. The discoverer was entitled to an additional 200 feet. In order to extend the claim to a sufficient length to access all the ore that might be expected to lie within the lode, a considerable number of claimants had to be involved. Indeed, in 1864 the average number of names included in a recorded claim was ten, with as many as fifteen names on some. When only a few men were in the prospecting party, and it was deemed necessary to claim a considerable length along the lode, they would enlist others who were not present, perhaps not even aware that the prospecting was taking place. For instance, Henry Munday appears as one of the claimants on quite a few of the 1864 claims. Munday, who came from Stockton, California, was a corporal in Company A of the Third Infantry Regiment of the California Volunteers at Camp Douglas. But on two of the claims he also included the names of one M. J. Munday as well as James and Owen Munday. Who were they? Possibly family members, perhaps wife and sons, but certainly not members of the California Volunteers. This common tactic often included naming the mining district recorder, financial supporters, or even fictitious names, and it introduced a considerable difficulty in determining ownership of claims in later years.

The bylaws of West Mountain and Wasatch Mountain Mining Districts both required that one day's work be done on each claim each month, lest the claim become "jumpable" or subject to relocation. But when the bylaws for the Mountain Lake Mining District were written, it was recognized that many of the prospectors were soldiers whose enlistment would end that year, 1864, and they would no longer be present to do their day's labor each month. A clause was added that a claim belonging to a soldier was not subject to relocation for six months after his discharge from the service. An additional clause stated that whenever three hundred dollars had been expended on a claim, it was deemed as belonging to the owners and their assigns and was not subject to relocation, a clause that was to cause much grief for miners in the years that followed. The problem of the multiplicity of claimants was resolved in 1872 in mining legislation passed by an act of Congress stating that a mining claim could be located by one or more persons and could be as large as 600 feet wide and 1,500 feet long.[25]

Although many of the soldiers may have left the area, 1865 saw an influx of miners from other parts of the country to take their places. During that year two groups were especially prominent, and both of them penetrated the upper reaches of Little Cottonwood Canyon. The first, led by Silas Brain and D. C. Nichols, was at work very early in the year; they recorded three claims as early as 15 May. Brain was to become very prominent in Little Cottonwood mining circles in the next few years. It has been suggested that he was one of the soldiers who came into Utah with Col. Albert Sidney Johnston following the Utah War, took a discharge while in the territory, and remained there.[26] While it is not known where he went or what he did in the years that immediately followed, he now surfaced to join in the mining activity in the Wasatch Mountains.

The second group, led by Dr. O. H. Congar, always included Joseph Wilde and W. S. and B. F. Dalton. Not much is known about Congar; he claimed to be superintendent of the New York and Utah Mining and Prospecting Company, a concern formed in New York to develop mining properties in Utah and Nevada. He sometimes signed his name as "O. H. Congar, M.D.," although the *Daily Union Vedette,* the newspaper published at Camp Douglas, referred to him as "one of the first thorough miners and geologists that ever visited this Territory."[27] After Congar arrived in Utah, Silas Brain supposedly brought some ore samples to him to assay. The quality of the ore attracted Congar to Little Cottonwood Canyon, where his first recorded claims there were dated 28 August 1865. On that same day Brain filed five more claims. Congar later said he bought Brain's early claims and employed him to continue locating additional lodes. A number of claims recorded in November of that year carried both their names.

During the winter of 1865–66 Congar traveled east and sold three of his claims—the North Star, American Eagle, and Morning Star—to James P. Bruner of Philadelphia. How Congar induced Bruner, a manufacturer of cotton and woolen goods, to get involved in Utah mining properties is not known, but there is some circumstantial evidence. In 1867, when Bruner was in Utah, several locations were recorded with the names James P., H. Naglee, Stanley, and Charles E. Bruner. H. Naglee and Stanley were sons of James P., who by that time had been taken into the family business, now named James P. Bruner & Sons. Charles E. Bruner dealt in "provisions" in Philadelphia. While it is not known where O. H. Congar came from, within one city block of Charles Bruner's residence in Philadelphia were four Congar families. Congar was not a very common name; the New York City directories for the late 1860s have no entries for that name, and the Philadelphia directories have only the four mentioned. One might well infer a connection between members of the two families.

Congar returned to Utah in the spring of 1866 as the representative of James P. Bruner, and he brought with him two other men in Bruner's employ. A. A. Hirst came to superintend the North Star Mine, while Dr. F. Davidis was to construct a cupeling furnace to reduce the ores and send to Philadelphia as quickly as possible the silver bricks it produced. On 12 May 1866 Congar filed a water claim for 750 feet of Little Cottonwood Creek to provide power, and then put in a flume, a large waterwheel, and a fan for the furnace.[28] He also built a road from the end of the logging road, near the White Pine slide, up to his camp at the smelter site, and he made a trail to the mine. He put up a cabin at the camp, located near the site of the future Central City. This was "the Gentile mining camp" where Samuel A. Woolley and Isaac Groo went to do some business in September 1866.[29] By the end of August Dr. Davidis reported that his furnace would process three tons of ore per day and would be operational within a week.[30] In spite of the fact that Davidis was principally involved, the furnace became known as the Congar furnace. The first references to it as such were in five mining claims filed about the same time as Davidis's report.[31] Coincidentally, Davidis was one of the claimants in three of them. A month later, Congar reported that his furnace was turning out metal in large quantities and that within a few days he would be turning out pure silver bricks. But that was not to be. A few weeks later he wrote to the editor of the *Vedette* explaining that the numerous adjustments and tests that were necessary had delayed the realization of a successful operation, and he gave a weak promise for the near future.[32] Possibly realizing that his project could not succeed, he went to the Pahranagat mines in Nevada and did not return to the Wasatch until 1871. James Wall, the district recorder, later said that Congar had smelted about two tons of bullion, but when they attempted to cupel it their result was only a small box of litharge, an oxide of lead.[33] Congar later wrote that the Davidis furnace was a total failure and that he, Congar, had built a replacement furnace in two weeks time and produced three thousand pounds of a crude silver-lead bullion that had a value of about $300 per ton, or fifteen cents a pound. Since freight charges to ship the product east were from twenty-five to thirty-five cents a pound, he sold it locally to Robert C. Sharkey to be used as lead.[34] Sharkey was a dealer in stoves and tinware, with a shop on the west side of East Temple (Main) Street between First and Second South Streets in Salt Lake City.[35]

In 1867 Bruner sent another man to take Congar's place, a German metallurgist named Reese. To prepare for his arrival, A. A. Hirst relocated Congar's water-power claim, this time in his, Hirst's, name as agent for James P. Bruner.[36] Davidis remained to work with Reese, and the two built a new furnace on the north side of the road a short distance below Congar's old furnace. By mid-August the furnace was completed, tested, and found to be a failure. The Bruner contingent then abandoned the operation and left the canyon. Reese remained in the area and in 1869 was reported to be making fire brick in Salt Lake City for use at the Stockton lead mines.[37]

When mining activity resumed two years later, Hirst returned and ran Bruner's mine, the North Star, for several years with but indifferent success. It was later stated that Bruner had expended $100,000 on the purchase and development of the mine. The mine was finally sold by the sheriff to the miners working there as a judgment for wages. They developed a new body of ore and sold the mine to Chicago interests for $75,000. In 1880 it was combined with several adjacent mines as the Alta Consolidated.[38] Toward the end of the century it was being worked by lessees and continued to produce ore well into the twentieth century. If for no other reason, the North Star is important for being one of the first, if not the first, productive silver-lead mines in the Wasatch. Congar's furnace lasted as a landmark for only a few years, the last mention of it being in 1870, but the Bruner furnace was more substantial and as late as 1882 remained as a monument to the first attempts to reduce Little Cottonwood's ores.[39] Congar returned to the Little Cottonwood mining scene briefly in 1871 and lived in Salt Lake City until 1874, when he moved to California.[40]

*

While there were a number of individuals and parties prospecting the local canyons during these early years, the Congar and Bruner operations were the most significant. Their departure in 1867 symbolized the state of the mining efforts at that time. As the accompanying table with the number of annual claims filed shows, 1868 was the low point. It was also the turning point, for that year a few of the prospectors who had been in the Wasatch in earlier years returned to further explore some of the potentially rich claims. Three of these men were James F. Woodman, Robert B. Chisholm, and Capt. James M. Day. Day and Woodman had met in the California and Nevada mining fields and had come to Utah together in 1865. While Woodman was not a party to any recorded claims at that time, Day contributed to three claims, two of them in concert with Congar.[41] Chisholm had worked in the Galena, Illinois, mines for many years before going to California. He prospected all over the West, including the West Mountain Mining District in Utah. While there in 1865 he made the acquaintance of Congar and saw some of the ore that Silas Brain brought into Congar's office for assay.

	MLMD					
	LCC	BCC	Other	Total	LCMD	BCMD
1864	10	19	1	30		
1865	36	27	4	67		
1866	70	22	4	96		
1867	45	4		49		
1868				0	14	
1869	19	1		20	19	
1870	72	2	5	79	208	130
1871	34		1	35	736	444
1872					531	399

It may have been the Congar association that induced the men to migrate to Pahranagat, Nevada, but now, in 1868, they returned. Day provided the resources for the three of them to visit the area, where they found very little activity. Most of the prospectors and miners, unable to support themselves in mining, had gone to work on the construction of the Union Pacific Railroad. The district recorder, James Wall, was among them. Some of the claims they left behind had considerable work done on them and a few had piles of ore surrounding their prospect shafts. While the mines may have appeared to have been abandoned, those owners who expected to return felt secure in their filings in the district records. Wall later said he had appointed a deputy recorder to take charge in his absence. Whether that person was present is not known, but this was the time when the Cottonwood Lakes Mining District was formed with

James F. Woodman as recorder. All claims filed in 1868 were recorded in the books of the new district, while none was recorded in the Mountain Lakes Mining District books. After the Little Cottonwood Mining District was formed, all Cottonwood Lakes records were transcribed into the new district's books. But from this time until the end of October 1871 the miners in Little Cottonwood Canyon had the unique distinction of being served by two recorders in two different mining districts.

While Woodman and Chisholm filed numerous claims in 1868, they concentrated on just one, working the prospect until winter snows and diminishing funds forced them to quit for the season. Convinced that this was the lode that would bring him fortune, Chisholm went back to Elgin, Illinois, to sell his farm so he could continue working the claim the following year. In the spring he sent his son William W. to Utah to work with Woodman. Later in the summer he sold the farm and returned to Utah with additional funds for the project. The men continued to push the development of the claim until about the middle of October when their shaft was down about ninety feet and they broke into a large chamber of solid ore.

It is a characteristic of this part of the Wasatch Range that most, although not all, rich ore bodies lie in a limestone formation or along the contact zone between limestone and quartzite. The limestone belt crosses most of the northern slopes above Alta and runs northwest, over the ridge into Days and Cardiff Forks and across Big Cottonwood Canyon. During the formation of the mountains, far in the geologic past, the rock layers suffered many fractures, or fissures, that often run for long distances. In the millennia that followed materials were deposited in the fissures, precipitating slowly out of gaseous or liquid solutions that oozed through the subterranean darkness, thus forming the veins of ore the miners were now seeking. Many of these veins were narrow, perhaps only the width of a pencil line, while others were generously wide. Some veins had ore of disappointing quality; others proved to be extremely rich. In some places, the fissures, when formed, opened into large cavities, very much like caves. However, unlike many limestone caves in many parts of the world, these were formed during cataclysmic upheavals rather than by erosion. When such a cavity fell along a fissure receiving a rich ore deposit and it too filled with ore, there resulted a lode such as the miners dreamed of but seldom found. Such was the lode struck by Woodman and Chisholm.

Further explorations in both vertical and horizontal directions showed the cavity to be very large. When the direction of the lode was established, the men filed a formal notice—not a difficult task since Woodman was the recorder of the Little Cottonwood district. The notice was dated 7 April 1870, although it carried the note "located by J. F. Woodman, June 10, 1869."[42] Fourteen names were attached to the notice; thus, with the extra 200 feet allowed for the discoverer, the claim was 3,000 feet long. Since in many cases other parties lent their names for the moment and then transferred their interest to the real owners, this was probably done with this claim as well, for some of the names appearing on the claim notice failed to appear in connection with the mine after that date. The prospect had been given a name before this time, but the notice made it official. It was the Emma. While all sorts of suggestions have been advanced for the source of this name, it was actually the name of the elder Chisholm's youngest child and only daughter.[43] Emma Chisholm later reflected that when she was only ten years old her brother William, who was in Utah working on the mine at the time, named the mine after her as a sort of peace offering, since she was constantly writing to him, begging him to come home to his "dear Emma." She did not see the mine herself until she visited the territory in 1873. She said she especially enjoyed the ride from the city to Alta, where she found the miners smoking a brand of cigars called Little Emma. She added, "They actually had my picture on the box."[44]

The Emma was not the only mine being worked at this time, nor was it the only one of promise, but the notoriety it gained in the months and years that followed placed the Wasatch on the mining map and did much for the mining boom that followed. The Emma and its story of international intrigue and scandal has been described and discussed in numerous articles and volumes.[45] It is not the intent to repeat that history here, but the early history of the mine, before it fell into British hands, has not been well documented and is interesting because it shows how most mines of that day, once discovered and producing, became pawns in the hands of promoters, schemers, and manipulators whose interest was far removed from the business of mining.

Perhaps the first big impetus for developing the Emma came in the spring of 1870, when Chisholm and Woodman sold a one-sixth interest in the mine to the Walker brothers of Salt Lake City for a reported $30,000—big money in those days.[46] This was an excellent move for all parties, for the four Walker brothers were prominent and very successful local businessmen and merchants. In the years before the railroad came to the territory, when their inventory had to be brought across the plains, they found they would lose far less material during the journey if they did the freighting themselves. Although they were known as merchants, they soon had a thriving freighting business serving not only themselves but other merchants as well. However, now that the railroad had arrived, their wagons were available for other loads, and they were soon used for hauling small lots of ore from Bingham and Little Cottonwood mines to the railroad at Uintah, at the mouth of Weber Canyon. As the output of the Emma increased, the Walkers were able to provide the necessary transportation. They also became the business managers for the newly founded Emma Silver Mining Company of Utah, making that concern a better-run business than it might have been without them.

After the Utah Central Railroad was completed from Ogden to Salt Lake City in January 1870, ore was hauled by wagon into town, loaded onto railroad cars, and taken to Ogden. From there some was shipped east to Newark, but most went to Reno and San Francisco on the Central Pacific Railroad. Then the Union Pacific Railroad, seeking more or all of the ore-hauling business, offered a one-year contract giving an excellent rate from Salt Lake City to New York. The contract was so good, in fact, that it became practical to ship the ore to Liverpool, England, for reduction. The ore shipped under the contract, which began on the first of August, was consigned half to Lewis & Son, of Liverpool, and half to Henry Bath & Son of London, the first sale to each being made at the end of October 1870.

Near the end of the year James F. Woodman sold his one-fourth interest in the mine to Warren Hussey, president of the First National Bank in Salt Lake City, for the sum of $110,000. While not a miner, Hussey had been associated with silver and gold mines for a number of years. In the early 1860s he was a land agent and gold broker in Colorado. In 1866 he and Charles L. Dahler, the Denver agent for Ben Holladay's Overland Stage line, formed a banking business—Hussey, Dahler and Company—with offices in Colorado, Montana, and Utah. During the latter part of that decade the company reaped great profits from its Montana operations; but when the gold mining business began to decline, Hussey withdrew from that northern territory and used his assets to buy the Miner's National Bank of Utah, the first national bank in the territory, and formed his own First National Bank of Utah. The new bank, with Hussey as president and Dahler as vice president, was an immediate success. Its advertisements boasted an astounding 50 percent dividend for the years 1871 and 1872.[47]

Hussey would not prove to be the asset to the Emma Silver Mining Company that the Walker brothers had been. His interests were to exploit the mine, increasing its value so he could reap a maximum return from his investment. Almost immediately the shipments of ore over the Union Pacific Railroad were increased to as much tonnage as could be taken out of the mine and hauled down the canyon, as much as 100 tons a day. The publicized excuse was to take advantage of the low shipping rates provided by the Union Pacific contract, to ship as much ore as possible before the one-year contract expired. To this end, all efforts at the mine were directed toward the removal of ore, while all other work, including the search for and development of new ore bodies, was halted. While not specifically stated, the increased traffic in rich ores was bound to give the Emma national recognition as a valuable mining property. In this unstated effort, Hussey was eminently successful. The relatively minor earlier shipments to San Francisco had alerted investors there, and almost immediately after Hussey bought into the Emma a group of California investors came to Salt Lake City to inspect the mine, although they declined to purchase at the asked-for $400,000 price. Yet almost immediately after that another California group took an option to purchase at $1.2 million. While they were unable to raise the cash, they did succeed in alerting New York investors to the importance of this new mine.

In March 1871 Warren Hussey and Joseph R. Walker, the eldest of the four brothers, went to New York and briefed Trenor W. Park, showing him maps and statements of earnings of the mine and explaining its brief history. Park was at that time president of a bank in North Bennington, Vermont, a director of a bank in New York, and president of the Panama Railroad Company. Park was convinced to visit Utah and inspect the mine, which he did at the end of March. He was accompanied by Gen. Hugh H. Baxter, a railroad magnate and Wall Street operator. Baxter was one of the few men in and around mining circles who had a legitimate claim to his title, for he was the Adjutant-General of the State of Vermont during the Civil War, superintending the fitting out of several Vermont regiments, and even personally assuming a share of the expense.[48]

After spending two days and part of a night examining the mine, Park and Baxter met with the owners who were present in Salt Lake City. Neither Day nor Chisholm was there, but Day's nephew and Chisholm's son were representing the absent owners and had their powers of attorney. Also present were James Smith, or his brother who had his power of attorney, and local minority owners Minerva Cunnington and Francis D. Clift. Cunnington was the wife of the Walker brothers' clerk, John Cunnington, who shortly thereafter went into business for himself and became a well-known and respected Salt Lake City merchant. Clift was another Main Street merchant at that time.

Although the owners had offered the mine to Park for $1.5 million, he made an offer to buy half of it for $375,000, an offer that, surprisingly, was accepted. Hussey was most likely the driving force behind this sale, for he later said, "My object in making the sale was to get Mr. Park and his associates interested with us. The parties interested with us were not willing to stock the mine and to place it where the returns would be known. They were simply working it together, and I thought that if the mine was understood, the half of the interest which I owned would be worth a good deal more to me than the whole of the interest, as it was."[49] All parties present at the meeting agreed to accept the offer, although Captain Day later said that the owners had agreed to accept nothing less than $1.5 million, and he was not happy with his nephew's acceptance in his name. Park later met with Day in Chicago and bought his remaining interest at the proportional rate of $1.5 million for the entire mine.[50]

Park's reason for offering half the asking price and buying out only half the interests was that the mine did not have a clear title. A patent application had been entered with the United States Land Office, but the government had not yet issued the patent due to pending litigation and a protest filed by one James E. Lyon, who claimed to be one of the original discoverers of the mine. Park contended that the original owners had to resolve the litigation difficulties and would be more inclined to do so if they were still holding an interest in the property. However, it was Park's intervention and negotiation with Lyon that caused the latter to withdraw his protest, allowing the patent to be issued in August 1871.

When the earlier option held by San Francisco parties expired, Park exercised his purchase agreement and on 26 April 1871 formed the Emma Silver Mining Company of New York, with a capital of $5 million, divided into 50,000 shares of $100 each. The new company took over the property and the operations at the mine, but activities there were unaffected and the Walker brothers continued as business managers. However, as soon as the patent was issued and the new company had title to the property, Park and Baxter went to England to dispose of the mine there. Because the Emma had received favorable recognition as a result of the well-publicized large shipments of rich ore, it did not take long to form a new company, the Emma Silver Mining Company, Ltd. Incorporated on 8 November, it was capitalized for one million pounds, divided into 50,000 shares with a par value of twenty pounds each. The stockholders in the New York company received an equal number of shares in the new company, but Park made a contract with them to sell their shares in England, for which they would receive the proceeds, not to exceed fifty dollars per share. The shares were sold and the Utah stockholders were bought out by the end of the year. Park did very well for himself, for at the time he disposed of the shares they were quoted at the U.S. equivalent of $115 each on the London market. After he paid the former owners their fifty dollars per share and covered his own expenses, he was still left with a tidy sum.[51]

With the mine now in British hands, the former owners were relieved of all their former responsibilities and duties—except for Warren Hussey. Surprisingly, he was appointed resident manager of the mine, reporting monthly to the British company. Although Hussey had signed the contract with Park, it is possible that he continued to hold his shares, for he later said that he had invested four to five hundred thousand dollars in the new company.[52] Hussey remained as manager for more than a year, but the mine was no longer yielding the large profits of the previous year. Transportation costs had risen after the favorable Union Pacific contract expired, and ore shipments fell far below the rates of the previous year. Then in May the mine suffered some flooding from the melting snows, followed by a caving of the tunnel. One miner died in this accident, a Cornish miner who had emigrated only one month earlier and had been working in the mine but thirteen days.[53] About a week later a massive cave-in occurred, also due to water seepage, but some reports also claimed there was insufficient timbering. This caused a considerable disruption to operations, for it was necessary to drive a new tunnel around the caved region to gain access to the rest of the mine.[54] It took much of the summer before large shipments could again be made. Then a conflict between the Emma and the Illinois Tunnel Company, in which the latter claimed rights to the Emma's ore bodies, resulted in reports and denials of violence and landed both companies in court.[55] While the Emma company won the case, the reports of events during the proceedings had not been reassuring to investors in England.

Late in the year it was announced that the company could no longer pay generous dividends to its stockholders, and the following March, after the mine had been reported in a miserable state, the board sent an English engineer to replace resident manager Hussey, of whom it was said that his administration partook too freely of an amateur character.[56] The Emma actually was suffering from the sins committed during the past two years when ore was indiscriminately removed and all other work ignored. Park later testified that when he first visited the mine, "The ore I cannot describe better than as a great sand-bank. A miner could break down, with a pick, 30 or 40 or 50 tons a day. There was five times as much work in putting up the timbers as there was in taking the ore out. In that respect it was the cheapest-worked mine in the world, perhaps costing only 2 per cent [of the value of the ore removed], while mines usually cost anywhere from 30 to 80 per cent to work them."[57] Small wonder it was very profitable during that period. When the deposit ran out, as it was unquestionably destined to do, there was still much rich ore remaining in the vein, but it had not been developed during the previous two years. Now, when the earnings were shrinking, the company had to face the expenses of the exploration that had been ignored. It also suffered continuously from the distance of a continent and an ocean separating management and the mine, as well as from a management that had unrealistic expectations and little mining expertise. The Emma was spiraling down to inevitable disaster, one that would cause an international scandal. While the mine continued to be worked by various companies and lessees until the mid-twentieth century, it would never again live up to the legend and output it had established during its first years of operation.

The Emma was not the only mine to be exploited by being promoted to British investors. There were a number of others throughout Utah Territory, and at least two more in Little Cottonwood Canyon: the Flagstaff and the Davenport. The lode that would become the Flagstaff Mine was said to have been discovered in June 1869 by Alexander Tarbet and John Snyder.[58] Snyder was probably one of the soldiers from Camp Douglas, since he had been actively prospecting in the Wasatch Mountains since the mid-1860s. But, like most prospectors, he was in need of financial backing, and for this latest discovery he found it in the person of Nicholas Groesbeck, a Salt Lake City merchant. As was the custom, the financial provider was rewarded with a portion of the claims that were filed. In Groesbeck's case, he not only placed his own name on record but those of a number of his sons as well. In the year following Groesbeck's sponsoring of Snyder, no fewer than fourteen claims were filed in the men's names, including the Flagstaff. In the latter case Groesbeck had four of his sons, a nephew, and an associate, Joseph S. Wing, who would become manager of the mine, included as claimants.[59] Almost immediately Groesbeck formed the Salt Lake Mining Company and began developing the mine. The Flagstaff, generally conceded to be on the same vein as the Emma, was less than a mile northwest of the latter mine, but it was much higher, very near the divide between Little and Big Cottonwood Canyons. Only a month after the claim was filed the Salt Lake Mining Company was preparing to ship ores from the Flagstaff and the nearby Lavinia, another of the Groesbeck claims.[60] Throughout the rest of the year regular shipments were made, all of them going west to San Francisco for processing.

Near the end of 1870 there were two events of significance to the Flagstaff. On 12 December the Salt Lake Mining Company filed a claim establishing a site for a tunnel to work the mine. Situated 400 feet south of and considerably lower than the existing shaft, it was to run in a northerly direction under the summit to allow the mine to be accessed at depth. The following summer a tramway was constructed to allow the ore to be carried from the tunnel some 1,640 feet down to where it could be reached by teams. When the winter's snows covered the tramway, the ropes and tackle were used to lower the ore on sleds.[61]

The other significant event was the construction of a smelter at the mouth of Little Cottonwood Canyon. While the Salt Lake Mining Company had nothing to do with this construction, it was to become very important to the Flagstaff Mine. The smelter was built by David E. Buel and J. C. Bateman. Buel, who used the questionable title of Colonel, had spent time in the Nevada mining fields during the 1860s and had built a reduction works to service the mines in the Reese River District.[62] In 1867 he went to the French Exposition in Paris as a representative of the mining district, and returned by way of London, where he made himself known in British financial circles. After his return to the West he joined Bateman in buying some claims and building a smelter in Bingham Canyon. Almost immediately the pair crossed the valley and began construction of the smelter at Little Cottonwood. It was located on the north side of Little Cottonwood Creek about three-fourths of a mile below the present-day fork of the highways at the mouth of the canyon. When the smelter was ready for operation in April 1871, Buel arranged for Flagstaff ore to supply one of the two furnaces, thereby starting a working relationship that was to last throughout the life of the smelter.[63]

After their smelter was operating Buel and Bateman redirected their attention to their operations in Bingham Canyon. After releasing glowing reports about the richness of the mine and upon being encouraged by Warren Hussey to market the property in London, Bateman traveled to England, where he floated the Utah Silver Mining Company, Limited, and disposed of his Bingham Canyon interests. This was to prove disastrous for the British investors, for the mine was highly overrated and failed to produce anything near expectations. On his return to Utah Bateman probably recognized the possibilities with the Little Cottonwood smelting plant and linked it up with the Flagstaff Mine. Meanwhile, Groes-beck got an option to buy out his partners in the mine for $200,000. Late in the year he traveled to England and disposed of his mine for $500,000.[64] Warren Hussey also has been credited with playing a significant role in the marketing of the Flagstaff property.[65] The newly formed Flagstaff Sil-ver Mining Company (Limited) took over not only the mine but also the Buel and Bateman smelter, which became known as the Flagstaff Smelter. After the sale of the Flagstaff, Nicholas Groesbeck left the mining scene and spent the rest of his life working in real estate. He was responsible for much early development in downtown Salt Lake City, which brought him even greater return than his brief but highly successful venture into mining.[66]

The Flagstaff was probably one of the better investments for the British, as it continued to run successfully for many years. It regularly employed a large contingent of teamsters who were kept busy hauling ore down to the smelter and materials back up to the mine. Often, after heavy snows closed the canyon roads, it was the Flagstaff that had the first wagons or sleighs moving again. When the railroad finally arrived at Alta, the com-pany built an improved tramway with a substantial ore bin at its lower terminal. It also installed a boiler and compressor at the base of the tramway to supply compressed air for the hoisting and pumping ma-chinery at the mine, the air being conveyed there through a 2,000-foot pipeline. To allow operations to continue throughout the winter, the tramway was covered with snowsheds over its entire length. And to han-dle larger quantities of ore, the company replaced the two-stack smelter at the mouth of the canyon with a larger four-stack facility at Sandy.[67] In spite of its successes, however, the Flagstaff, too, suffered from various le-gal conflicts and management problems, some of them due to the dis-tance between the mine and its owners. By the end of the 1870s it was said that there was enough room in the mine's underground workings to build a good-sized town. Yet, in spite of the millions of dollars of ore that had been taken out of the property, it was a mystery where all the money had gone.[68] The mine remained in British hands through numer-ous reorganizations and continued to operate sporadically until the col-lapse of mining operations at the end of the century.

The common thread in these three British ventures was the banker Warren Hussey. He played a strong role in the marketing of other Utah mining properties in England, and through his First National Bank of Utah he invested heavily in many mining ventures. During the financial panic of 1873 his bank, finding itself unable to meet the demands of its depositors who wanted to withdraw their money, was forced into receivership and involuntary liquidation. Final settlements amounted to twenty-five cents on the dollar. Hussey was forced to leave Utah, but he continued in the fields of banking and investments. In 1886 he was living in Spokane, where he became involved in the Spokane National Bank, only to have it fail during the Panic of 1893.[69] While he may have been the bane of his investors and ruthless in his business ventures, he left behind a legacy for the gentile population of Utah. During his period of residence in Salt Lake City he was instrumental in establishing an Episcopal mission in the city and securing a building lot and arranging a loan for the construction of St. Mark's Cathedral. He also was one of the principals in the launching of St. Mark's Hospital in 1872.[70]

*

The third British venture in Little Cottonwood Canyon was the Davenport Mine. The original discovery, in 1870, was on the ridge dividing Big and Little Cottonwood Canyons, just west of the pass between Grizzly Gulch and Silver Fork. Lying on the dividing line between two mining districts, the claim was filed in both.[71] The mine was worked for a year, yielding some rich ore, but was so located that transportation was quite difficult. The owners planned a tunnel on the Little Cottonwood side and a tramway to take ore from the portal to the canyon road, but before any action was taken the mine was sold in England.[72] While the circumstances of the sale are not clear, the mine was in the charge of John H. Ely before the sale, and he remained in charge for the next year and a half. Ely was one of the early prospectors in the canyon, having filed claims with Dr. Congar, Silas Brain, and James M. Day in 1865; however, he was not one of the original locators of the Davenport.[73]

Immediately following its sale the Davenport enjoyed a flurry of activity and lavish expenditures. The tunnel was started and pushed rapidly. As soon as the snow disappeared, a wagon road was built from Alta to the mine. During the summer of 1872 a tramway was built, extending some 750 feet from the tunnel down to Grizzly Flat, where a steam sawmill was placed into operation and an ore bin was constructed to receive the ore. By the end of the year the community at Grizzly contained an outfitting store, two boardinghouses, two saloons, and an office, in addition to the sawmill and ore house.[74] The company also undertook the construction of a smelter at the mouth of the canyon. Situated south-southeast of the Flagstaff Smelter, on the south side of today's Little Cottonwood road,

Utah State Road 209, at about 3800 East, the installation had two cupola blast furnaces, each provided with a water-powered blower. Although the smelter was completed during the winter, it was not set in operation until the following April.[75]

During this period the mine was producing some ore, but it failed to meet its current expenses, especially in view of the elaborate construction that took place. In May 1873 the superintendent, W. B. Smith, resigned and left for New York. It was suggested that Smith had fled to South American gold fields.[76] The company had at least three superintendents in the next two months, one of them, James H. Hyter, dying after being injured by a falling rock in Superior Gulch.[77] While operations continued throughout the year, the financial situation was not good. In December the miners, dissatisfied because they had not been paid for two months, placed an attachment on the company's property at the mine and smelter.[78] The mine remained idle for four months before the property was sold to satisfy the attachments. Under new management it began operations again, but seemingly without direction or goal. While ore was shipped throughout the decade, the smelter was leased both that year and the next, after which it was shut down for good.

These three mines were but a few of the many that blossomed during the early 1870s. What all of them had in common were the workers—miners, laborers, teamsters, and others who directly or indirectly supported the mining operations. And all of those people needed a place to live, to rest, to socialize, and to house families. As a result, many communities appeared, most of them small and many of them short-lived. The biggest and best known of them all was Alta in Little Cottonwood Canyon.

CHAPTER 6

Alta: A Location Being So Near Heaven

THE FIRST MINING COMMUNITY in Little Cottonwood Canyon was Dr. Congar's camp, located at the extreme west end of the flat where today's Alta is located. It was the site of the first smelter and became the base of operations for most early mining operations. In June 1870 it was known as Galena City and comprised some four log shanties, a blacksmith shop, and the indispensable whiskey shop. It was the principal mining camp in the canyon, housing some of James F. Woodman's men as well as the owners and proprietors of the Flagstaff and Lavinia mines.[1] As the camp grew, it became known as Central City. The first documented use of the new name was in March 1871, when the telegraph line the Deseret Telegraph Company was building to the Emma Mine reached Central City.[2] The telegraph office was installed in Alexander Daft's store, which had been built since the previous summer. By the end of April T. F. Fuller and A. Sullivan had announced the opening of their Central House hotel and barroom.[3] Clasbey and Read opened a general store with a large stock, and a Mr. Donnell started a brewery. By mid-June Central City could boast two stores, one hotel, and three or four eating houses, in addition to the new brewery.[4] But the city's growth was limited by lack of space, for it was located on a bench where there was room for only one street. The fact that most of the mines were located farther up the canyon caused some to relocate and settle at the confluence of the North and South Forks, where the creek from Grizzly Gulch joins Little Cottonwood Creek. Gradually others followed, and soon Central City was in its decline. By mid-November Aaron Sullivan was offering the Central Hotel for sale at a bargain price.[5]

The new community took the name of Alta, a common name for mining communities, one whose root is the Latin *altus*, meaning high. The first documented use of the name was in the *Salt Lake Herald* on 23 May 1871, at a time when Central City was still growing. Alta was very much influenced by the Emma Silver Mining Company. It was at the site of Samuel Woolley's steam sawmill, which had been sold to the Emma company in January 1871, and the principal street on which most business houses were being built was named Walker Street, certainly a tribute to the Walker brothers, who managed the mine and, as will be seen, had much to do with the creation of the town. The presence of a named street may suggest planning and orderly growth, but that was not the case, at least not so far. The first structures were randomly located, and some of them now found themselves in the middle of the new street. In fact, a meeting was called on 19 June to consider removing all obstructions upon Walker Street for the benefit of miners and property holders. About that same time miners were bemoaning the fact that Alta did not have a good hotel or general merchandise store, forcing them to walk to Central City for every article, and many had to go there for their meals and lodging.[6] But that situation did not persist very long; L. S. Fuller, who had already opened a wholesale liquor house in the new town, built a new hotel, the Alta House, at the west end of Walker Street. By the end of the year Alta society was taking form; at the end of the first week in November the very first dance party was held in the Alta House, hosted by a Mrs. Bradner to celebrate her birthday. It was promised to be only the first of a series of social entertainments.[7]

Throughout the next season Alta continued to grow while Central City continued in its decline. By November 1872 the latter place was referred to as "the old town site of Central."[8] Samuel Woolley, who made frequent trips to the head of the canyon, never made mention of Central City in his diaries after November 1872, although he did business in Alta as long as he ran his sawmills. This is not to suggest that the city disappeared, however, for some miners working claims in that area remained there. Alexander Daft kept his store open and, located as it was at the top of the long grade up the canyon, it continued to do a good business.[9] The site also continued to be used as a reference for nearby claims. As late as 1902 Albert Potter, while making a survey of public forest lands, wrote: "Central City is a mining camp and consists of just a few houses."[10]

The owners of the Clasbey & Read store at Central City built a new one in Alta one block east of the Alta House at the end of 1871, but it was crushed by the weight of the winter snows. The next spring they rebuilt the store and moved in. Merchants Street and Ralph opened a general merchandise store and found their business growing so fast that they immediately built a larger stone building, countering the common belief that Alta was a town of lumber. The hotels were filled to overflowing, making it difficult for visitors to find accommodations. Near the end of the season Nick Dramer opened his Grand Hotel with a grand ball. In a swell of exaggeration, the new building was reported to have accommodations for 120 persons.[11] One correspondent reported that during the summer Alta doubled its size, while the number of buildings on mining claims increased by a factor of three.[12] Another reporter thought Alta had grown tenfold since the last spring.[13]

New construction was also apparent on the sides of the hills, as many structures were being erected at the various mines. The most significant and impressive new additions were the tramways, however, built to bring ore down from the mines and haul materials and supplies back up. The first was installed by the Savage and Montezuma mines, located well above the Emma. It was some 1,500 feet in length, connecting the mines to the road near the Emma Mine. If it was an aerial tramway, it was not built very high, for winter's snows rendered it inoperable.[14] The Flagstaff was the next to construct a tramway, but it used rails instead of aerial cables. This was some 1,640 feet long, connecting the mine to a point on a road on the mountainside between Alta and Central City. This tramway was unique in that it continued to operate during the winter by replacing the railcars with sleds that were raised and lowered on the snow surface by the tramway cable.[15] The Davenport company built an aerial tramway from their tunnel, somewhat below their mine, to an ore bin on Grizzly Flat, up the North Fork from Alta. It was only 700 feet long but was high enough to allow operations throughout the winter.[16] However, the most noticeable and impressive tramway was built by the Vallejo company. It was more than 2,300 feet long, running between the Vallejo Tunnel, high on the mountain, down to an ore bin on the north side of Alta's Walker Street, about 500 feet below the Alta House hotel. Since it was an aerial tramway with towers some thirty feet high, the wire cables and the buckets they carried remained above the deepest winter snows.[17] It was one of the first things a visitor noticed upon approaching Alta, and therefore attracted considerable attention. As with most tramways of that period, it was gravity powered, meaning the weight of the descending buckets and their contents caused the ascending buckets to be pulled up.

As soon as this tramway had been placed in operation it received more than its share of publicity. An adventurous young lady, at first identified only as "Miss S——," on being challenged by her male companion, climbed into one of the tramway's ore buckets and soared high above Alta's north slopes for nearly nine minutes before reaching the upper terminal. When the *Salt Lake Herald* reported this event a number of ladies in the city made enquiries as to who this woman might be, but the editors chose not to reveal her name, only saying she had "scaled on foot the highest peaks of the Cottonwoods and explored the cavernous recesses of the deepest mines." While she was later revealed to be a Miss Simons, or Simmons, her identity remains a mystery today, although she may have been a daughter of Alta's Dan H. Simmons, who, with his wife, ran a restaurant on the north side of Walker Street between First and Second West.[18]

Although a surprisingly large number of inhabitants remained in the mining camp over the winter, it was as if the town went into hibernation. As the weather turned cold and the snow piled deep, Alta society became more respectable, as the gamblers, prostitutes, and loafers left for more congenial climates. When a New Year's Eve dance was scheduled, it too was affected by the severe weather. Tommy Dobbs, in a letter to the *Herald*, wrote, "in point of numbers it was not a success, there being but three ladies present. If I am not misinformed it resulted in a 'stag dance.'"[19] As the spring sun grew warmer, however, the town came back to life. In April 1873, Standish Rood, writing for the *Herald* under the pen name of Archibald, penned one of the best surviving descriptions of the town. He started at the corner of Third West and Walker Streets, walked up the north side of Walker Street to the end of town, then crossed and walked back down the south side, describing every business along the way.

On the northeast corner of Third West and Walker Streets was the Alta House, run by Stewart Fuller, and soon to be known as the Fuller House. It could accommodate seventy-five lodgers and feed three times that many. Next door to the east was the saloon and billiard room run by Behrman and Fitzgerald, offering ales, wines, liquors, and cigars, and featuring two billiard tables. Alexander Bernay had a saloon called the Miner's Exchange in the next building. He dealt almost exclusively in lager beer and his place featured a bagatelle table. Next was the Kanyon House, sometimes written as the Cañon House, a boardinghouse for laboring men run by Thomas C. Thomas. In his large second-floor room he could accommodate fifty lodgers. Archibald claimed that Thomas made no charge for lodging, only requiring that his guests provide their own bedding. While not stated, it was also likely that Thomas required his guests to take and pay for their meals at his place. On the uphill side of the Kanyon House was Street and Ralph's general merchandise store. They had a large inventory worth about $15,000, including mining supplies, hardware, crockery, footware, clothing, liquors, and tobacco. On one side of their store was the office of Wells, Fargo & Co., James E. Weigle in attendance, and the office of the district mining recorder, who at this time was John McDonald. McDonald also acted in the capacity of notary public. The post office was also in the Street and Ralph store, with the junior member of the firm, Mr. Street, serving as postmaster. One door farther up the street was one of the pioneer merchants of Alta, William Gill Mills, merchant and judge. His store offered hardware, clothing, and groceries. Next to Mills's store was the small, neat, and new bar of Ed Egan. His was a "perpendicular saloon," in that it had no chairs. Next door was the Tivoli saloon, restaurant, and shooting gallery operated by Fritz Adolphy and his wife. The ceiling of the barroom was covered with muslin in stripes of red, white, and blue. Mrs. Adolphy was reported to be an excellent shot— she could beat all the men at shooting. And the last business on the block, on the corner of Second West Street, was the El Dorado saloon and first-class chophouse, run by Thom J. Masters, where one could buy a square meal for "four bits." This building was supposed to be one of the pioneer frame houses of Alta, and it had become slightly warped by the annual heavy accumulation of snow.

Second West Street may not have looked much like a street, but nonetheless it was known as Second West. Crossing the street, on the opposite corner was Harlow and King's saloon and billiard hall, the largest in the city. It had four carom tables, as well as a new bar. The proprietors gave a new twist to saloonkeeping on Sunday, 4 May 1873, when they converted their saloon into a church and hosted the first documented church services in Alta, Rev. J. P. Schell, a minister of the Presbyterian Church, officiating.[20]

Next door was Dan H. Simmons's restaurant. His wife's kitchen was reported to have been the paradigm of cleanliness. It had, through some unexplained arrangement, clear running water, an escape pipe to carry off waste water, and a cellar for the storage of fruits and vegetables. Adjoining the restaurant was Nick Dramer's Grand Hotel, with sleeping accommodations for fifty guests. The hotel, run on the European plan, also had a first-class restaurant run by Cy Iba. However, Iba's lease expired that May, at which time Dramer took over the restaurant and Iba moved down canyon to Tannersville, where he took over and ran the Mountain House. Dramer also left the hotel business soon thereafter when the Grand Hotel was taken over by James A. Varnes and Lafayette Seaman. Such change of personnel and ownership was quite common in Alta, where businesses were established and failed with astonishing rapidity.

Continuing up Walker Street, the next establishment was the California Brewery, William Nischwitz, proprietor. He specialized in lager beer and had a bagatelle table in his saloon. The final business house on this side of the street was Abe Cohen's Pioneer Cheap John store. While he specialized in mattresses and pillows, he also kept in stock a variety of other things. On Sundays he would move out on the street and auction off anything available, from a prospect to a developed mine. Cohen was soon to relocate to Second West, off Walker Street, where he would continue his variety-store business.

Crossing to the south side of Walker Street, the first business was Joseph Brandy's Miner's Restaurant. A neat and homelike establishment, it was patronized almost exclusively by miners. In spite of the proprietor's name, neither liquor nor cigars were to be found there. If a miner wanted a drink he could go next door to McTiernay's saloon. Next was a two-story building housing a bar and billiard room run by Ewen and Rafferty. On the second floor, furnished rooms provided lodging accommodations for twenty guests. Continuing west, the small saloon of Murphy and Hughes was found. At that time it was the only building in Alta that had a house number—No. 10. Private parties could be held in a room at the rear. The next building, occupying the corner at Second West Street, was the big general store of Clasbey and Read. Their large stock was said to contain everything from toothpicks to mining cables. Of course, that included giant powder, much used in the mines. As Alta continued to grow there were concerns about storing powder in the center of the town, so the following summer, in 1874, a powder house was built on the south side of Little Cottonwood Creek, behind Clasbey and Read's store, connected to the store by a small rail tramway.[21]

On the southwest corner of the intersection was another store, the clothing house of D. Auerbach & Bro. This business did not last long, for the building was soon being used as a restaurant by C. O. Dean. The next building to the west was the Alta Meat Market, run by John A. Quick, whose motto was "Quick sales and small profit." Quick had been the butcher in Central City and moved up to Alta when it became apparent that the population was shifting. But he would soon leave Alta as well, and his Alta Meat Market would be taken over by J. C. Armstrong and R. R. Munro. Quick's next-door neighbor was the Mountain Saloon, run by James M. Halford but soon to be taken over by J. Gunther. It was claimed that this place was run on strictly puritanical principles, with nothing but faro and draw poker allowed. The saloon had butchers on both sides, for its western neighbor was William P. Rowe, who kept a well-stocked market and was always ready to furnish calves or hogs to customers.

The next two establishments were both grocery stores, the first being the California Store, the next the house of Ed Forney, dealer in groceries and provisions. One corner of Forney's store housed the camp barber, where miners and male visitors could go for a shave. The last place on the street, across from the Alta House hotel, was the Alta Lodging House, run by Charles H. Collins. Behind Collins's place, at the foot of Third West Street, was Parlin and Thompson's California Feed and Livery Stables, where travelers could stable their mounts and where the owners, who were shortly to start their own stage line from Alta to Granite City, also housed their animals.

This, then, was the Alta City business section in the spring of 1873.[22] The first impression one might get is that there was an excessive number of saloons and bars. However, if one considers the plight of the off-duty miner in this tiny hamlet, at a time when there were none of the modern entertainment features like radio, television, motion picture theaters, and the like, what was a man to do but visit the saloons and socialize with his comrades to while away the evening hours? The next impression might be that there were a number of hotels and lodging houses that accommodated a large number of people. However, photographs of Alta from this period do not show the type of buildings we would associate with large hotels. What was offered was pretty bleak by today's standards: probably cramped dormitories, with few amenities other than a roof overhead, certainly nothing like a private room with bath. In fact, there is no evidence that Alta had any bathing facilities until the spring of 1874 when Bob Bauman, the town barber, fitted up a "suite of bath rooms," complete with hot and cold water, where the miners could "wash off the chlorides" and visitors could wash off the dust.[23]

Dust, or mud, was always present in Alta. Although a few of the buildings had wooden sidewalks out front, the streets were not paved and the dirt was constantly churned by passing horses and wagons. There were a number of large trees in town, three of them standing prominently in the middle of Walker Street. A *Salt Lake Tribune* correspondent attributed the trees' survival to the "proprieties of taste" shown by Alta residents. He noted that "even in the main street there are several towering monarchs, whose lofty proportions and evergreen foliage are in relief to the plain white or unpainted buildings of the town."[24]

An interesting aside to Archibald's description of the town is the fact that he failed to make mention of the Chinese laundry that was on the south side of Walker Street at the upper end of town. A photograph of Alta taken during that period clearly shows a sign, "Sam Gee, Washing &

Ironing," next to the Miner's Restaurant. Many Chinese workers came into the territory during the building of the transcontinental railroad. Upon completion of that task some of them remained to take up residence in various communities, including mining camps, where they performed important and necessary domestic services, such as washing and ironing clothes or cooking meals. But they were commonly treated as nonpersons, invisible except when their services were needed or they were involved in disturbances. In July 1873 Archibald reported that eight Chinese men and four women were charged with keeping disorderly houses, gambling establishments, being a regular nuisance, and making indecent exposures on the public streets. They were tried and charged $120 in fines and the same amount in costs. Whether the charges had any merit is not known, but the contempt felt toward them was shown in his statement that "we would advise the 'Heathen Chinee' to quit breathing rarified air, and emigrate to climes more congenial."[25] But they remained, never in large numbers, yet always present. They suffered in the various calamities that struck the camp, helped in the rescues, and were among the first to rebuild. They were present as late as 1909, when a Chinese cook was killed in a snowslide that demolished the kitchen at the South Columbus Mine.[26]

As Alta grew it also matured, as evidenced by attempts to develop local social and cultural interests. At the beginning of May 1873 Alta's first newspaper, the *Alta Independent,* was published. However, the town wasn't ready for it, and the paper died before the end of the month.[27] Within two weeks another newspaper was announced, though it didn't appear until early July; it was the *Cottonwood Observer,* issued on Wednesday and Sunday of each week.[28] It did better than its predecessor but still lasted only three months. In October the *Salt Lake Herald* announced, "The Cottonwood Observer is defunct."[29] In its short life, however, it did manage to stir some excitement.

In its first issue the *Cottonwood Observer* revealed that the land on which the town was located had been secured by private parties. The Walker brothers or some of their associates, it charged, had acquired a patent on the land, had formed the Alta Town Site Company, and would sell lots to the occupants for prices ranging from $50 to $200.[30] Needless to say, this report raised considerable excitement among the residents. But it was not the first time allegations like this had been made. Back in 1871, when the name of Alta City first appeared in the newspapers, it was suggested that the real estate of both Central City and Alta was owned by a company in Salt Lake City that was prepared to sell lots for as much as $300 each.[31] While this initial report was false, it appeared that some unnamed party was attempting to preempt a quarter section, 160 acres, as agricultural land. Federal statutes regulating preemption specifically exclude mineral lands, and since the land in question contained numerous mining claims, protests against the preemption were filed with the U. S. Land Office.[32] Nothing more was heard about the matter, so the protests apparently had the desired result. Now, however, two years later, the matter arose again. Amid the resulting debate it was revealed that the earlier situation revolved about one Robert Nagler, who had been prospecting in Little Cottonwood as early as July 1869.[33] He had a log house near the future site of Central City and after living there for nearly two years decided to apply for a government title to the land. He was then approached by Salt Lake businessmen Joseph R. Walker, Henry W. Lawrence, and William S. Godbe, all of whom were interested in Little Cottonwood mines, about making a town site of his location. When he agreed, the group had the site platted and chose the name of Alta. About that time the protests were entered and the town site plan appeared to have been dropped.[34] While Lawrence and Godbe's connection with the Alta town site ended at this time, Walker, for reasons known only to him, continued to pursue it.

Title to public lands could be secured by individuals through a number of methods. The best known is the homestead claim, where a person lived on the land and improved it for a number of years. Some tracts could be purchased, while others could be preempted for agricultural or mining purposes. There was also a method using scrip, which was issued by the government for a variety of reasons, one being in the nature of rewards or compensation for military service. Scrip was also issued in some cases to Indians who did not qualify for land on Indian reservations, such as Sioux half-breeds who could not draw on reservation lands in Minnesota Territory.[35] Unknown to the residents of Alta, one of these scrip certificates would directly affect them. On 3 January 1873 the United States government granted to one Amanda Brown, who surrendered her Sioux half-breed certificate No. 595D, 160 acres of land in Little Cottonwood Canyon, land that encompassed the sites of Central City and Alta.[36] Miss Brown had no special interest in this specific plot of land other than its value, for on the last day of January she sold it to Samuel J. Lees of Salt Lake County for the sum of $2,500. Brown was not present for this transaction; she was represented by her attorney in fact, Robert Nagler.[37] At this time Samuel J. Lees was the manager of Walker Brothers store in downtown Salt Lake City.

Even though these two transactions were now a matter of public record, no one in Alta appeared to be aware that anything was going on. It was not until a new plat for the Alta town site was prepared that the *Cottonwood Observer* printed its exposé. The plat that had been drawn in 1871 included both Central City and Alta, and had their blocks numbered contiguously. In 1873 B. A. M. Froiseth published a map of the Little Cottonwood Mining District and its vicinity on which he showed the combined plat.[38] Shown in Figure 14, it had 116 blocks, with streets platted well up the sides of the hills on both the north and south sides of the canyon. The new plat, drawn in 1873, contained only thirty blocks, taking in Alta City proper, as it existed at that time. The street names remained the same as on the earlier plat, as did the assigned block numbers. Hence the blocks had an irregular and noncontiguous numbering system.[39]

In retrospect, it is apparent that Miss Brown, with her scrip certificate, was involved at the time the 1871 townsite plat was prepared. The application for patent probably had been made at that time, causing the initial flurry of concern among the residents. In fact, in March 1871 Robert Nagler had sold a lot in the proposed town site, but he did so as "Robert Nagler, Attorney in fact for Amanda Brown."[40] He must have sold a number of other lots as well, since deed records show that more than twenty lots were in the possession of other parties during the years 1871 and 1872, although the original transactions involving Nagler were not recorded.

Technically, most of the Alta residents were squatters, in that they held no claim to their land except by possession. Hence, the reaction to the *Cottonwood Observer*'s article was immediate and intense. Amid accusations of fraud in acquiring the patent, a meeting was held to determine what action could be taken to protect the residents. A committee was appointed to contact the Walker brothers and to consult with attorneys to determine if the patent could be revoked. The editors of both the *Salt Lake Tribune* and *Salt Lake Herald* thought the Alta residents were overreacting and that the reputation of the Walker brothers was such as to discount any charge of fraud.[41] The committee met with the Walkers' representative, and J. R. Walker himself visited Alta to assure residents that they would settle the matter reasonably with those parties who had made improvements on their property. But the residents did not want to give in to the patent holders. The committee found two attorneys who thought the patent could be overturned and offered to take on the task. At a meeting of residents one offer was rejected as being too expensive. The second offer was acceptable, but when they went to find the attorney, who had been in Alta that day, it was found he had left the canyon. Some believed that he had second thoughts and was afraid he would lose either the friendship or patronage of the Walker brothers if he took the job.[42]

With this setback the residents' indignation began to simmer, and it finally cooled when General George Maxwell and O. H. McKee made statements about the matter. Maxwell was the United States Marshal in Salt Lake City but had been the register of the Salt Lake Land Office when the Alta townsite patent request was initially filed; McKee was with a San Francisco law firm but formerly had been chief of the mining bureau in Washington. Under instructions of the commissioner of the U. S. Land Office he had personally visited Alta, examined the ground, and filed a report of his findings. At least two earlier on-site examinations had been made and reports filed. These two gentlemen assured the residents that the patent was granted on the basis of the evidence before the general Land Office, implying that it could not have been fraudulently issued.[43] With this explanation the fervor died, and little, if anything, more was said about the matter.

As it turned out, the editors of the *Tribune* and *Herald* were right; there was no great need for concern. There is no evidence that anyone was evicted from their land during the years that followed. Samuel J. Lees became the agent to handle sales of town lots. Those who bought the lots on which they settled obtained them for more reasonable prices than anyone expected, the prices noted in the deed records varying from about $25 to $350, most being well under one hundred dollars. But while matters turned out well for the Alta residents, the holders of the townsite patent still had some legal problems. Sometime during the period between the application for and issuance of the patent, the Sioux half-breed woman Amanda Brown married one John W. Heines, and on 4 October 1873 they sold the land in question to Samuel J. Brown of Minnesota for two hundred dollars. It is not known who Samuel Brown was, but the name suggests he may have been a relative of Amanda's. At any rate, on that same day, he turned around and sold half of the land to Alexander F. Bell of Gource, Michigan, for two hundred dollars. Two months later, on 17 December 1873, Samuel Brown sold another one-third interest in the land to Samuel G. Anderson of McLeod County, Minnesota, for $9,500. It appears that Alta townsite land had become much more valuable in Minnesota than it was in Alta. Since the land was in Salt Lake County, these transactions were recorded in the local county recorder's office.[44]

While the Walker brothers may have held control of the patent on the land, these transactions cast a shadow over their title, so they sent their agent Samuel J. Lees to Michigan, where he tracked down Alexander F. Bell and on 31 December 1873 bought Bell's interest in the land for five hundred dollars.[45] Lees then went to Minnesota, where he found Samuel J. Brown and Samuel G. Anderson and bought their combined half interest for another five hundred dollars.[46] One wonders how Samuel Anderson took his big loss, having bought his one-third interest for $9,500, but most likely no money had passed hands in his purchase transaction. Now Samuel Lees again held clear title to the entire property. That he was acting for the Walker brothers was made clear about ten years later after he had relocated to Walkerville, Montana, to manage Walker interests there. At that time he transferred the land to Benjamin G. Raybould, trustee.[47] Raybould was a long-time employee of the Walkers, acting first as book-keeper in their store, later becoming cashier of their bank, and always being their most trusted and dependable associate.

*

With the townsite question behind them, the Alta-ites, or Altitudinarians, as one correspondent called them, returned to their daily lives and entered two of Alta's most golden years.[48] Popular legend has given Alta a reputation of having a rough-and-tough lifestyle that featured frequent fights and shootings, equally frequent burials in its boot hill, and an overly generous serving of ruffians, prostitutes, and the like. There certainly was some of that; as the reporter Archibald once reflected, "it is true we have had a few drunken skirmishes. One man has had his nose masticated, a girl has been shot at and missed, two or three individuals have had their heads raised; but the boys up here do not mind these little things and would prefer not to have it mentioned."[49]

What popular legend ignores is the fact that Alta also had its more sensitive side, especially during the winter months when the transients moved on to warmer climes and left the town to the cadre of permanent residents. Grand balls were then a regular event. The New Year's Eve Ball of 1872 that drew only three ladies has already been mentioned; but the following winter attendance grew, starting with a grand ball at Wright's new saloon in September. In November Judge Varnes hosted a ball and dinner at his Grand Hotel. With music furnished by the Alta string band, under the leadership of J. W. Murphy, it was attended by at least eleven ladies and a host of gentlemen. Murphy's band played again at the New Year's Eve Ball and dinner.[50] Social life hit its peak in March at a ball and dinner at the Fuller House. There were some thirty-five ladies present, most residents of the snowy locale but a few from neighboring communities of Sandy, Granite, and Salt Lake City. The dance began at nine o'clock and continued until six o'clock in the morning, when the party finally broke up.[51] That winter's season finally wound down with the Strawberry Ball at the Grand Hotel late in May. Sixteen ladies and twenty-four gentlemen were in attendance. Dancing began at nine o'clock in the evening and continued until three in the morning, with a one-hour intermission for a refreshment of strawberries and cream.[52]

The following September the winter social season began with a ball at the Grand Hotel attended by twenty ladies and the same number of men. Dancing began at nine o'clock and continued until midnight, when the group adjourned to Ryan and Barratt's Alta Restaurant for supper. After the meal they returned to the hotel, where they danced until four o'clock in the morning.[53] Dances continued throughout the winter, and by February they were "as common as snowstorms." The Grand Hotel had a "magnificent ball" on the fifteenth, followed by a masquerade ball at the Fuller House on the twenty-second and a dance at the Grand Hotel the same evening. The following month a St. Patrick's Ball was given at Johnson's Hall. Late in April another dance party was given at Dan Simmons's hall, with supper at Barratt's restaurant.[54]

In addition to dances, intellectual interests were nourished by the occasional visits of entertainers or speakers. The year 1874 was especially rich in this respect. In July a group of five actors, part of a stock acting company in Salt Lake City, traveled to Alta and performed selected scenes from plays in Fuller's hotel on three successive evenings. They were followed at the end of the month by Ann Eliza Young, who had gained notoriety through her highly publicized divorce from Brigham Young the previous year. She drew standing-room-only audiences for her lectures on each of two nights, her topics being "My Life in Bondage" and "Polygamy as It Is." In December a Mrs. Stenhouse gave a lecture on Brigham Young in the Alta church. In spite of the cold season, there was a full house, including a great number of ladies. Alta had even managed to assemble a choir, and after the lecture the choir "sang several choice pieces in their best style."[55]

Spiritual and intellectual interests overlapped at the Alta church. Rev. T. P. Schell has been mentioned when he held services in Harlow and King's saloon. Schell was a young minister, just out of college, when he came to Utah as a missionary. His visit to Alta must have inspired him, perhaps upon seeing all that fertile ground awaiting his work, for he returned repeatedly to conduct services. By the end of his first summer at Alta he had inspired people to subscribe toward the building of a church. The Walker brothers, through their agent Samuel Lees, deeded three lots on the southeast corner of Belinda and Second West Streets to Rev. Schell as a site for the church.[56] Ground was broken on 1 September 1873; one month later the edifice was dedicated. The building was twenty-two by forty-four feet, with one large room on the main floor and two in the basement. It cost $1,800, two-thirds of which had been provided before construction began. The building had a belfry, for which a twenty-eight-inch brass bell had been provided, a gift of Capt. Thomas Goldsworthy, a tall Cornishman who had arrived in Alta in 1872 to superintend the Flagstaff Mine. The bell arrived late in October, was hoisted into position and began to add its echoing tones to the pulse of Alta life.[57]

While the church building was a welcome addition to Alta City, it apparently did little to increase the size of Rev. Schell's congregation. Archibald wrote, "We are called, but few choose to go." Another correspondent, "John T.," said of Schell, "His converts are few and his audiences thin."[58] During the following winter the church building became a meeting place for musical and literary entertainment. Lectures and debates were given on such diverse and esoteric subjects as the distinctive elements of alcohol, the discouragement of Chinese immigration, labor vs. capital, and the philanthropical principles involved in the words "Jesus Wept."[59] The following spring Miss May Belle Mosby started a school in the church's basement rooms. A year earlier, before the church was built, one Spencer Perser had tried without success to establish a school. Miss Mosby had better luck, for it was soon reported that she had about twenty pupils who were making fine progress.[60] She was destined to teach for only one season, however, for the following March, 1875, she married Mr. C. H. Enos, Alta store owner, justice of the peace, and correspondent for the *Salt Lake Tribune.* Unfortunately, the new Mrs. Enos's married life was to last no longer than her teaching career. She died at the beginning of March 1876, about three months after the birth of her first child. Enos remained in Alta for the rest of that summer but then resigned his appointment as justice of the peace and left both Alta and the territory.[61]

Miss Mosby had a brother, Wade, and a sister and brother-in-law, Allie G. and C. L. Torbitt, living in Alta, which probably explained her presence there. Mrs. Torbitt cared for the newborn babe immediately following May Belle Mosby's death, but Wade Mosby and his wife took the child, named May Belle after its mother, and raised it as their own. In January 1900 the young May Belle Mosby died in Salt Lake City of complications following an appendicitis operation. She was buried next to her mother in City Cemetery. When a monument was erected for the two May Belle Mosbys, the dates of the mother's death and the daughter's birth were in error by two years, probably due to faulty memory or recollection; the younger May Belle actually was two years older than the reported twenty-three years at the time of her death.[62]

In September 1874 Rev. Schell departed Alta to embark on a new mission assignment, leaving the local church without a spiritual leader. During the following two years William Gill Mills, local merchant and justice of the peace, and a Mrs. Davis, "an admirable teacher and lady of erudite attainments" who had taken over the school in the church basement after May Belle Mosby married, attempted to keep a Sunday School running. Occasionally a minister would visit and preach a sermon, and for a brief period in 1876 both Presbyterian and Episcopal clergymen traveled to Alta on alternate Sundays to preach, but the church never again realized the activity that had taken place during the tenure of Rev. Schell.[63]

The events just described did not involve the majority of Alta's residents. There were many whose tastes ran toward the more mundane activities like playing nine pins, billiards, or faro, or in chicken and pigeon shooting. While these were daily events, occasionally formal matches were held between the better players or shooters, such as a billiard match between experts named Donnelly and Atwood at Harlow and King's saloon. It was reported that about three hundred dollars changed hands after Donnelly won the match. In a similar contest between two sharpshooters named Swan and Masters, the former won the $300 prize when he shot all twelve pigeons released as targets. In a sweepstake shooting that followed, another 150 birds were shot. Although barroom debates often escalated to an exchange of blows, such minor scuffles never created as much interest as sparring exhibitions that were arranged and announced in advance.[64] Occasionally such matches featured pugilists from neighboring mining districts, adding a higher level of competitive interest. Wintertime activities included snowshoe races, some of them causing more than a few dollars to change hands. Impromptu races were held between miners from the Flagstaff, Vallejo, and other mines high above Alta, who would get astride a scoop shovel and go sliding down the slopes. Since their ability to control their mounts decreased as their speed increased, spectacular crashes were the usual result.[65]

An attempt to bring together all of Alta society in one event was made for the Fourth of July celebration of 1875. The year before, the Fourth had been celebrated by a ball at the Fuller House on the evening of 3 July. But that was the typical "high society" event, attended by ladies from Alta and Salt Lake City and a "fair proportion of gents." It followed the usual pattern: dancing started at 9:30 and continued until 12:30, when a supper was served. Dancing then resumed until 3:00 in the morning. But there were no organized events to occupy the residents' attention the next day. Since the residents were left to their own resources and fueled by liquid refreshments, several fistfights broke out and at least one shooting took place.[66] Although there were no serious injuries, a more detailed program was planned for the following year. It included a parade around several of the city's blocks, an oration at a grove of evergreens below the city, fireworks in the evening, and several balls that night. Because the Fourth fell on Sunday, the celebration was held the following day. Spectators and participants came from surrounding mines; a few even came from Salt Lake City. The procession formed and the parade commenced. The *Salt Lake Tribune* correspondent estimated that 2,500 persons took part, certainly an inflation of fact.

After the parade the crowd moved down to the grove, where a speakers' stand had been erected for the occasion. After an oration by Judge Tilford, the special event of the afternoon was to take place. A symbolic bird, a great American eagle, was to be given its freedom to fly overhead, among the lofty peaks. The bird had been in captivity in Salt Lake City for several months before General Maxwell purchased it for this occasion. The *Salt Lake Tribune* was certainly one of the promoters, for it provided a silver medal to be attached to the bird's leg. On the medal was the engraving: "SALT LAKE TRIBUNE EAGLE, Alta, Utah, July 4th, 1875. From the chain gang of Zion to the free air of Alta." The latter referred to the fact that the bird had been in chained captivity for some time. However, that confinement was to thwart the plans for the day, for when the bird was released it was unable to fly. It flapped its wings, but instead of soaring overhead it crashed into the crowd. The high spirits of the crowd were manifest in its forgiveness, and the bird was taken back into Alta where it was given good care worthy of its rank as the national symbol. That evening large bonfires lit the slopes of Emerald Hill southwest of town, while rockets were fired from adjacent peaks and Roman candles ignited near town. The day's festivities ended with balls at the Fuller House and Joe Barratt's Grand Hotel. The symbolic American eagle lost a bit of its symbolic stature several days later when it was found feasting on the "carion and filth" at the local slaughter yard. The *Salt Lake Herald,* probably still smarting from publicity the *Tribune* had gained by sponsoring the event, claimed of the bird: "So mean has he shown himself to be that it is now said he is a base, ignoble buzzard, and not a drop of eagle blood flows in his veins."[67]

*

Before leaving the subject of social life in Alta, it is necessary to note that it also had a dark side. Still, much of what has been written about the rough-and-tumble life at Alta, the shootings, killings, and overpopulated "boot hill," has been exaggerated. While the population was predominantly transient males whose social lives centered upon the saloons, there were surprisingly few deaths due to shootings or other violence. There were many fights and skirmishes, but that sort of thing happened in downtown Salt Lake City as well. While many an argument escalated to the point of extreme violence, only twenty-five such instances were reported in the newspapers in the decade of the 1870s. All but two involved gunplay; the exceptions made use of clubs, drills, axes, and teeth. And, in all these, only six deaths resulted.

The first killing, the result of a dispute over the title of a claim, took place at a mining camp in Little Cottonwood Canyon. On 13 August 1870 Henry J. Woodhull was shot in the abdomen by one Nathan Springer and died the following evening. Woodhull was the eldest of at least three brothers who had come to Utah from Michigan to take part in the budding mining scene. They did some prospecting in Little Cottonwood Canyon and filed numerous claims in association with other prominent old-timers such as Silas Brain, J. F. Woodman, R. B. Chisholm, James Wall, and A. A. Hirst. Early in April 1870 the brothers announced the opening of an office in the Elephant Store on East Temple (Main) Street in Salt Lake City where they would purchase silver and lead ores and perform assays on ore samples.[68] Initially they shipped these ores to West Coast smelters via the Utah Central and the Central Pacific railroads, but by early June they had started the construction of two furnaces of their own on the State Road at Little Cottonwood Creek. By the end of July the new smelter was in operation, primarily working on ores from the Little Cottonwood district.[69] With the amount of activity at their smelter the Woodhull brothers were becoming well known in the Utah business community, so the news of the killing drew a great deal of interest. The body was brought into the city, where an inquest was held, and then was shipped to Michigan for burial. Springer was taken into custody and denied bail pending the next term of the District Court.[70] The surviving brothers, William S. and Sereno D. Woodhull, continued running the smelter and remained active in the Little Cottonwood mining scene until the middle of the 1870s, when they drifted into obscurity.

The next killing took place nearby about one year later, after the camp had become known as Central City. On election day in Little Cottonwood, Monday, 7 August 1871, men gathered at a saloon and after partaking of much good cheer moved outside for some athletic exercises. Patsey Marley and Mike Gibbons led off by engaging in a friendly sparring match. After some skipping about, dodging and feinting, Marley planted a solid blow and Gibbons found himself on the ground. He got up and went after Patsey with a vengeance. The boxing became furious fighting. At that point, Gibbons's friend Tom McLaughlin entered the fray by planting a rock upon the head of Marley, whereupon the popular Marley fell to the earth. Someone fired a challenging shot into the air, which was followed by shots from several quarters. When it ended one Capt. Carey had been shot in the leg and a Mr. McCarty suffered a bullet through the bowels, an injury that proved fatal, as he died early Wednesday morning. McLaughlin and Gibbons beat a hasty retreat; however, since the coroner ruled the death was due to a pistol shot from an unknown source, no charges were filed.[71]

Another shooting death took place in November 1872. Bill Hawes, a former foreman at the Flagstaff Mine, broke into Fritz Adolphy's saloon and shooting gallery in the early hours of the morning and demanded something to eat. H. C. Miller, night watchman and employee at the saloon, refused to serve Hawes and threw him out. That evening the two men met on the street and after a few words began shooting. Miller was shot in the hand, but Hawes, who was said to have been a dangerous man when drinking, received four bullets in his body and died almost immediately. He had come into the territory from Montana via Austin, Nevada.[72]

Many of the fights and shootings revolved around women, specifically the prostitutes who frequented the mining camps. One of the better known women was Annie E. Miller, a native of Maine who came to Utah by way of California. In 1871 she appeared in Central City, where she opened a saloon, the scene of several shootings, albeit none of them fatal. When the population migrated to Alta City, Madam Miller moved too and opened a bagnio in the new city. Although many of the "soiled doves" left for warmer localities during the winter, Miller and a few of her "associates" remained. In February 1873 she had one of her "lovers" arrested for assault and battery, but at the trial her wrath cooled and she withdrew the charge. Judge Varnes thereby fined her twenty-five dollars for contempt of court, but, after consultation, the fine was remitted. "Such is justice in Alta," mused the newspaperman Archibald.[73]

Only two months later Madam Miller was in the news again when she was shot at by another "lover," a "Stalwart Missourian" who took after her with a double-barreled shotgun. He was arrested but soon was released without a trial. "The madam always relents after her first passion," wrote Archibald.[74] That same week, one of the other ladies of the evening, Mattie Ritticker, shot a Flagstaff miner through the cheek with a revolver. Through a little "judicious petting" by Ritticker, charges and arrests were avoided.

Madam Miller met her end on Sunday, 30 November 1873, when her house was discovered to be on fire. An alarm was sounded, but the fire had such headway that nothing could be done except to prevent it from spreading. When the flames died two bodies were found in the debris. They belonged to the madam and her companion of several months, a Captain Hart. Earlier in the day Hart had been arrested on a charge made by one of Miller's female lodgers. Later Hart, now intoxicated, found Miller in a saloon, where they began to quarrel. They returned to the house where Hart shot Miller, set the house afire, and shot himself.[75]

Only two months later, in February 1874, another of the demimonde met her end in her own house of ill repute. Two visiting miners, Archelans Barratt and Stephen Jeffries, got into an argument. Barratt took a rifle and shot at Jeffries, but the bullet missed its intended target and struck the hostess, Lizzie Sanders, in the abdomen. Dr. Bevan was called to attend to the woman, but he pronounced the wound a fatal one. Justice William Gill Mills was called to take a deposition from the dying woman, who also dictated her will, leaving her house and lot in Alta, a gold watch and chain, and over $500 in currency to her nephew in Canada. Only a few weeks earlier she had married a miner from the Emma Mine and on their wedding night robbed him of his money and turned him out of the house. Yet, in her deposition she evinced no animosity against Barratt and claimed she believed the shooting was accidental. Many shared in the belief that Barratt was innocent, but he was later sent to the penitentiary for the shooting, only to be pardoned by the governor in September 1879.[76]

There were as many or more people in and around Alta who died by their own hand as died through shootings. At one point a *Salt Lake Tribune* editor wrote, "suicides are becoming epidemic at Alta, it is said on account of the location being so near heaven."[77]

*

In September 1875 Alta realized a significant milestone in its development when a long-awaited railroad connection with the Salt Lake Valley was completed. The miners had often hung their hopes on promises of more efficient, less expensive means of transporting their product down the canyon, only to see one scheme after another vanish. The railroad headed their way, but ever so slowly. Now, on the twelfth day of the month, the first cars came up the canyon carrying passengers, freight, Wells Fargo express, and United States mails. They then returned with a full load of ore, twenty tons coming from the Vallejo Mine alone. The impact of rail service was seen immediately, when the passenger fare from Alta to Fairfield Flat near the mouth of the canyon dropped to $1.50 from the $2.50 charged by the stage operators.[78]

After the completion of the Utah Central Railroad from Ogden to Salt Lake City, Brigham Young wanted to continue the line to the south to link the many communities in that direction, but he realized that he could build only as far as present business traffic would warrant.[79] Since a considerable part of the Utah Central's traffic was made up of ore shipments from the various mining districts, he chose to capitalize on that market. The new road, the Utah Southern, was incorporated on 17 January 1871. Initially it was planned to be narrow gauge, with the rails three feet apart instead of the standard four feet eight and one-half inches, because there was considered to be "too much dead weight, involving useless expense, in the wide gauge, with its cumbrous rolling stock."[80] However, by the time ground was broken, on 1 May, the preference for narrow gauge was fading. A week later the *Salt Lake Tribune* noted, "The gauge is changed and revelation has again given place to common sense."[81] The first spike was driven on 5 June, and the railroad slowly pushed southward. By the beginning of August, about twelve miles south of the capital city, a station had been located where ore and freight from Bingham and Big and Little Cottonwood Canyons would be received. Because of the nature of the soil at that place, it was called Sandy Station. Service between the city and that station was begun on 6 September.[82]

While it was expected that branch lines would be built to both Little Cottonwood and Bingham Canyons, all track laying was halted at Sandy Station due to a lack of rails. Meanwhile, grading continued southward for the main line, as well as to the east for the Little Cottonwood branch. There can be little doubt that this branch line was doubly important to Young, for it would serve the granite quarries, the source of the stone for the Salt Lake LDS Temple, as well as the mines. So important was the first consideration that on 13 August the Salt Lake City Council approved the construction of a track on South Temple Street to link the railroad depot with Temple Square. About five weeks later Brigham Young wrote, "It is a novel sight to see the locomotive steaming into the Temple Block with its five car loads of granite."[83]

The stone was brought to the Sandy railhead by teams and wagons, although other means of transportation were under consideration. In May 1871 a "road steamer," which had been purchased in the east by Brigham's son John W., arrived in Salt Lake City. It was fired up and run on the city streets, much to the amusement of the local residents. John W. Young later used the road steamer to haul his steamboat, *Lady of the Lake,* from the Utah Central depot to the Jordan River, where it was launched to cruise down to the lake.[84] About this same time Brigham Young revealed the real purpose of the road steamer when he wrote, "The railroad will prove of immense advantage in freighting rock for the Temple, & to further facilitate this work we contemplate putting a Road Steamer, which is now here, on the road between the quarry & the station."[85] Although at least one personal history made mention of the steam road engine running between Little Cottonwood and Sandy, D. McKenzie, Brigham Young's secretary, wrote that they had two road steamers, but because of the general unevenness of the roads and the frequency of irrigation ditches they were not yet in use.[86] It is likely they never were placed in general use, because in the fall of 1871 the Utah Southern Railroad began grading a roadbed toward Little Cottonwood Canyon. Before ties and rails could be laid, however, a new plan was afoot which would take the Utah Southern out of the branch line business so it could push its main line southward and leave the branch lines to others.[87]

On 24 October 1872 an independent company, the Wasatch and Jordan Valley Railroad, was incorporated to build a railroad line from Sandy to Alta in Little Cottonwood Canyon. Most of the officers of the new company were Salt Lake businessmen and staunch members of the Mormon Church. Even though Brigham Young's Utah Southern Railroad no longer controlled this important branch line, Young still held effective de facto control. As the *Salt Lake Tribune* once acidly commented, Brigham "is fond of projects and is really the 'Boss' engineer in all such matters. It is he who determines the grades and curves for home railroads and the composition for whitewash for the Theatre . . ."[88] It is interesting to note that the new company chose to build a narrow gauge line. While no explanations were published for this choice, several reasons can be offered. With its rolling stock and rails being smaller in size and lighter in weight than their standard gauge counterparts, a narrow gauge railroad required less capital to construct and operate, and thereby became the gauge of choice for small branch railroads. It would not be the first narrow gauge railroad in the territory. For the past year John W. Young had been building the narrow gauge Utah Northern Railroad running north from Ogden toward Idaho and Montana. In November 1871 the narrow gauge Summit County Railroad Company was formed by Brigham Young's eldest son, Joseph A., to link the Grass Creek coal mines above Coalville with the Union Pacific Railroad at Echo Station. Its close link with John W.

Young's railroad operations was shown by the fact that John W. acted as eastern agent for the company. Also, the Summit County Railroad began operations with a locomotive borrowed from the Utah Northern. In April 1872 the narrow gauge American Fork Railroad was organized by principals in the Miller Mining Company to link that company's Sultana Smelter in American Fork Canyon with the Utah Southern Railroad at American Fork. And finally, in September 1872 the Bingham Canyon and Camp Floyd Railroad was formed by Bingham mining interests to build a narrow gauge connection between the Bingham mines and the Utah Southern at Sandy Station. This last event may well have precipitated the decision to form a separate company to build the Little Cottonwood branch line.

The Wasatch and Jordan Valley company took over from the Utah Southern the three and a half miles of roadbed prepared the previous year, then wasted no time getting construction under way. Within a week a preliminary survey was completed, and on 5 November a groundbreaking ceremony was held at the end of the completed grade.[89] The immediate goal was to get the line built and operating to the mouth of the canyon. In spite of severe winter weather the road progressed rapidly. On the last day of February 1873 the company's first locomotive arrived from the east and was placed on the track, and at the beginning of April it was announced that the road was completed to the first granite quarry opposite the Davenport Smelter, about three-quarters of a mile west of today's fork in the highway at the mouth of the canyon.[90] On 4 April, on the invitation of Wasatch and Jordan Valley president William Jennings, Brigham Young and a party of guests left the city and traveled to Sandy Station on the Utah Southern, then transferred to the narrow gauge train and rode to the quarry. There they witnessed the first loading of granite onto a flatcar, which was brought down on their return trip.[91] Shortly thereafter the track was extended another three-quarters of a mile to Granite, the quarrymen's camp. Although the railroad still had no passenger cars, two more excursion trips were hosted in April, the participants riding on flatcars or in the mail and baggage car.[92] On 28 April, as soon as the railroad's two passenger cars arrived, it began regular freight and passenger service, scheduled to meet the Utah Southern trains at Sandy Station. Fare was seventy-five cents each way for passengers and one dollar per ton for freight.[93]

While the new rail connection made travel to and from Alta much more convenient, especially for passengers, Alta residents looked forward eagerly to the extension of the line to their city. The railroad did continue construction farther up the canyon to Fairfield Flat, where Wasatch Resort would later be located; on 22 September it opened the additional section for business and established its terminus there for the winter.[94] The following season there were great expectations that the railroad would be continued up the canyon, but nothing happened. It is possible that with service established to the granite quarries the company's management had achieved its goal and had no further interest in going farther, even though a good part of its traffic, and income, was a result of mining business. Matters remained at a standstill throughout 1874 and into the spring of 1875, when the railroad company changed hands.

In July 1873 a group of eastern investors headed by Charles W. Schofield had taken control of the Bingham Canyon and Camp Floyd Railroad, before that company had laid any track. The new management moved quickly, and by the end of the summer it had trains running between Sandy Station and Bingham Canyon. Now, in June 1875, the same group took over the Wasatch and Jordan Valley Railroad and immediately put a force of graders to work on the roadbed from Fairfield Flat to Alta.[95] While the Little Cottonwood wagon road was in the canyon bottom, running close to the creek, the line of the railroad stayed well up on the north slope, following an alignment much like that of today's highway. With construction crews working from both ends, the grading and track laying was completed in the brief period of two months. The road was opened on Sunday, 12 September 1875. The grade up the canyon was so steep that it was considered impractical to use steam locomotives, so the new section used horse and mule power to take the cars up the canyon, and gravity power to bring them back down. As a result, the line was commonly known as the Alta tramway, or simply the tramway. The animals were harnessed in tandem to haul the cars up and were either driven down the road or turned loose to find their own way back to the stables at Fairfield Flat. As many as seventy horses and mules were used. The cars were much smaller than those used on the adjoining steam line, these being small four-wheel cars, the freight cars appearing as a box on wheels, and passenger cars being flat with three bench seats, carrying nine passengers. Although the freight cars could have held bulk ore, the ore was usually sacked for convenience, the sacks weighing seventy to one hundred pounds, since the freight had to be moved into the narrow gauge cars at Fairfield Flat, and again into standard gauge cars at Sandy Station. Ore was always sacked for handling by teamsters.

On the trip down the canyon a brakeman would ride the car, keeping its speed under control. It must have been an exciting ride, as the grade was very steep. Within a few weeks the line suffered its first accident when a brakeman panicked during the descent and imagined his brake was broken. He jumped off to save himself, leaving his car to career down the track unrestrained until it crashed into several other cars, knocking them off the track and killing four mules.[96] Surprisingly, however, accidents were infrequent after that incident.

The new management realized that there was little chance of keeping the tramway open throughout the winter, so it proposed to cover the tracks with snowsheds. They started immediately about a half mile above the lower terminus and got the track covered with wooden sheds as far as Tannersville before winter's snows halted construction. In 1876 they expected to complete the sheds all the way to Alta but got only as far as Superior Gulch, a mile and a half below their destination. But the tramway ran that far all winter, and the following spring the sheds were completed up to Alta. It had been a phenomenal task, costing some $105,000, well above the initial estimate of $60,000 to $70,000. The uphill side of the track was protected by a substantial stone wall that supported the butt end of heavy timbers sloping down over the track at the same slope as the hill. At places having the most severe avalanche danger the timbers were placed only two feet apart.[97] The entire structure was then covered with heavy planks.

While the sheds were advantageous to the railroad company, they drew mixed reactions from passengers. One correspondent wrote that the unbroken snowshed gave one the impression of riding through a log cabin.[98] Another wrote, "The natural feeling of a person going up for the first time is that the tunnel is without end. It is narrow and far from straight, and in places, where the worst snowslides occur, the massive granite walls and the darkness make it look like a huge dungeon."[99] Another writer noted that while riding up through the sheds, "Some one mentioning the snow sheds as being seven miles in length, Charly [a young rider] being somewhat fatigued answered, 'I thought it was two hundred miles long.' A remark which any one could not fail to appreciate who had experienced the tedium of the journey."[100] Charley Cunningham, the good-natured conductor during the years immediately following completion of the tramway, did his best to make the ride through the dismal snowsheds as pleasant as possible. But when one woman reportedly told him she didn't like to ride in the sheds in the summertime because the occasional patches of sunshine coming through knotholes and cracks made her face freckle, he didn't, or couldn't, say a word. He just released the brake and let the car slide gently out of Alta.[101]

CHAPTER 7

Alta Disasters

OVER THE YEARS the city of Alta had its economic ups and downs. Mining companies came and went, miners went on strike, and prices paid for lead and silver fluctuated—all of which affected the well-being of the city and its occupants. Nevertheless, the mining camp generally remained a safe and secure place to live. But all that changed in the early evening hours of Tuesday, 19 January 1875, when a snow avalanche, starting near the top of the mountain on the north side of the canyon, came roaring down the slopes and crashed into the upper part of the city, smashing several houses and claiming six lives. Never before had the mining camp been assaulted in this way, and the tragedy left residents of both Alta and Salt Lake City stunned and reeling in disbelief.

Snow avalanches are the natural result of heavy snowfalls on steep mountain slopes, a situation realized annually in the Wasatch Mountains. The early Salt Lake City settlers were not particularly affected by this phenomenon since they did not frequent the higher mountains during the winter. Avalanches soon became a recognized fact of life, however. In January 1850 a group of men from the city took their teams and sleighs into Dry Canyon, east of the city, to gather firewood. While they were working on the slopes an avalanche came down, filling the bottom of the canyon to a depth of some fifty feet and burying their sleighs. Fortunately neither the oxen, having been taken off the sleighs, nor the men were injured, but they returned to the city without their firewood.[1] As loggers and miners increasingly worked the higher mountains during the winter, avalanche incidents became more frequent, and by the time of the first Alta tragedy at least twenty-seven lives had been lost to this cause, ten of them in Little Cottonwood Canyon alone. A serious avalanche, still vivid in the memories of Alta residents, happened two years earlier, on the day after Christmas in 1872. On that day, teams were moving up and down the canyon road when a large avalanche came down the slopes of Mount Superior or one of the gullies east of it. It smashed into the wagons, scattering them, their teams, teamsters, and contents along the slide path, carrying some of them across the creek onto the slopes on the south side. Rescue parties worked several days looking for survivors. Six men, plus numerous mules, died in that accident, the body of one of the men not being found until the following June.[2]

During all this time, although avalanches came down all around them, the residents of Alta were spared. However, during this same time drastic changes could be seen on the slopes above the camp. The trees that had served to hold and stabilize the heavy snows were being cut at an ever increasing rate to satisfy the growing appetite of the community and the mines in the vicinity. Without this vegetative support the snows had a greater tendency to slide, especially in conditions of very heavy snowfall in a short period of time. That situation occurred in January 1875 when a storm of unprecedented strength and duration struck the area. Starting on Monday, 4 January, it continued for two weeks. As the snow accumulated on the slopes, the inevitable avalanches began. The following Monday four men were killed when an avalanche swept down over the Annie Tunnel in Honeycomb Fork, while the superintendent of the Cooper Mine lost his life en route from Alta to his mine.[3] That same day a slide came down Emerald Hill, south of Alta, trapping two miners who took refuge in their shaft; it then crossed the creek before coming to a halt only 150 feet from Parlin and Thompson's stables, very close to Alta's business district.[4]

The next day a slide demolished the boarding and ore houses at the Vallejo Mine, killing the black cook.[5] Still the storm raged on. Another week later, on the evening of 19 January, the dreaded avalanche finally entered Alta city. It actually came down in two portions, the first striking and destroying the ore house at the Emma Mine, the other sliding into town, smashing several houses, and coming to a halt in front of William P. Rowe's house on the east end of Main Street. The snowy mass was soon covered with rescuers armed with lanterns, shovels, and poles. A faint cry for help was followed until the men uncovered the cabin in which Michael Kelly and George Tomlinson were trapped. Kelly had suffered a broken back and was dead, but Tomlinson, though severely bruised, survived. At another place of avalanche debris men were digging for the broken remnants of James Carey's house. When they reached it they found a heartbreaking scene: Carey; his wife; their nine-month-old child; Laura Wilson, the six-year-old daughter of Clara Wilson, who boarded with the Carey family; and John Vanderleen, a young man who was in the house at that fatal moment, all were dead. Mrs. Carey was found in a chair, grasping her child to her breast. C. H. Enos wrote, "It was a terrible sight, one which I hope I may never be called upon to witness again. Two men and one woman, in the prime of life, and two children hurled into eternity without a moment's notice."[6] This event became an Alta legend, one that was told and retold in the years that followed. Fifteen years later Alfred Lambourne wove this story into his short novel *Jo—A Christmas Tale of the Wasatch.*[7]

The bodies of the victims were taken to the house of local businessman Fritz Adolphy, where they were dressed for burial. The next morning they were taken down the canyon and into Salt Lake City, where all but Laura Wilson were buried in City Cemetery. Mrs. Carey and her child were buried in the same coffin, "the child resting on the mother's breast, as sweet and life-like as on the night of the terrible calamity," according to a newspaper account. The body of the Wilson girl was sent to Virginia City, Nevada, for burial.[8]

The storm raged for two more days before it finally broke, allowing the sun to brighten the scene. It had lasted seventeen days and left seventeen people dead. While the avalanche tragedies may have shaken the residents of Salt Lake City, they left Alta citizens in a state of near panic. Before the storm ended, some thirty or forty of them left their homes and went down to Salt Lake City to spend the rest of the winter. For those who remained, the huge mass of snow studded with boulders, timbers, trees, and other detritus was a constant reminder of their narrow escape from injury or death. They tried to resume a normal life, even organizing more dances and balls than ever before, until Enos reported "dances are becoming as common as snow storms."[9] Then, in March, after barely enough time to have allowed the resident's anxieties to subside, another storm blew in and hammered the Wasatch Mountains again, precipitating new avalanches in many quarters. This storm also caused the destruction of Mill A and the buildings surrounding it in Big Cottonwood Canyon. On that same day, 17 March, another avalanche plowed into the upper part of Alta, wrecking several houses and killing John Strong and his wife. Several others were buried but were rescued uninjured. The danger was so great that miners from surrounding mines went to Alta for safety, but even there everyone moved into the lower part of town, fearing another slide might wipe them out of existence.[10]

After the spring sun melted the snows, the surrounding hills turned green and the flowers bloomed in their alpine profusion; those who had fled Alta during the winter returned and life in the camp began to return to normal. Soon the horrors of the previous winter were but a faint memory. Then, after three years of tranquil existence, disaster struck again. This time, however, it came not with a dreaded avalanche but from another equally feared source. For many years Alta residents had been concerned about the danger of fire, perhaps more so than about snow slides. In July 1873 the stables of Messrs. Lingo and Anderson caught fire and burned to the ground. Out of twelve horses in the stable at the time, four perished. Although a bucket brigade was formed, it was recognized that it was only the unusual direction of the wind that allowed the firefighters to save an adjacent house. The intensity of the flames caused one correspondent to remark he had no idea "that Hades was so near the clouds as the altitude of Alta."[11]

At the end of November of that same year the house of Madam Miller burned, the result of the murder-suicide already mentioned. By the time the fire was discovered, it had gained such headway that there could have been no thought of saving the house. It was only through the intense exertions of the firefighters that the fire was confined to the one house, and it was recognized that the entire town had been in danger of burning.[12] While the winter snows reduced the fire threat, many businessmen took preventive measures the following spring. Some of them, including Clasbey and Read, Street and Ralph, James Tucker, and William Gill Mills, built platforms on the roofs of their buildings on which were placed barrels filled with water. The first mentioned used a 430-gallon barrel. Correspondent Archibald estimated the weight of that water to be two and a half tons, then joked that "more danger is to be apprehended from a flood than a fire." But he applauded the action. The town was built of pine lumber, he noted, and if a fire started when there was a fair wind, the waters of Little Cottonwood Creek would not be sufficient to arrest its progress.[13] This also was the time when the merchants, realizing the danger of storing explosives in their stores, got together to build a powder house on the south side of the creek, connecting it with Clasbey and Read's store by means of a tramway.

In the summer of 1874 a public meeting was held in the courthouse to organize a "hook and ladder company" and to appoint a committee to examine stove pipes and take other measures that might be advisable to prevent the town's destruction by fire.[14] In November of that year a fire started in the back room of Armstrong and Monroe's meat market, caused by heating tar in a lard can on the stove. Prompt use of a fire extinguisher and the formation of a bucket brigade brought the fire under control, but only after everything in the room was destroyed. The *Salt Lake Tribune* correspondent was very concerned about this. "People cannot use too much caution about fire," he wrote, "for it certainly would be very unpleasant for us to be burned out and have to camp out some of these nights in six feet of snow."[15] His concern was still strong the following spring when he recommended that the previous year's fire committee be formed again. The recommendation was taken and the committee commenced their duties in earnest.[16] Their efforts must have been successful, for three years passed without another major fire incident. Then in the afternoon of Thursday, 1 August 1878, came the dreaded call: "Fire!"

As people ran out into the street they saw smoke boiling up from the Swan Hotel at the lower end of town. But as they rushed there they were beaten back by the intensity of the flames. Within five minutes the fire had spread to a nearby blacksmith shop and to the Fuller Hotel. The strong wind blowing up the canyon fanned the flames until they jumped the street, setting fire to Tucker's store. The fire had reached such proportions that the people were rendered helpless. As flames marched up Walker Street, spreading from one building to the next, loud reports of giant powder exploding were heard, convincing everyone to retreat to the slopes above town. Within forty minutes the entire town was a boiling mass of flame, the fire scorching Patsey Marley Hill to the east and sweeping up the road to Grizzly Flat, consuming most cabins on the way. When the flames finally subsided, the town was in ashes. Out of about 140 buildings that stood that morning, only fifteen remained, including the courthouse, Mrs. Sanford's lodging house, the California Stables, and an odd assortment of homes and cabins. The flagstaff on Walker Street still stood, charred and blackened, but of the trees that had graced the center of town there remained only stumps.

The fire had moved up the north side of town, burning the ore house and tramway of the Alta Consolidated, formerly Vallejo, and the church that Rev. Schell had worked so hard to build, but it spared the Wasatch and Jordan Valley tramway station. Incredibly, no one was killed or physically injured. But the townspeople stood over the smoking ashes in stunned disbelief. Very little had been saved and most people had nothing but the clothes on their backs. Superintendents of mines in the vicinity, such as the Flagstaff, Lavinia, Prince of Wales, and Emma, responded by sending such supplies and food as they could spare, while those of the

Iris and Emma tunnel companies opened their boardinghouses for the survivors, as did the few people whose homes had not been destroyed.

The next day a considerable number of Alta residents left for Salt Lake City, while others remained to sift through the ashes. Many of the merchants, though ruined, took steps to continue their businesses. James Tucker made arrangements to bring in a new stock of goods and set up shop in the California Stables, John Strickley proposed to open a new store in a building in old Central City, Ed Williamson opened a new drug store in the courthouse, S. Fuller made plans to build a new hotel, the post office was moved into the office of the Wasatch and Jordan Valley tramway, J. J. O'Riley reopened his clothing store in a tent, and the California Brewery reopened in a tunnel on the south side of the canyon where they had stored a quantity of beer. Meanwhile, citizens of Salt Lake City responded by raising about $600 and sending a carload of provisions to the beleaguered mining camp. Additional contributions were sent by citizens of Bingham and Park City.[17] At this same time contributions were being solicited for a fund to aid people in cities in the southeastern part of the United States who were suffering from a yellow fever epidemic. In spite of their own problems, the Alta masons sent $94.50 for the yellow fever fund, a noble act of charity.[18]

Before anyone could start to rebuild, the citizens held a meeting at which they discussed changing the site of Alta. It was announced that the Walker brothers had offered to deed lots to those who had lost buildings and intended to rebuild. It was then resolved that the new Alta should have its main streets running north and south instead of east and west; that the principal business street should be no farther east than Third West Street on the old plat, that being the lower end of the burnt-out town; and that there should be a space of at least twenty-five feet between buildings. The new location also was thought to be much safer from snowslides.[19] With this resolution the reconstruction of Alta began.

Within six weeks it was reported that ten business houses had been completed and occupied. The new Main Street was made wider than its predecessor and the buildings were separated so water could be poured on them from all sides in case of fire. To ensure an adequate supply of water for both fire and culinary use a pipeline was planned to bring water from the Emerald Hill Mining Company's reservoir, about a half mile southeast of the new town. It would, however, be another year before the waterpipes were installed and operating.[20] By mid-October about one hundred houses had been built and nine saloons were in operation. With that report, the correspondent remarked that another fire would make every citizen a saloon keeper.[21] Within another week the Alta Consolidated (Vallejo) had its tramway rebuilt and running, allowing that company to resume ore shipments. The town was ready and secure for the winter.

Not all those who were burnt out chose to rebuild, however. Dan H. Simmons had offered his restaurant for sale in 1875, but, apparently unable to find a buyer, opened it again the following spring. Now, with his business in ashes he abandoned Alta and moved to Salt Lake City, where he opened an oyster parlor on First South Street.[22]

By the beginning of the new decade Alta had settled into its new appearance and lifestyle. The fact that the main street ran across the canyon rather than along the canyon bottom gave the new Alta a strikingly different appearance from the former camp. A *Tribune* correspondent using the pen name of Amos described the main street of the new town. At the northwest end of the street, adjacent to the Wasatch and Jordan Valley Tramway depot, was John Strickley's large fireproof Miner's and Outfitting Store, built of stone and iron. It also housed the express office, which was run by H. C. Wallace. Next door to the south was Martin Wright's saloon and barber shop. "Old Joe's" variety store offered candies, nuts, underwear, and knick-knacks, and shared its building with a saloon run by Fritz Rettich, popularly known as "Baldy Fritz." Below Fritz was one of the few residences on Main Street, this one occupied by George W. Case, a local teamster. Next door was J. J. O'Riley's clothing store and saloon. Then came Albert Thomas's boarding and lodging house, known as the Canyon House. Next was William P. Rowe's Altamarket, a butcher shop, and then Charles H. Collin's Colorado House, a lodging house and saloon. The last building was the office and drug store of Alta's physician, F. H. Simmons. His building also housed the post office, with the doctor acting as deputy postmaster. Two buildings faced north at the lower end of Main Street. One was James Tucker's California Store, well stocked and busy competing with Strickley at the other end of the street. Henry C. Wallace would soon leave Strickley and join Tucker to operate the store as Tucker & Wallace, a partnership that would continue throughout the remainder of the century. Adjacent to Tucker's store was the large and well-appointed billiard hall of Charles Sickler, who also was the Alta outlet for malt beverages from the Henry Wagener breweries in Salt Lake City.

Going back up the hill on the east side of the street was the general merchandise store of A. J. Fitzgerald, followed by A. R. Devers's Alta Hotel, soon to be renamed the Hotel de Fritz. It boasted a 20-by-40-foot dining room and sleeping accommodations for about fifty persons. Farther up the street was the shoe and boot manufactory run by B. C. Sandberg, and at the top of the hill was a brewery run by Peter Smith. Most of the houses and cabins were located west of Main Street, but at least two Chinese laundries had been built to the east, on the ashes of the old town.[23] Situated as it was, the new Alta was considered to be much safer from snowslides than the old town.[24] It was a false sense of security.

At the end of December 1880 an exceptionally heavy storm descended upon the Wasatch and continued for many days, causing the usual flurry of avalanches. As they grew in intensity the snow slides smashed large sections of tramway snowsheds and tore up considerable track, rendering the tramway inoperable. As the storm continued, numerous mine buildings on the sides of the mountains were covered or crushed by avalanches, but without loss of life. Then, at thirty minutes past midnight on Wednesday, 12 January 1881, a large avalanche came down from the top of Davenport Hill, sweeping away the Grizzly boardinghouse on Grizzly Flat. In the house were three men; one woman, Mrs. Jonathan Hoskins; and four Hoskins children, three boys and a girl. One man and a boy, who were sleeping on the second floor, were thrown out of the wreckage and survived. The rest were buried in the snow. When rescuers arrived, the remaining two boys were dug out alive, but the other four, Mrs. Hoskins, her daughter, Jennie, and the two men, Evan Morris and John Howarth, were dead. The husband and father, Jonathan Hoskins, was at the Prince of Wales Mine and thereby avoided the disaster.

Although Alta residents might have been somewhat anxious as they heard avalanches rumbling down the slopes and heard the news of the nearby Grizzly disaster, they still felt relatively secure in their new location. But their confidence began to ebb the next day when another avalanche crushed the Toledo compressor house, killing two men, Charles Burbridge and Frank LaPorte. Some Alta residents now were very frightened. Many wanted to leave to seek safety in the valley, but the crushed tramway sheds and the huge snowbanks covering them discouraged all but the most determined. The few men who did abandon the camp were later branded as cowards for leaving the women and children behind. Many miners moved into Alta from their mines, and the residents relocated to those buildings they felt most safe. Then, shortly before midnight on Friday, 14 January, an enormous slide came down the north slopes above Alta. It started near the ridge and was nearly a half mile wide. When it struck a small flat above town it split and continued as two separate slides. The western portion passed over the tramway sheds and depot building, leaving them undamaged, but it then slammed into Strickley's store, crushing it and a few adjacent buildings. The force of the wreckage also caused the next building below the store to be tipped over. The other half of the slide passed east of the town, taking with it a few Chinese houses reported to be "of trifling value." The fact that the avalanche split as it did saved most of the town.

As residents turned out to rescue their less-fortunate neighbors, they discovered the wreckage of Strickley's store to be on fire. All those who had taken refuge in the stone store building and those who were in the adjacent structures were recovered safely, except for three men who had been in the basement of the store. The fire and the fear of explosions from powder stored in the store drove the rescuers back. Although there were reports of explosions, the explosive powder had been stored in a separate vault, separated from the basement by a stone wall and iron door. The three missing men were John Fitzgerald, T. Burt Lee, and Will Hollingshead. Earlier that day, Fitzgerald, the telegraph operator and railroad agent, had attempted to leave Alta with a group of miners who were fleeing to the valley, but he was unable to keep up and returned to Alta. He then chose to spend the night in the store, where he felt safer than in his room in the tramway office. Ironically, his office survived whereas the store did not. Hollingshead was John Strickley's nephew and was in charge of the store. Lee had for some time been in the employ of the Walker brothers in Salt Lake City but had lately been working with Hollingshead in Strickley's store. The three men decided it would be safer in the basement of the building. They moved down there, but took along a shovel full of coals from the stove on the main floor to give them some warmth. This was the source of the fire after the avalanche struck, and may well have been their fatal decision. Had it not been for the fire, they might have survived; but the fire drove the rescuers away, and when the bodies were recovered several days later, they were horribly burned.

The avalanche had also swept away two boardinghouses, two blacksmith shops, two ore houses, and every station on the aerial tramway of the Alta Consolidated mines, by this time known as the Joab Lawrence. Men at the mine had been ordered to quit work and abandon the premises, but two men, John Washington and Richard Williams, had stayed to look after the property. They were ordered to stay in the tunnel, but light was seen in the boardinghouse window shortly before the avalanche struck. The two men were presumed lost.

With this avalanche and the destruction of Strickley's store, many of the remaining residents decided they had had enough. A large party, including women and children, started down the canyon on foot. They used the tramway, walking inside the sheds until they would reach a section that had been smashed. Then they climbed out above the wreckage and waded across the avalanche debris to the intact sheds on the other side. Many times they had to use axes to escape from or gain access to the sheds and had to dig their way through the snow to reach the surface. It was an exhausting journey, many people carrying children or helping the weary. About fifty persons reached the city that night, but others didn't get down in time to catch the train that had been sent to bring them to the city and arrived the following day. It was claimed that Alta was practically deserted, with only twenty-five people remaining after the large group left. Among them was James Tucker, manager of the Emma Mine, who had taken his children from school in Salt Lake City so they could spend Christmas at Tucker's Alta home. When they were unable to return to the city by way of the tramway, Tucker chose not to send them down the canyon through the snow. Instead, he moved his family into the Tucker Hotel. The children certainly had an exciting Christmas vacation.

As if satisfied with the destruction it had wrought, the storm abated, giving rescuers better conditions for recovery of the victims. On 19 January nine bodies were taken to Salt Lake City for burial.[25] It was not until late in May that the body of Richard Williams, killed by the slide at the Joab Lawrence boardinghouse, was found. The body of his companion, John Washington, was found in mid-June. The remains of each were taken to the city for burial.[26]

John Strickley resolutely rebuilt his store much in the same style of the original building, except with a roof pitched to allow slides to pass over it should it be threatened again. He announced the opening of his new store the following May.[27] Alta continued to flourish, albeit not in the manner of its prosperous days of the early 1870s. During this period Alta had but four major producing mines: the Emma, City Rocks, Flagstaff, and Joab Lawrence, formerly Vallejo. One correspondent who visited the camp for the first time in twelve years thought it had a deserted appearance and was very quiet.[28] It probably was quieter, since many of the miners lived in boardinghouses at the mines, while the town of Alta became the business center and housed those with families. There were enough families to encourage the miners and businessmen to organize another school. Miss Maud Crosby, a graduate of St. Mark's School in Salt Lake City, became its teacher.[29] Alta had not had a school since the fire of 1878 destroyed the town, including the church that had served as its schoolhouse. In 1880 the correspondent writing under the pen name of

Amos wrote, "Alta needs a school building and a public free school. . . . No matter about a church, or a courthouse, . . . whether you have one wife or twenty, or none at all, is a matter of minor importance; but give the children a school."[30] Now it had one.

In February 1884 winter staged another massive attack upon the Wasatch Mountains. As the snow depth increased, numerous slides came down, some of them carrying away the mine works around some of the idle mines, others demolishing most of what was left of the tramway snow-sheds, again isolating Alta.[31] As the storms continued, residents of the mining camp, remembering the tragedies of 1881, became very anxious about conditions surrounding them, especially the threat hanging over them on the slopes. They began to seek places of safety to spend the nights, and one of the best places was the Bay City Tunnel, located a short distance above Alta and through which the Emma Mine was being worked. So it was on the evening of Friday, 7 March 1884, when a group of people made their way to the tunnel to spend the night. It was a stormy night, and the tunnel was cold, so the eleven people—eight men, two women, and one boy—joined the Emma fireman in the boiler room where it was warm. It was later speculated that they were heating tea or coffee for their supper, believing the boiler room to be perfectly safe. But it was not, for at that moment a huge snow avalanche a half mile wide came roaring down the north slopes above Alta, carrying away the Emma boiler house, blacksmith shop, boardinghouse, and concentrating works, instantly killing the twelve people in the boiler house. In a stroke of irony, the avalanche failed to enter the town of Alta—the people would have been safe had they remained in their homes. Again the residents were faced with a rescue mission, although in this case they searched only for bodies. As the corpses were removed from the snow, they were carried to the courthouse, where they were prepared for burial. But Sunday morning, when the miners planned to take the bodies down the canyon on specially prepared sleds, the storm raged so wildly that they abandoned the attempt and left the bodies to be preserved by the frost. Still, between twenty and thirty people did make the trip to Wasatch at the mouth of the canyon. When they reached the city they brought the first news of the disaster, since the slides that had wrecked the snowsheds had also taken down the telegraph line.

A rescue party of fifty-seven men was organized in the city to travel to Alta to bring the bodies down, but it was Thursday, 13 March, before the storm allowed them to make the attempt. When they reached Alta that evening they found a gloomy atmosphere, some people so panic-stricken that they would not venture out of the Emma Mine tunnel. The next day they took the victims to the city for burial. The dead included five employees of the Emma Mine, although only one, 48-year-old fireman Edward Crockett, was on duty. The others were 44-year-old Gus Lybecker, foreman; 35-year-old D. D. Wasson, machinist; 38-year-old Sam Prethero, engineer; and Willard Stephenson, fireman. The remaining victims were Peter A. Carlgren, a miner, and his wife, Mary; 21-year-old Byron F. Wasson, miner and brother of D. D. Wasson; 18-year-old Lottie Olsen, whose two brothers, a sister, and a brother-in-law lived in Alta; N. S. Delano, reported as an "old '49-er," although he was only about 50 years old; O. J. Johnson, 37-year-old brother of J. S. "Regulator" Johnson; and a 15-year-old boy, John Richardson.

As is often the case in such tragedies, there were a number of curious ironies. Three people, two men and a woman, were en route from Alta to the Emma Mine just before the disaster. The woman, being exhausted, had to stop to rest, so the men stayed with her. The avalanche came down, crushed the surface buildings, then split and left the three untouched but undoubtedly badly frightened. There also was an old man, John Burrows, and his son who were in their small, frail cabin just below the Emma works when the avalanche struck. The cabin was covered with spray from the slide, adding to the snow that was falling from the storm, but, other than that, neither they nor their cabin were harmed. Also, Willard Stephenson, one of the victims of this avalanche, had a friend who was killed in the avalanche three years earlier. That person left some personal property that was divided between his best friends. As a result, Stephenson received the watch his friend was wearing on the night he was killed. Before Stephenson's body was found his friends expressed a superstitious fear of the watch and said they would destroy it if it were to be found on Stephenson's person. Records, unfortunately, do not tell us if that was done.[32]

The new Emma company spent most of the next summer clearing away the debris, rebuilding structures, and installing new machinery. It was October before the works was back in operation and the pumps were again pumping water out of the mine. To ensure that the disaster could not be repeated, the machinery was set well into the hill and the roofs were pitched, with the slope such that any future avalanches would pass over them without causing damage.[33] It was a precaution that proved well worth while, for it was only a few months later that the buildings were put to the test.

In February 1885 the Wasatch was hit with another intense storm. The residents of Alta, still remembering the disaster of the previous winter, kept a wary watch on the snow depths on the slopes above the city. In the evening of Friday, the thirteenth, after eight days of continuous snowfall, a group of men in the Tucker and Wallace store were speculating upon the possibility of a snowslide. Henry C. Wallace, one of the owners of the store, felt the danger to be so severe that he urged the men to retire to the Bay City Tunnel at the Emma works. He left the group and went there. No sooner had he entered the tunnel, at about 8:15 p.m., than an avalanche came down and passed over the Emma works, doing no damage other than sweeping the smokestack away. The slide continued down the slope, grinding through the center of the town of Alta before it came to a halt. When the dazed residents struggled out to survey the damage they were amazed to find very few buildings still standing. The snow had passed over the top of Strickley's store, demonstrating the wisdom of rebuilding into the hill with the roof sloping with the grade, but it destroyed his adjacent warehouse and most buildings below until its energy was spent, sparing four buildings on the west side of lower Main Street, two on the east side, and a few surrounding houses. In addition, the railroad office and sheds were destroyed, as were all the buildings, ore houses, and tramway of the Joab Lawrence company, as well as other unoccupied mine buildings on Emma Hill.

Some of the many people who were buried were able to extricate themselves, while others were dug out by rescuers. At least one was not found until late the next morning. The body of James Watson, 46, was one of the first to be dug out. He had started out with Henry Wallace to spend the night in the Emma Mine but had stopped at Albert Thomas's hotel, where he stayed and met his death. In rapid succession others were found. Jeremy Reagan, 27; Barney Gilson, 50; and Mattie Hickey also were found in the wreckage of Thomas's hotel. They had been in Fritz Rettich's saloon up the street from the hotel, but their bodies were carried through the wreckage of two other buildings and deposited in front of the bar in the hotel. Timothy Madden, 47, was dug out alive but badly injured. He spoke to his friends and told them he could not survive. He died several hours later. Ed Ballou lived in a building opposite the hotel on Main Street where he ran a brewery. His building was crushed, but he managed to get out by himself. With the aid of others, he dug his wife and one child out of the wreckage, but four of his other children were later found dead.

At the time of the avalanche John Ford was in his cabin with his wife, child, and a neighbor, a Mrs. Keist. Although badly hurt, Ford was able to dig his way out. Mrs. Keist was rescued some five hours later, severely injured but alive. She and Mrs. Ford were found together, but Mrs. Ford was dead, presumably killed at the instant the slide hit the house. The child also was found dead. In the house next door Andrew White was rescued alive, some twelve hours after he was buried. Two others were reported killed, one being a man by the name of McDaniels, who lived in a house on the south side of Alta, the other being a Chinese cook named Charlie Foh.[34] James Cullinan, who was in the Thomas hotel, also was listed as one of the dead in the first reports of the disaster, much to the grief of his family in the city; however, he was dug out, alive although injured, late the next morning.

Word of the avalanche was not received in Salt Lake City until late the following day. Henry Wallace and two other men went down the canyon, telegraphed the news from Wasatch, and then took the train into the city. A relief party was organized and, despite stormy weather and advice against starting, traveled to Wasatch and started breaking a track up the canyon. Meanwhile, a party of thirty to forty men and one woman from Alta had started down with ten sleds, eight of them carrying twelve of the dead, the other two carrying John Ford and James Cullinan, who were being brought down for medical attention. The men in the party were completely fatigued, both from anxiety and the rescue efforts of the past several days and from breaking trail while dragging the sleds. Near Tanners Flat they were considering leaving some of the bodies and continuing with the injured when they were met by the relief party. The combined party continued to Wasatch where a train waited to return them to the city. The dead were taken to the sexton's office to await burial, while the injured were taken to the hospital.

At least four women and five children remained at Alta, preferring the dangers of the camp to the rigors of the canyon. But two days later another relief party went all the way to Alta and brought down several men, three women, and four children. Alta was now nearly abandoned. Out of nearly 150 people in and around Alta before the avalanche, only ten men, one woman, and her child remained. The woman, a Mrs. Bennett, refused to leave her husband, who was working in the City Rock Mine in Grizzly Gulch.[35]

This last disaster nearly spelled the doom of Alta. Although the camp continued to exist in later years, it was a mining center inhabited only by hardened miners. It never again boasted the small-town atmosphere and amenities that it had enjoyed for nearly fifteen years. George Cullen, superintendent of the Emma Mine, gave his opinion that Alta would never be rebuilt, but some reconstruction took place the following summer.[36] John Strickly built an addition to his store to serve as a warehouse, replacing the one destroyed in the snowslide. The railroad company erected a new depot—good, strong, and safe from snowslides—to serve as headquarters for its agent, Mr. J. W. King.[37] Albert Thomas, whose Cañon House hotel was destroyed, vowed he would never return to Alta. After his first hotel had been destroyed in the fire of 1878, he had built a new one on the relocated Main Street. In June 1881, while out searching for the remains of John Washington, who was lost at the Vallejo Mine when the avalanche struck in January 1881, Thomas fell into a vertical shaft that was covered with snow. He was lucky, however, for he suffered only a broken foot. When the avalanche of 1885 destroyed his business, he said he would return only to retrieve his belongings. He also noted that the last three avalanches to strike Alta all happened on a Friday, although only the last was on Friday the thirteenth.[38] But the lure of Alta was too strong; he returned to start a new hotel in a building belonging to James Tucker, probably the old Fitzgerald saloon, across the street from the wreckage of his old hotel, again calling it the Cañon House.[39] Later that summer, however, he sold his business to F. H. Grice, a black man and long-time resident of Alta, who renamed it the Alta Hotel. Thomas remained to work his mining claims for another year before disappearing from the Wasatch scene.

When winter arrived again most people packed up and left. Superintendent Cullen shut down the Emma Mine for the winter, saying he did not wish to furnish victims for snowslides. By midwinter there were scarcely ten people remaining in the district.[40]

In the spring of 1886 Alta enjoyed a brief revival. A new flagpole, seventy-two feet high, was erected in the street in front of Grice's Alta Hotel, where a huge flag, ten by twenty feet, was flown on the Fourth of July. There were enough families with children that summer to warrant the starting of another school, with Maria Cullinan as teacher.[41] Alta also became a way-station for travelers to the increasingly popular hotel and summer camp run by William Brighton across the divide at Silver Lake in Big Cottonwood Canyon. Travelers would take the train from the city to Wasatch, then ride the tramway cars to Alta, where many of them would take a meal at the Alta Hotel or at Charles Collins's Colorado House before taking saddle horses or a wagon over to Silver Lake. Following the same route on their return to the city, they created a sense of activity and probably did more than a little for the economy of businesses in the mining camp.

Also in 1886 it was announced that a telephone line would be built to Alta, although it would be mid-1887 before it became a reality. The wires came to Alta from Park City, by way of Silver Lake. Alta telephones continued to be served by that communication link for over a century until microwave technology rendered the wires unnecessary.

For the rest of the nineteenth century Alta suffered a desultory existence, its activities increasing and decreasing with the fluctuations in silver and lead prices. In 1876, for instance, silver was quoted at $1.20 per ounce, while in the 1890s decade it was only half that value. Enthusiastic predictions for the future would be made each time prices rose, only to have them crushed when prices fell again. And with each cycle Alta seemed to shrink a little more. During a surge of prosperity at the beginning of the 1890s the Rio Grande Western Railroad company rebuilt the line from Sandy to Wasatch, making it standard gauge, replacing the old narrow gauge tracks that had been used for nearly twenty years.[42] But the branch line didn't generate enough business to warrant much service, and by the end of the decade the service was so poor that Jake Smith, who by this time was leasing the tramway to Alta, took steps to provide rail service between Wasatch and Sandy. With the consent of the railroad company, he got an obsolete horse-drawn streetcar from the Salt Lake City Railroad and began passenger service to Wasatch. Horses pulled the car up to the mouth of the canyon and gravity provided the motive power on the way down.[43]

As for the tramway, its business had fallen off too. After the railroad was built from Salt Lake City to Park City by way of Parleys Canyon and began providing passenger service in the spring of 1890, tourists and pleasure seekers who had traveled to Silver Lake by way of Alta began taking the train to Park City and a stage across Scotts Pass to the resort, thereby cutting down on the passenger traffic on the Alta tramway. The new resort at Wasatch, in the mouth of Little Cottonwood Canyon, provided an almost inconsequential amount of traffic in its tramway parties to Alta. During most years the tramway was used to transport a small amount of the ore that was shipped to market; but some years, such as 1895, it did not run at all.[44] It was still being used for limited passenger service in July 1902, when Albert Potter rode a tram car to Alta when he was making his survey of public forest lands.[45] But in 1903 the tracks were said to be in bad shape, the ties having rotted so terribly that the line was unsafe for any heavy traffic. While still a local landmark, the tramway was for all intents and purposes out of business.[46]

The few businessmen who remained in Alta were not there for the sole purpose of attending to their business; all of them either had mines that they were working or else they were working for one of the mining companies. Charles M. Sickler continued to run his saloon at least through 1891, but he remained in Alta working his Jack Mine through the middle of the decade. There is no evidence that John Strickley, who also had a general store in Bingham, kept his store in Alta open after 1886. F. H. Grice, who took over Thomas's hotel in 1885, remained through 1888 but then moved to Park City, where he opened a restaurant on that town's Main Street.[47] Henry Wallace, who ran the Tucker and Wallace store, had numerous mining interests. He was the Little Cottonwood Mining District recorder for at least ten years, was manager of the Emma Mine for several years, and at various times had interest in the Highland Chief, Vallejo, Sells, and Darlington Mines. He also was a director and the manager of the Albion Mining Company at the turn of the century. By that time he was one of the few real old-timers, his association with the camp dating back into the 1870s. He disappeared from the Alta scene in 1903.

Another old-timer who stuck it out into the twentieth century was Charles H. Collins, who with his wife had run boardinghouses and saloons since the early 1870s. He also prospected the area and for a time was a foreman at the Emma Mine. In 1900 he struck some rich ore in his prospects on the east side of Peruvian Hill. After three more years of development work, he sold the prospects, then known as the Collins Group, to Park City investors.[48] There is a story that had him depositing the cash from the sale in the bank, then returning to the mine to collect his tools and have a last look around. When he failed to return, his wife sent someone to look for him. He was found in his mine, dead of a heart attack.[49] While nothing has been found to substantiate this story, the terrain surrounding the Collins Group is known to this day as Collins Gulch.

As the nineteenth century approached its close, Nature had one more surprise in store for Alta. On 1 March 1899 another avalanche swept down over the townsite, crushing the Emma ore bins and all the better buildings in the camp. By this time the Salt Lake City newspapers had little interest in Alta, and since no one was hurt the event was not extensively reported.[50] There were only about twenty-five men in the area and they had taken refuge in the mines. Apparently not all buildings were destroyed, however, because as late as July 1904 a mining claim mentioned Sickler's old saloon and Tucker and Wallace's old store.[51]

Fatalities Due to Snow Avalanches in the Nineteenth Century

(BCC = Big Cottonwood Canyon, *DN* = *Deseret News,*
DNW = *Deseret News Weekly,* LCC = Little Cottonwood Canyon,
MCC = Mill Creek Canyon, PC = Parleys Canyon, *SLH* = *Salt Lake Herald, SLT* = *Salt Lake Tribune.*)

1852	19 Dec.	MCC	A Mr. Grover, at Gardner's mill (*DNW,* 25 December 1852)
1860	8 Mar.	PC	Oliver Ogilvie (*DNW,* 14 March 1860, 25 April 1860)
1861	28 Jan.	BCC	Francis J. Stokes (*DNW,* 10:396)
1864	1 Apr.	MCC	Thomas Pierce, Robert Spurgeon, at Ellsworth's mill (*DNW,* 6 April 1864, 11 May 1864)
1869	7 Apr.	MCC	Wright, Stewart, John McDonald, a Mrs. Robbins, at Gardner's upper mill. (*DNW,* 14 April 1869, *Weekly Telegram,* 15 April 1869)
1870	5 Dec.	BCC	J.C. Smith, at Richmond Mine, Silver Fork. (*SLH,* 8 December 1870)
1871	21 Feb.	BCC	Nephi B. Fretwell, at Davenport Mine (*SLH,* 25 February 1871)
	23 Nov.	LCC	Charles Morrison, near Frederick Mine (*SLH,* 25 Nov.1871, 19 June 1872; *SLT,* 20 June 1872)
	23 Nov.	LCC	Unnamed, at Ohio Mine on Emma Hill (*SLT,* 25 November 1871)
1872	5 Feb.	BCC	Edward Samuels, William Hampton (*SLH,* 7 February 1872; *SLT,* 8 February 1872)
	11 Apr.	LCC	H. H. Murray, at Wellington Mine in Albion Basin (*SLH,* 13 April 1872; *SLT,* 13 April 1872)
	26 Dec.	LCC	Peter "Dutch Pete" Kahn, LeRoy Dibble, Thomas Triplett, Peter Elliott, Francis Brown, and an unnamed teamster, on road below Brunner furnace (*SLH,* 27–30 December 1872, 14 June 1873, 3 July 1873; *SLT,* 31 December 1872)
1874	17 Jan.	BCC	George D. Lee, at Teresa Mine in Silver Fork (*SLH,* 22 January 1874; *SLT,* 22–23 January 1874)

1875	11 Jan.	BCC	John Cox, John Tremberth, James Gleason, James Renfrey, at Annie Tunnel in Honeycomb Fork (*SLH*, 13–14 January 1875; *SLT*, 13–14 January 1875)
	11 Jan.	BCC	William Slensby, in Honeycomb Fork (*SLT*, 19 January 1875; *SLH*, 20 June 1875)
	12 Jan.	LCC	Jackson, cook at Vallejo Mine above Alta (*SLH*, 13 January 1875; *SLT*, 13 January 1875, 19 January 1875)
	19 Jan.	LCC	James Carey, wife, and nine-month-old daughter, six-year-old Laura Wilson, John Vanderleen, Michael Kelly, in Alta (*SLH*, 21 January 1875; *SLT*, 21 January 1875)
	20 Jan.	BCC	Thomas H. Broderick, William Ritter, Charles Drabble, James Breeze, Reuben Moor, at the Richmond Mine in Silver Fork. The last three were buried at Silver Springs. (*SLH*, 21, 24 January 1875, 20 June 1875, 14 July 1875; *SLT*, 21, 23 January 1875, 8 June 1875, 8 July 1875)
	3 Mar.	BCC	William G. Thomas, near Little Giant Mine on Kesler's Peak (*SLH*, 6 March 1875; *SLT*, 6 March 1875, 1 April 1875, 22 May 1875, 25 May 1875)
	17 Mar.	LCC	John Strong and wife, at Alta (*SLT* 18 March 1875; *SLH*, 20 March 1875)
	6 Apr.	BCC	J. A. Johnson, at Prince of Wales Mine (*SLH*, 8 April 1875; *SLT*, 10 April 1875, 28 May 1875)
1876	28 Dec.	LCC	J. B. White, John W. Parks, wife, and child, J. W. Brown, at Wellington Mine in Albion Basin (*SLH*, 3–5 January 1877; *SLT*, 3–5 January 1877, 9 January 1877, 24 July 1877)
	29 Dec.	LCC	Charles Stantini, Louis Labrie, at Bald Mountain (*SLH*, 31 December 1876; *SLT*, 31 December 1876)
1877	11 Mar.	BCC	Matthew Ingram, Jared Pratt, at Butte Mine in Honeycomb Fork (*SLH*, 13–15 March 1877; *SLT*, 13–14 March 1877)
	10/11 Nov.	LCC	William Deuser, Paul Ritter, at Bald Mountain (*SLH*, 17 November 1877, 27 November 1877; *SLT*, 17 November 1877)

1879	20 Jan.	LCC	Richard Duncan, at Emily Mine (*SLH*, 21 January 1879; *SLT*, 21–22 January 1879)
1881	12 Jan.	LCC	Mrs. Jonathan Hoskins and daughter Jennie, Evan Morris, John Howarth, at Grizzly boardinghouse (*DN*, 14 January 1881; *SLH*, 15 January 1881; *SLT*, 15 January 1881, 19 January 1881)
	13 Jan.	LCC	Charles Burbridge, Frank Laporte, at Toledo compressor house (*DN*, 14 January 1881, 20 January 1881; *SLH*, 15 January 1881, 20 January 1881; *SLT*, 15 January 1881, 20 January 1881)
	14 Jan.	LCC	John Fitzgerald, T. Burt Lee, Will Hollings-head, in Strickley's store, Alta (*SLH*, 16 January 1881, 20 January 1881; *SLT*, 16 January 1881, 19 January 1881, 20 January 1881)
	14 Jan.	LCC	John Washington, Richard Williams, at Vallejo Mine (*SLH*, 16 January 1881, 19 January 1881, 20 January 1881, 19 June 1881; *SLT*, 16 January 1881, 19 January 1881, 20 January 1881, 27 May 1881, 19 June 1881)
	? Jan.	LCC	Frank Darby, at Wellington Mine (*DN*, 20 June 1881; *SLH*, 28 June 1881; *SLT*, 19 January 1881)
1882	22 Feb.	BCC	Charles Taggart, wife, and five children, in South Fork (*DN*, 24–25 February 1882; *SLH*, 25–26 February 1882; *SLT*, 25–26 February 1882)
	25 May	BCC	Robert Greaves, at Richmond & Teresa Mine in Silver Fork (*DN*, 26 May 1882; *SLH*, 28 May 1882; *SLT*, 27 May 1882, 14 July 1882)
1884	7 Mar.	LCC	Gus Lybecker, D. D. Wasson, Samuel Prethero, Edward Crockett, Peter A. Carlgren and wife, B. F. Wasson, Miss Lottie Olsen, N. S. Delano, O. J. Johnson, John Richardson, Willard Stephenson, at Emma Mine (*DN*, 12 March 1884; *SLH*, 11 March 1884, 6 June 1884; *SLT*, 11 March 1884)
1885	13 Feb.	LCC	James Watson, Mrs. John Ford and child, Timothy Madden, Jerry Reagan, Barney

			Gilson, David P. Evans, Mattie Hickey, four children of Ed Ballou, Charlie Foh, McDaniels, in Alta (*SLH*, 15 February 1885, 18 February 1885; *SLT*, 15 February 1885, 18 February 1885)
	14. Feb.	LCC	Sam Trescott, R. Angove, John White, at Superior tunnel (*SLT*, 18 February 1885)
1889	? ?	MCC	Indian Pete, in Porter Fork (*SLT*, 19 June 1889)
1891	4 Mar.	LCC	Barney Cast, Hans Olson, at Emily Gulch below Alta (*SLT*, 6 March 1891)
	? ?	BCC	Bob Nelson, at Prince of Wales Mine (*SLT*, 19 May 1891, 7 July 1891)
1892	5 Dec.	LCC	Peter L. Mattson, at Greely Mine (*SLT*, 8 December 1892)
1893	late Mar.	LCC	L. E. Eckman, near Alta (*SLT*, 16 April 1893, 19 April 1893)
	late Nov.	BCC	John DeSteffani, at Prince of Wales Mine (*SLT*, 9 April 1894, 10 June 1894). He was thought at first to have fallen into an abandoned shaft; see *SLT*, 15 December 1893.

CHAPTER 8

Mining in Little Cottonwood Canyon

INTO THE MINING DOLDRUMS that persisted in Little Cottonwood Canyon around the turn of the century came a new generation of optimistic prospectors and miners. Among this group of men were the brothers Anton (Tony) J. and Alfred O. Jacobson. The elder brother, Tony, was born in Salt Lake City on 2 September 1869, about the time the original mining boom was beginning. He showed up at Alta and began filing claims in the late summer of 1899. He briefly worked the Columbus Mine under lease from the owner, Fritz Rettich, one of the true old-timers. Encouraged by what he found in that mine, he filed several claims in the vicinity, and in the spring of 1902 he brought them together in his newly formed Columbus Consolidated Mining Company. While he installed an air compressor at the mine to aid in driving a tunnel, he recognized that the water flowing down the canyon offered a much less expensive source of power than hauling coal up the canyon to supply a steam plant. He also recognized that the mining operations could be more profitable if the ore could be concentrated before it was hauled down to the smelters. But the concentrating process also required power; so, in the spring of 1903, he filed a notice of appropriation of the water of Little Cottonwood Creek and announced his intention to build a concentrating mill and an electric power plant.[1]

This was not the first hydroelectric project proposed for Little Cottonwood Canyon, but in this case the proposal was quickly followed by action. Surveys and plans were completed by midsummer; a contractor, J. J. Burke of Salt Lake City, was chosen; and by the first of October construction was under way.[2] The site of the new power plant was about four and a half miles below Alta, less than a mile below Tanners Flat, and only a few hundred yards above the old site of Woolley's sawmill. A substantial and stately stone building was built on the south side of the stream, with a bridge providing access from the canyon road. A transmission line was constructed to carry electricity from the plant to the mine at Alta. A dam at Tanners Flat directed the water into a twenty-two-inch-diameter pipeline approximately 4,500 feet long to drive the Pelton waterwheels. The plant was designed to have two 300-kilowatt generators, but only one was installed, it being a Westinghouse product.[3] After numerous delays, the plant finally went into operation, and on the evening of the Fourth of July 1904 what remained of the old mining camp of Alta found itself illuminated by electric lights. Within a few days electric motors were driving air compressors, providing compressed air for drills in the Columbus Consolidated Mine as well as the neighboring Alta-Quincy and South Columbus Mines, all being Jacobson enterprises.[4]

The construction of the concentrating mill was carried on at the same time the power plant was being built. Originally conceived as a 100-ton mill, its size was increased to 150 tons during construction. It was located below the mine, on the north side of the canyon a short distance below Alta, near the site of old Central City. It was a large building that dominated the scene as one approached Alta. The machinery was installed and the mill began operating in November 1904. The very first tests demonstrated its value: when ore valued at only fifteen dollars per ton was concentrated, the product shipped to the smelters had a value of from forty-five to sixty dollars per ton.[5] The mill was powered by electric motors driven by electricity from the company's power plant. Electric power quickly established its value in the Alta mining community, so much so that late in the 1905 season a two-stage turbine pump driven by an electric motor was installed in the Columbus Mine to take care of the ever-present water problem. An elaborate new three-story boarding-house also was built, with accommodations for one hundred men, illuminated by electric lights, of course.[6] The rapid expansion of the company's operations made it necessary to install a second generator in the power plant near the end of 1905, this one purchased from the General Electric Company.[7]

The Columbus Consolidated Mining Company was riding high at this point. It had started the year 1905 about $60,000 in debt but finished the year free of all encumbrances. The Jacobson brothers became heroes of the day. In February 1907 they pleased their investors and the financial community at large by paying a twenty-cent dividend on their stock, followed by three more similar dividends during the year. The following January they suspended their dividend payment, however, blaming severe winter weather that restricted the delivery of ores to the smelters. But when the snows melted the following spring they failed to restore the dividend. The problem of water in the mine was causing many difficulties. When a pump in the 400-foot level broke, the water rose rapidly, flooding the lowest one hundred feet of the mine. New, bigger pumps were installed and the water levels slowly subsided, but the Columbus Mine seemed unable to return to its former productive state. However, the concentration mill was kept busy with ores from the nearby South Columbus Mine. In March 1910, when heavy snows again prevented the shipment of ores to the smelters and the company was unable to make payments on notes that became due, a twenty-five-cent-per-share assessment was levied. When that was followed by another ten-cent assessment in October the stockholders became very unhappy, so much so that at the next stockholders' meeting, in June 1911, a completely new management team backed by Michigan interests was elected to take control of the badly flooded mine.[8] But the new management had no better control over the water. The situation had become so bad that in November 1912 they pulled the pumps and closed both mine and mill for the winter.[9]

Perhaps the closure was part of a larger plan: there had been rumors of a merger involving the Columbus Consolidated. Within months after the closure the rumor became fact; the merger would include the Columbus Consolidated, Columbus Extension, Flagstaff, and Superior-Alta, the new company to be known as the Wasatch Mines Company. It was incorporated in February 1913. The stockholders of the Columbus Extension failed to approve the merger, however, so that company remained as a separate entity. In the new organization the old Columbus Power Plant was made a subsidiary under the name of the Wasatch Power Company. By combining the assets of the individual companies with new financing, the Wasatch Mines Company was able to meet the present indebtedness, with enough left over to drive a tunnel to drain the mines.[10]

The concept of a drain tunnel was not new; almost since the first days of mining in the Wasatch there had been thoughts and attempts to drive tunnels well below the most productive mines to drain the water from them and to provide access at or below the lowest mining levels. It was a good idea, for it would be much easier and less expensive to take the ore out of a long tunnel than to hoist it to the top of a shaft, and to allow water to run out a tunnel rather than have to pump it to the surface. But in practice very few of the drain tunnel enterprises were successful, even if the companies did manage to find the resources to drive a tunnel. In the case of the Wasatch Mines Company, the tunnel did achieve its goals and became one of the more successful drain tunnels in the district. The portal was located at the southwest end of the old Columbus Consolidated property, on Hawleys Flat where one of Woolley's sawmills had been located years before. Work began in June 1916 on the construction of buildings to house machinery and workers. For three years the crews worked until in October 1919, at a distance of 5,000 feet from the portal, the tunnel was taking out so much water that the water level in the old Columbus shaft was dropping about three feet per day. By the following August the Columbus Mine was free of water and could be worked through the tunnel rather than the shaft.[11]

The Wasatch Power Company plant was an important part of the tunnel effort, for it provided the power required to accomplish the task. In addition, it provided power in the form of electricity and compressed air for nearly every other mining company in the Alta area. To meet the growing demand for power, in 1915 the company installed another pipeline, this one with nearly 3,000 feet of twelve-inch pipe, bringing more water down to the power plant to drive the generators.[12] But occasionally that was not enough and the plant was unable to supply the demands of the users, especially in cold weather when the stream and/or the compressed air lines would freeze, or when avalanches would fill the streambed and restrict the flow of water. To supplement the Wasatch Power Company's plant, Utah Power and Light Company in 1917 built a power line from the Cardiff Mine in the South Fork of Big Cottonwood Canyon over the pass and down into Alta.[13] Remnants of this power line caused Cardiff Pass to be popularly known as Pole Line Pass in later years.

In 1923 the Wasatch Mines Company fell upon hard times and went into receivership. In September 1925 a lease agreement was signed between the receiver and the Mineral Veins Coalition Mines Company, whereby the latter became the majority stockholder in the Wasatch Mines Company and assumed control.[14] Curiously, the general manager of the new company was A. O. Jacobson, one of the brothers who started the company's ancestor firm a quarter of a century earlier. Under the new management the drain tunnel was continued in a northerly direction, running under the divide and eventually connecting with the Cardiff Mine workings in the South Fork of Big Cottonwood Canyon. Operations in this tunnel continued past the middle of the twentieth century. The Jacobson association continued to the end, for in the 1940s the operator in the Wasatch Drain Tunnel was A. O. Jacobson's son Tony.[15] The tunnel was worked by lessees well into the 1960s, when it was finally closed.[16] Today its flow of water supplies the needs of the Snowbird Resort.

The Wasatch Power Company continued to operate until 8 January 1929, when it was absorbed into the Utah Power and Light Company. By that time most other remnants of the Columbus Consolidated Mining Company had disappeared, for the large bunkhouse burned in January 1913 and the huge concentrating mill was destroyed by a fire that started in the early morning hours of 7 September 1914. It is not known when the power plant ceased operations, but it probably was in the early 1940s.[17] Today the walls of the power plant, long since abandoned, still stand, though the roof has collapsed, on the south side of Little Cottonwood Creek about midway between Hogum Fork and Maybird Gulch. Bits and pieces of the pipeline may be found along its route up the canyon, and the diversion dam, now breached, may be seen at the southwestern corner of the Tanners Flat campground.

*

In 1903, about a year after the Jacobson brothers started the Columbus Consolidated, another promoter, Henry M. Crowther, gained control of a number of mines in Grizzly Gulch, including the Grizzly, Lavinia, Regulator, and Darlington. They were absorbed into the newly founded Continental Mines and Smelting Company, a New York corporation. The group of mines at Alta was known as the Continental-Alta.[18] Immediately a new three-story boardinghouse was constructed and a force of fifty men was employed to do extensive development work in the mines.[19] To process the ores, a mill was planned and construction was started at Tanners Flat. Recognizing the need for better transportation to carry ores to the smelters, Crowther negotiated with the Denver & Rio Grande Railroad to obtain a lease on the old horse tramway between Wasatch and Alta as well as the remnants of the railroad between Wasatch and Sandy. After six months of fruitless negotiation, he gave up on the rail tramway and announced that the company would build an aerial tramway between the mines and mill. Construction was started and was carried on simultaneously on both the mill and the tramway. At the end of October 1904 the mill building was completed and the machinery was installed. It was a 100-ton mill, somewhat smaller than the Columbus mill at Alta, and was water powered. A pipeline about 4,000 feet long and fourteen inches in diameter was installed to bring water from Little Cottonwood Creek. After passing through the Pelton waterwheel, the water was returned to the creek just above the Columbus power plant's diversion dam. The mill was completed, and a trial run was made in mid-January 1905; but operations had to wait for three months before the tramway was completed and both could be placed in successful operation.[20]

The Continental-Alta aerial tramway was a magnificent accomplishment. From its upper terminal at Grizzly Flat it ran for 25,000 feet, nearly five miles, with one angle station designed to accommodate the slight bend in the canyon without having to transfer the loads. The lower terminal was at the mill on Tanners Flat. There were sixty-two wood-frame towers supporting the cables and a private telephone line. The installation of the cables had to be the most impressive task; the first cable shipment contained a single reel with 6,350 feet of steel cable, weighing eight tons.[21] It was supposed that the tramway would be less susceptible to snow avalanche damage because the cables were well above the surface. However, even before all the cable had arrived, much less been installed, a small snowslide in the vicinity of Alta took out two of the towers. They were quickly repaired, and as the remaining cable arrived it was installed. Finally, on 10 March 1905, the cables and their suspended buckets began to move up and down the canyon. During the trials it was found that the tramway had too few buckets for successful operation, and it was the end of April before additional buckets were shipped in, placed into service, and enough ore was being moved to supply the mill.[22]

The tramway served the company well that year, but the following January as many as twenty-five towers were knocked down by snowslides. They were not repaired and back into operation until the end of April, by which time some people must have been wondering whether the aerial tramway was indeed the solution to their transportation problems.[23] Yet, when it worked, the tramway worked well. In 1907 the Continental-Alta was reorganized as the Unity Mines Corporation, with Crowther still at its head. But the company had seen its glory years and would soon disappear. In the summer of 1908 Tony Jacobson announced he had leased both the aerial tramway and the mill for use by his Columbus Consolidated and South Columbus Mines.[24] As has been seen, however, the following years were not good ones for either him or his companies, and the big tramway fell into disuse.

During the years from 1908 to 1910 a group of Michigan investors bought the City Rocks Mine, adjacent to the Continental-Alta ground. They built a short tramway to link to the Continental-Alta tram and used it to ship their ores down canyon. They also continued development well under the divide between Big and Little Cottonwood Canyons, either very close to or even into the ground of the Utah Mines Coalition Company. The latter company had the tunnel at Lake Solitude, at the head of Mill F South Fork in Big Cottonwood Canyon, which was driven under the upper end of Honeycomb Fork and toward the divide between Honeycomb and Grizzly Gulch. When the City Rocks development became known to the Utah Mines Coalition Company, the latter took the former company to court. Although the alleged encroachment may well have taken place before the Michigan group took control, the plaintiff had in effect tweaked the tail of the defendant tiger. At about that same time, the Michigan interests had proposed a merger of a number of properties on the Little Cottonwood side of the divide, a merger that took over a year to become fact; however, when it did, a new company, the Michigan-Utah Consolidated Mining Company, was formed to take over not only the proposed properties but the Utah Mines Coalition as well. As a result, the litigation was settled out of court to the satisfaction of both parties.[25] The old Continental-Alta properties were part of the merger, so the long tramway running down to Tanners Flat became the property of the new Michigan-Utah company. However, the tramway did the company little good, for there is no indication that it ran after March 1910, when it was damaged by snowslides.

The management immediately addressed the company's transportation needs and decided to build a new tramway instead of repairing the old one. That was done during the summer of 1912, and by the end of September the new tramway was carrying ore to Tanners Flat.[26] It served well for many years, suffering many more service interruptions for maintenance purposes than from avalanche damage. It was somewhat realigned from the old one to do away with the angle station, and many of its towers were of steel and concrete construction. The upper terminal of the original tramway was on the east side of a deep gully coming down from the north slopes, a gully that channeled avalanches that repeatedly threatened the tramway. The upper terminal of the new tramway was relocated to the west side of the gully, removing it from the avalanche danger. A suspension bridge was built across the void, allowing mine cars to be pushed from the mine to the loading station. Today some detritus from the upper terminal remains on the site next to the road that runs up Grizzly Gulch, and remnants of the bridge can be seen scattered across the bottom of the gully.

In spite of the tramway there still was a serious transportation problem. It did little good to get the ore to Tanners Flat when the smelters were many more miles away in the center of the valley. The mining operators longed for the convenience of the old horse tramway and the railroad from Sandy to Wasatch, and many attempts were made to replace both. It was mentioned that Henry Crowther built the first aerial tramway after he failed to get a timely lease of the old horse tramway; but he did not drop the goal of rail transportation, for he continued negotiations with the Denver & Rio Grande Railroad company and managed to get a lease on the roadbed from Bingham Junction in Sandy to Wasatch at the mouth of the canyon. In June 1906 he formed the Continental Transportation Company to construct and operate the proposed railroad, but by the end of that year it was reported that his scheme had fallen through.[27] Shortly after that, Crowther left Alta, but others who remained continued to press for rail transportation. Yet nothing happened until the State of Utah started to build its capitol building, which was to use granite from the quarries at the mouth of Little Cottonwood Canyon. Possibly in anticipation of the traffic the granite would create, J. G. Jacobs, a longtime railroad promoter who had built the Salt Lake & Mercur Railroad, took a lease on the D&RG branch line from Sandy to Wasatch in May 1911. However, nothing further was done until June 1913 when it was formally announced that the capitol building would use Utah granite and contracts were awarded for the stone. Jacobs announced that $40,000 would be spent to build the standard gauge railroad, and work began immediately.

The old roadbed was in good condition and some rails were still on the ground.[28] In August Jacobs incorporated the Salt Lake and Alta Railroad Company and transferred his lease to the new company. Incidentally, one of the incorporators was Norman W. Haire, managing director of the Michigan-Utah company.[29] Construction work was pushed, and by mid-November 1913 the railroad was completed and trainloads of granite were passing over the new tracks. The railroad generally followed the old Wasatch and Jordan Valley Railroad grade and ran into the mouth of the canyon to the doors of the hotel at the Wasatch Resort, where a switchback took cars up to the granite quarry.[30] Loading platforms were erected at the railhead so teamsters could transfer their loads from the ore wagons to the railroad cars. There was talk of extending the tracks to Tanners Flat or continuing the aerial tramway to Wasatch, but that essentially was wishful thinking. Teams were still used to carry ore the few intervening miles to the railroad, or, in the case of some mining companies, teams carried the ore all the way down the canyon.

In 1914 the flow of ore down the canyon to the railroad was limited only by the condition of the roads and the availability of teamsters. Early in 1915 Jacobs claimed that the amount of supplies and freight going up to the mining camp was three times greater than the year before. The time was ripe for better transportation in the canyon. It came in September of that year when Walter K. Yorston, a Salt Lake City general contractor, formed the Alta-Cottonwood Railway Company to build a narrow gauge railway from Wasatch to Alta. He planned to use the old horse tramway right-of-way on the north side of the canyon, but he proposed to use steam locomotives—geared Shay locomotives that had proven themselves on many railroads with exceptionally steep grades. Surveys were started immediately and graders were active in the months that followed.[31] However, Yorston faced some formidable problems in building the railroad, the first being the financing of his venture. He found that through the Pearson Engineering Company of Pennsylvania. Upon returning from an extended trip to the east, Yorston reported that the financing was completed and the company would be reorganized to accommodate the new eastern partners. In May 1916 a representative of the Pearson company made an inspection visit and helped plan the construction. Pearson continued to be involved for the next year and Yorston became its western manager. Shortly after that company's involvement the railroad became known as the Little Cottonwood Transportation Company.[32]

The next problem was the acquisition of rails, since these were war years and steel was not readily available. Yorston managed to find enough rails for four miles of track, but they did not arrive at Wasatch until mid-September. The remaining rails were found close to home; they were salvaged from the Emigration Canyon Railroad when it was dismantled the following summer. Although a locomotive had been ordered, there was no motive power on hand to help in track laying. Yorston solved that problem by purchasing a Ford automobile, fitting it with flanged wheels, removing the body, and extending the frame. With this vehicle he was able to haul ties and rails as fast as the track-laying gang could use them.[33] Even with the motorized assistance, however, the track moved up the canyon slowly, and it was not until February 1917 that it reached Tanners Flat. By that time a Shay locomotive had arrived at Wasatch, but it was not placed in service. Instead, the Ford "locomotive" was used briefly to take light ore cars to Tanners Flat; later, teams were placed into service to take cars up. Once loaded, they came down under gravity, restrained by brakes, just as the old horse tramway had operated forty years before.[34] The railroad obviously was not providing the transportation relief that had been expected of it, nor was the company receiving the revenues it had anticipated. In July one of the company directors reported that Yorston had been dismissed and that H. A. Dunn, a construction engineer who had been brought in from a job in Spain, was replacing him. Dunn remained about a year before being replaced by Shand Smith, a Utah mining man. Under new management the railroad slowly pushed toward

Alta, arriving at the Columbus Rexall Tunnel late in October 1917.[35] Passenger service was started about that time, the first run using the Shay locomotive, but the long-awaited freight service did not come with it. It was found that the freight cars were too heavy for the steep grades, sharp curves, and light rails. Two new, light Shay locomotives and forty smaller freight cars were purchased and were on hand when the snow left the tracks the following spring. The company, now ready for serious business, posted its rates for the season and had its first train arrive at Alta on 18 June 1918.[36]

Finally able to transport large quantities of ore down the canyon, the railroad added extra track at the upper end to provide service to individual mines. A major spur crossed to the south side of the canyon to the South Hecla works at the bottom of Collins Gulch. A short spur was built to the Wasatch Drain Tunnel, and another 850-foot spur went to the ore bins at the base of the Sells tramway. This was a single-span, gravity-powered aerial tramway that ran from the Lexington Tunnel high on the east slopes of Peruvian Gulch to the ore bins in the canyon bottom just east of the Wasatch Drain Tunnel, about a mile below Alta. The 2,350-foot-long tramway was built in November and December 1917, about the time the railroad reached that vicinity.[37]

It had been planned that trains would use from ten to fifteen of the six-ton cars, carrying sixty to ninety tons per trip. Indeed, at the end of the 1918 season a train took down 131 tons in a single trip, about twenty-two cars. But that was a dangerous stunt—the grades were very steep, as much as 11 percent, and averaged 7.5 percent.[38] Even though the locomotives and all cars were equipped with air brakes, the engine crew had to be extremely careful. Air for the brakes was provided by a steam-driven air compressor on the engine, but if the engineer used the brakes too often he could use the air faster than it could be compressed and find himself without brakes. Also, steel wheels on steel rails do not provide the ultimate in traction; if the rails were wet or covered with weeds, the wheels could slide on the rails and deny the train some, if not all, of its braking abilities. Elbert G. Despain, who with his father ran motor-car mail and passenger service over the tracks for many years, claimed the railroad had trouble finding engineers who would stay on the job because of the danger. It was just too steep, he said.[39]

In spite of the grades, the railroad had an unexpectedly good safety record. It ran for two years before suffering a major accident. On 5 May 1920 one of the small locomotives, equipped with a snow plow, was used to force its way through the snow and open the road to Alta. After bucking drifts for more than four miles, the engine left the track. Through the efforts of the ten men on board, it was placed back on the rails and continued its work. Then it quivered to a halt in a deep drift, the wheels began to spin, and the engine started to slide backwards down the grade. The air brakes were applied, locking the wheels and making the locomotive slide even faster. The engineer, recognizing the danger, shouted for the others to jump. As they did, he followed; but two men, new itinerant employees, failed to muster the courage to go overboard. Fear stricken, they stayed with the locomotive as it gained speed down the steep grade, then left the rails and turned over on its side. The rest of the crew picked themselves up and ran to the wreck, where they found the two men, Isaac Urtilla, 45, and Thomas Jordan, 46, crushed to death under the locomotive. After the engine was raised slightly with jacks, the bodies were removed and taken to the city for burial; but the engine remained on its side next to the track for another dozen years before it was cut up for scrap.[40]

The Little Cottonwood Transportation Company operated with indifferent success for three years. While it did provide rail service and was able to haul ore down the canyon at less cost than before, it never replaced the teamsters. When winter snows covered the tracks the company shut down until the spring thaws allowed trains to operate again. During those periods, the mine operators had to depend upon the teamsters, often finding them in short supply because they knew that their work would be seasonal at best. As a result, some companies continued to use teams during the summer, dividing their shipping between the wagons and the railroad, thereby cutting down on the latter's summer profits. In 1921 when the railroad company applied to the Public Utilities Commission for permission to increase its rates 25 percent, claiming it was operating at a $30–50 daily loss, it narrowed the cost difference between rail and wagons to such a small margin that the railroad no longer held a significant advantage. Wagons could haul five tons as compared to a railroad car's six and one-half tons, and when the county improved the canyon road in the spring of 1921, the teamster's trip was made considerably easier. Also, for the first time, the road was opened to motorized travel. Before that time it was the boulders that stood above the road surface, rather than the steepness of the grade, that discouraged the use of motor vehicles.[41] This is not to suggest that the road became a good one by today's standards. As late as 1931 George H. Watson, in encouraging visitors to drive to Alta, wrote, "Alta can easily be reached in less than two hours from Salt Lake. . . . Most of the distance in the canyon can be driven in high and second gear. Several 'dugways,' however, require low gear and slow driving, as the road is narrow and rocky."[42] But back in 1921 the road was good enough to cause the days of the railroad to be numbered.

In the spring of 1922 manager Shand Smith received instructions from the company's New York offices not to operate the railroad that season. Smith noted that the company's share of the ore transportation market was too small to allow it to operate at a profit. He applied to the Public Utilities Commission (PUC) for permission to suspend operations that season, implying the railroad might operate again in later years; but near the end of August he addressed the PUC again, stating that conditions did not warrant reopening the road and that he would retire as manager on the first of September.[43] Steam locomotives never ran again in Little Cottonwood Canyon.

Passenger service continued over the rail line between Wasatch and Alta for another half-dozen years. Almost as soon as the railroad was built Lewis E. Despain and his son, Elbert G., took over the passenger and mail service using converted motor vehicles called "jitneys." Like the original Ford that had been converted to haul construction supplies, the jitneys were automobiles equipped with railroad wheels and multiple bench seats. Several different vehicles were used over the years, some larger ones proving unsuitable and being retired immediately. When the railroad ceased operations, the Despains continued running the passenger service at least until 1928. Meanwhile, George H. Watson, manager of the South Hecla Mine, gained control of the railroad and renamed it the Alta Scenic Railroad.[44] Watson, a character to the core and later to assume the unofficial title of Mayor of Alta, encouraged the use of the jitney for tourist visits to the camp and often was seen riding or even driving the vehicle himself. Elbert Despain later recalled Watson standing in the jitney while he told the tourists all sorts of tales and pointed out local landmarks. He would tell them that back up there, pointing toward the high mountains to the south, was country never seen by white men, and inhabited by bears, wolves, and unknown animals.[45]

The use of the railroad came to an end on 2 June 1928 when the jitney, while coming down the canyon, ran away, left the tracks, and crashed into the mountainside. That day four men had ridden up to visit the Peruvian Mine. Elbert Despain dropped them off at the Wasatch Drain Tunnel and continued up to Alta. In the afternoon he picked them up again. Since there were no facilities to turn the vehicle around, the jitney had to go down backwards. Because of the very low gear ratio in reverse, the usual practice was to put the transmission in neutral and depend upon the brakes to control the speed on the way down. That afternoon, however, a hailstorm had knocked weeds down over the rails and the wheels began to slide. Unable to control the jitney's speed, Despain called for his passengers to jump. Everyone did, and the vehicle continued on to its destruction. Of the five men on board, only two suffered significant injuries: Edwin Hillier, of Nampa, Idaho, suffered a fractured clavicle and John P. Clays, manager of the Utah Peruvian Lead Company, suffered a compound fracture of the right leg, and fractures of the right shoulder and left arm, in addition to numerous bruises and contusions. Despain and T. A. Jacobson, son of long-time Alta miner A. O. Jacobson, walked to the Wasatch Power Plant, where they called for help. After the injured were taken into the city, Clays's leg was amputated below the knee. Ten days later he died from his injuries.[46]

The tracks remained in the canyon, unused and gathering rust, for another half-dozen years. Then, in the midst of the Great Depression, at a time when the federal government was pumping money into local economies for public projects, the Little Cottonwood road found itself a recipient of this federal largess. The county, being responsible for the project, chose to move the road from the bottom of the canyon to the north slopes, following the right-of-way of the abandoned railroad. Accordingly, Watson, still the owner of the tracks through his Alta Scenic Railroad, had his men cut the rails and haul them out of the canyon to be used as smelter flux.[47] The two locomotives still stored in the sheds at the mouth of the canyon were sold, the larger one going to Pioche, Nevada, and the smaller one to the Utah Iron Ore Corporation at Iron Springs, Utah. The other small engine was still on its side next to the tracks where it had turned over years before. It was sold for scrap and dismantled on the site.[48] After the railroad was removed the county built the new road following the railroad roadbed, very nearly the same alignment the road has today.[49] About this same time, the Sells tramway, which had been badly damaged by a snow avalanche some time before, was taken down by the same people who dismantled the railroad and shipped to Cedar City, where it was used to build a coal mine tramway near Kanarraville. By this time the Michigan-Utah tramway had been shortened so that its upper terminal was near the Wasatch Drain Tunnel; the cable that had been taken down was left lying about on wooden reels. That cable was also sent to Cedar City to extend the Sells unit at that site.[50]

The Salt Lake and Alta line from Sandy to Wasatch was officially abandoned in 1934, although some track remained until 1943.[51] Today little can be found of either railroad. The dirt road from the forks at the mouth of the canyon up to Wasatch Resort, used as a bicycle trail today, was the route of the railroad. The road that doubles back to the northwest to go up to the canyon highway was the first leg of the switchback to the granite quarries. At several places along the canyon highway stone walls may be seen on the uphill side, remnants not of the Little Cottonwood Transportation Company's railroad but of the old 1870s horse tramway and its snowsheds.

*

With the railroad gone and the country in the depths of the Depression, little of significance was happening in Alta. Some mining activity continued, much of it by lessees and most through the Wasatch Drain Tunnel or on properties controlled by George H. Watson, who remained and became an icon of mid-century Alta. Watson had come to Utah from Michigan in 1902, a young man of nineteen years. He spent several years working as a bartender in Salt Lake City. He may have dabbled in mining during that time, but his name did not appear in the press until 1907. Then he became secretary, treasurer, and one of the incorporators of the Alta and Hecla Mining Company, formed around five claims located about a mile southeast of Alta.[52] Watson rapidly learned about corporate wheeling, dealing, and promotion. By May 1908, when the Alta-Hecla company absorbed the property of the Lilburn Mining Company, Watson was the former's manager.[53] Two years later he maneuvered to increase his influence and power in the area's operations and corporate politics. At that time the South Columbus Mining Company, a 1906 consolidation of the South Columbus and Alta-Quincy Mines, operated by the Jacobson brothers, had fallen on hard times and found it necessary to levy assessments on its stock. The stockholders retaliated at their meeting in February 1910 by electing an entire new directorate, which included George H. Watson. He was soon elected president of that company and orchestrated its merger with Alta-Hecla, to become the South Hecla Mining Company, with himself as president and general manager.

By the end of 1916 Watson was managing the South Hecla, South Hecla Extension, Alta Germania, and Alta Michigan companies in addition to his own brokerage firm, G. H. Watson & Co.[54] In the years that followed he formed a growing succession of companies, some of them taking in claims from older companies, others being a restructuring of companies he had previously formed. With much impressive juggling, negotiating, and some sleight-of-hand, Watson amassed an empire of mines and claims covering most of the land south of Alta, and some to the north and east. In 1923 he consolidated most of his holdings in the

Alta Merger Mines Company, which received ninety-four surveyed and five unsurveyed claims, aggregating some 1,400 acres. In 1927 his Alta Michigan Mines Company, which had been formed in June 1916 to take a five-year lease on the Kate Hayes vein from the South Hecla company, took an option on a controlling amount of stock of the Emma Silver, Alta Merger, and Alta Consolidated companies. This gave the Alta Michigan company control of 138 mining claims and about 1,600 acres of mineral ground.[55] Notice that Alta Merger already was one of Watson's companies; thus did he weave a web of organizational chaos, but even chaos has its own intricate, if incomprehensible, organization.

Watson still had two more tricks up his sleeve: in 1932 he formed the Alta United Mines Company to consolidate various of his interests, including the Alta Merger, Alta Consolidated, and Alta Michigan companies. Alta United, however, was not a new corporation, it was the old Emma Silver Mines Company, incorporated on 3 July 1919, which had fallen under Watson's control. In 1931 he was president of Emma Silver Mines, while his associate Robert F. Marvin was secretary. At a stockholders' meeting on 26 October 1931 a resolution was passed changing the name of the company to the Alta United Mines Company.[56] It is interesting to try to unravel the relationships of the various companies involved, for in 1930 Emma Silver Mines was placed under control of Alta Michigan. Now, in 1932, Emma Silver Mines' name was changed to Alta United and it was given control of Alta Michigan, a corporate example of being one's own grandpa.

In 1933 Watson did it again. This time he formed the Alta Champion Mining Company to revive the Alta ore zone. As with Alta United, this was a new name for an old company, in this case the Alta Michigan. At a stockholders' meeting on 5 May 1933 the articles of incorporation of Alta Michigan, which had been absorbed into Alta United, were amended to change the name to the Alta Champion Mining Company. Leases were taken on properties of thirty-four companies, consisting of 164 mining claims covering 2,100 acres of mineral land.[57]

One of Watson's associates in his Alta adventures was Robert F. Marvin. As a youth of twenty years, Marvin had gone to work in Watson's brokerage firm, the result of his parents' financial interest in Watson's enterprises, where they lost their proverbial shirts. He remained with Watson until the mid-1930s. He said that Watson's tactics to acquire companies were to buy big blocks of stock at assessment sales, where stock was very cheap, and to promise great things and drum up proxies for the next annual meeting. Watson, he said, was a genial, likable, and persuasive rascal who could talk the Devil out of his pitchfork. He had a certain amount of success in mining and liked to play Pied Piper to the stockholders of properties under development. Mergers, consolidations, and reorganizations were his standard techniques. He thereby built up the South Hecla holding from about 330 acres and a couple of paying ore bodies to some 2,400 acres with no ore at all. "And so he departed this life in slippery possession of an empty empire."[58]

Although some may consider George H. Watson to be Alta's worst scoundrel, it is necessary to note that in 1937 and 1938 he deeded to the Wasatch National Forest the surface rights to some 1,800 acres of land around Alta. Whatever his reasons may have been, the land allowed for the development of the ski industry there. And when the first ski lift was built, parts of it were taken from the old Michigan-Utah aerial tramway, another gift from the mining industry that allowed the new ski industry to grow. Watson remained, playing the role of the bumbling old prospector, charming skiing visitors and regaling them with stories—some true, but most of them a product of his fertile imagination. He described his inventions, such as a home cremation kit, talking tombstones, and the "Christmas Club" (a piece of a tree limb). Visitors called him the Mayor of Alta, but most likely the title was one he bestowed upon himself. He liked to call the place "Romantic Alta," and himself the Mayor of Romantic Alta. The "romantic" adjective arose in 1920 when Robert Marvin was courting his future wife. One day the couple climbed to the top of Mount Superior, where the future Mrs. Marvin found a tobacco tin containing an engagement ring carefully placed there in advance by her suitor. When they returned to Alta where the Watsons were waiting, they celebrated "Romantic Alta" and the appellation was born.[59]

Watson remained at Alta until 1952, living in a cabin that once was the station for the Little Cottonwood Transportation Company. On the last day of March a forest ranger and an Alta Lodge employee were checking local residents to see if they had enough food after being snowbound for several days. When they got no response at Watson's cabin they entered and found him dead, presumably of a heart attack. The next day the U. S. Forest Service sent a snow tractor to bring the body down to the city. George H. Watson was buried in Mt. Calvary Cemetery.[60] Today his name lives on at the Watson Shelter, now known as Watson's Cafe, on the ski runs in Collins Gulch.

Figure 30. Columbus Consolidated mill at Alta, 1906. (SLMR, 30 May 1906)

Figure 31. The Continental-Alta aerial tramway going down Little Cottonwood Canyon from Grizzly Gulch in 1905. The Columbus Consolidated mill is on the right. (SLMR, 15 October 1905)

Figure 32. Columbus Consolidated compressor building and boardinghouse in 1903. (SLMR, 30 November 1903)

Figure 33. Mine and loading station of the City Rocks company in 1907. This mine was located in Grizzly Gulch. (SLMR, 30 August 1907)

Figure 34. Columbus Consolidated mill and loading station; view looking up the canyon toward Alta in 1907. (SLMR, 30 August 1907)

Figure 35. The Columbus Consolidated power plant as it appears today. It is located on the south side of Little Cottonwood Creek between Hogum Fork and Maybird Gulch. (Author's photo)

Figure 36. Remains of the boiler and winch at the Prince of Wales Mine at the head of Silver Fork. This equipment was installed in 1875. (Author's photo)

Figure 37. Boiler and steam engine remains at the Wellington Mine in Silver Fork. The boiler was supplied with water from the Prince of Wales Mine located directly above. (Author's photo)

Figure 38. Boiler and pump at the Prince of Wales pumphouse. Located high in Grizzly Gulch just below Twin Lakes Pass, it was installed in 1875. Water was taken from three springs at the site and pumped through a pipeline of more than a mile in length to supply the boilers at the mine. (Author's photo)

Figure 39. Boiler and pump remains at the Eclipse Mine in Days Fork. Two boilers were installed in 1880; one of them has all but disappeared as it sinks into the shaft. (Author's photo)

Figure 40. Large cast gear at the Eclipse Mine in Days Fork. (Author's photo)

Figure 41. Cardiff Mine workings in the South Fork of Big Cottonwood Canyon as they appeared in 1911. (SLMR, 30 July 1911)

Figure 42. Sacking ore at the Cardiff Mine in 1911. Ore was sorted and placed in sacks weighing between 75 and 100 pounds, allowing easier handling during transportation to valley smelters. (SLMR, 30 July 1911)

Figure 43. Cardiff Tunnel in the South Fork, located a short distance below the earlier Cardiff Mine. The large ore bins are in the center of the photograph, with the mine buildings on the left at the same level. The bunkhouse is above the mine at the extreme left. The Howell mine is above the large dump on the right. This is from a Shipler photo taken on 23 July 1923. (UHS)

Figure 44. Cardiff Mine bunkhouse, located above the tunnel, as it appeared in 1916. The Beefsteak raise was later driven to connect the tunnel with the bunkhouse, allowing miners to go directly to and from their work in the wintertime. (SLMR, 15 April 1916)

Figure 45. Cardiff ore bins located in Big Cottonwood Canyon at the mouth of South Fork. The company built these bins in 1916 to be used for temporary storage when motor vehicles replaced teams to haul ore down the canyon to the smelters. Company teams, and later trucks, brought ore from the mine to these bins. (UHS)

Figure 46. Foundation and basement walls of the Cardiff Mine bunkhouse as it appears today. The remains are located on a flat above the tunnel. (Author's photo)

Figure 47. Remains of the ore bin at the Cardiff Tunnel in Cardiff Fork. When the mine was active there was a trestle extending from the dump in the foreground to the top of the ore bin, allowing mine cars to dump ore directly into the bin. The trestle was destroyed in a spring flood in the early 1970s. The Howell Mine dump is in the background. (Author's photo)

Figure 48. A Knox tractor with ore trailers on the new road through the cut at the Stairs power station reservoir. The tractors were introduced in 1916 to replace teams of horses in the canyon. The new road through the cut was built to remove steep grades over Jones' Hump at the same site. (SLMR, 15 October 1916)

Figure 49. The Maxfield Mine in Big Cottonwood Canyon, circa 1916. This view looking up canyon shows the road passing between the mine buildings and the dump. A covered trestle between the tunnel and dump crosses the road in the distance. (SLMR, 15 October 1916)

Figure 50. Maxfield Mine sleeping quarters for the night shift. It was located some distance up canyon, on the other side of Mill A Gulch, to give the workers a quiet environment in which to sleep. (SLMR, 15 October 1916)

Figure 51. The Maxfield Mine. The tunnel is on the left with a covered trestle passing over the canyon road to the dump. Blacksmith and machine shops are on the left. (SLMR, 15 October 1916)

Figure 52. William S. Brighton's first hotel, built at Silver Lake in 1873. (UHS)

Figure 53. Brighton's first hotel after a front porch was added. This side of the hotel faced Silver Lake. (UHS)

Figure 54. View of Brighton's resort from across Silver Lake before the new hotel was built. The old hotel is the large building with individual cottages to its right. C. R. Savage photo. (UHS)

Figure 55. William S. Brighton's second hotel, built in 1893. (CHO)

Figure 56. A familiar view of Brighton from the road above Silver Lake. The new hotel is seen above the center of the meadow. This C. R. Savage photo was probably taken shortly after the new hotel was completed; the old hotel is still standing to the left of the new one. (CHO)

Figure 57. The Balsam Inn, built by T. C. Davis in Brighton about 1911, was popular with both hikers and skiers. It was destroyed by fire on 31 March 1937. (UHS)

Figure 58. The Brighton hotel being razed in 1945. (UHS)

Figure 59. The Alpine Rose Lodge in the late 1940s after it was newly enlarged using materials salvaged from the Brighton Hotel. This building burned to the ground on 4 June 1965. (UHS)

Figure 60. The "Parlor" at Youngs Peak Lodge, circa 1888. Located on Mill C Flat in Big Cottonwood Canyon, its site was covered by the Big Cottonwood Power Company's reservoir in 1895. (CHO)

Figure 61. Youngs Peak Lodge. Youngs Peak, known today as Storm Mountain, looms overhead. (UHS)

Figure 62. The hotel at Wasatch Resort in the mouth of Little Cottonwood Canyon as it appeared in 1920. (SLT, 15 August 1920)

CHAPTER 9

Mining in Big Cottonwood Canyon

Much of the early prospecting in the Wasatch Mountains took place in Big Cottonwood Canyon because of the presence of a road into that canyon and the relative ease of access. Most of the early prospectors were soldiers from Camp Douglas. As they left the area and were replaced by a different group of miners, the newcomers were drawn to the rich deposits that were becoming known in Little Cottonwood. Hence the mining activities in Big Cottonwood diminished until about 1870, when it seemed that the excitement flowed over the divide between the two canyons and flooded the Big Cottonwood side with a mining frenzy.

One of the first things that happened as the miners moved into the canyon was the formation of the Big Cottonwood Mining District on 17 March 1870. This was done at Slate Springs, which was in the lower part of the canyon, a short distance above today's water treatment plant. Subsequent meetings were held at the mouth of the canyon and at Mill F to pass bylaws. In August a meeting was held at the Wellington Mine in Silver Fork where it was agreed to build a wagon road from Algiers, near the head of the fork, down to the canyon road. The building parties were authorized to charge one dollar for each ton of ore taken out over the road.[1] Algiers was a mining camp located on the flat on the north side of the east bowl, at the head of Silver Fork. This name may have been given by Hugh L. Rice, who was developing the Maggie and Mabel lodes located nearby. In the years that followed the Algiers name was dropped, and the site became known as Rice's Camp and Rice's Flat.[2]

The road that was built crossed Big Cottonwood Creek near its confluence with the Silver Fork stream, climbed the slope to the south, staying east of the stream until it reached the site of the much later Alta Tunnel. From that point the trail that exists today follows the route of the original road. Much of the lower part of this road can still be followed, although the lower end becomes somewhat obscured as it reaches the Silver Fork home sites. Stone retaining walls may be seen at several places below the Alta Tunnel. This road was the only route into Silver Fork until well into the twentieth century.

In the spring of 1871 the county court recognized the activity in Silver Fork and saw fit to designate all that part of Big Cottonwood Canyon lying east of Mill B as the Silver Precinct. Since the recorder for the Big Cottonwood Mining District, Henry C. Hullinger, had chosen the mouth of Silver Fork as his headquarters, that place was designated as the site for elections and he was appointed the precinct's justice of the peace. By the end of June the place had been named Silver Springs and Hullinger was offering meals and accommodations in a building he had erected there.[3] Meanwhile, there were at least two parties engaged in burning charcoal for smelting purposes, presumably for the smelter Buel and Bateman had erected at the mouth of Little Cottonwood Canyon. Stephen G. Sewell was burning coal at Silver Springs, while William Howard was operating about a half mile down the canyon. But they soon had a market for their product right in Silver Springs, for Col. Isaac Weightman, president of the Hawkeye Milling and Smelting Company, moved in and started the construction of a smelter for ores produced in Silver Fork. The company bonded the Richmond Mine, located at the head of Silver Fork, and by the end of the summer the smelter, operating with Sewell's charcoal, was reducing Richmond ores. It was reported to be producing three tons of bullion per day.[4] The smelter very nearly was destroyed by fire soon after it began operating, for it was at this time that a great forest fire raged through Silver Fork and the main canyon. The conflagration was blamed on loggers setting fire to some brush to destroy a yellow jackets' nest. The smelter was saved by workers building backfires, but much timber was lost and Howard lost some seven thousand bushels of charcoal. The fire also threatened Mill F, a short distance up the canyon, before it was quenched by a heavy rainstorm.[5]

In the late summer of 1871, about the time the smelter was under construction, the fast-growing community took on a new name: Belleville, named after Sewell's wife, Isabelle. It even boasted a Sunday School, sponsored by the Hawkeye company. But its prosperity was short-lived, for the next spring the smelter did not resume operations, due to financial embarrassments of the company, it was said. In July 1872 the smelting works and its ore passed into the hands of William Howard, who intended to move it to Camp Carr, soon to be renamed Argenta.[6] While there is no evidence that the smelter was moved, the mining district recorder did relocate his office to Argenta, moving the focal point of the mining community away from Belleville. At the end of July 1872 an ad appeared in the newspaper for the Belleville House, opposite the Hawkeye smelter.[7] But that was the last time the Belleville name appeared, and the community again became known as Silver Springs. While Isabelle Sewell no longer had a town named after her, she could console herself that the land on which the town rested belonged to her and her husband, for on 1 May 1872 both Sewell and Howard received patents from the Land Office for the land on which they were living, in both cases a quarter section— 160 acres. Sewell's patent took in much of the land encompassing today's Silver Fork community, while Howard's patent ran from Beartrap Fork down to Greens Basin.[8]

Directly above Algiers was the Wellington Claim, belonging to J. Albert and Stephen Groesbeck of Flagstaff Mine fame, but named after fellow claimant Wellington Sprouse. Above the Wellington a chain of claims was located, all following the same fault and most of them overlapping one another. They included the Congress; the Antelope; the Highland Chief, located by Woodman and Chisholm of Emma fame; the Prince of Wales Mine, located and owned by Thomas E. Owen; and the Wandering Boy.[9] When these claims were located they were described and measured relative to their discovery point. It was up to the miners to verify that other claims did not conflict with theirs. It was only after the Land Office began making mineral surveys to support patent applications that the enormity of the location conflicts became apparent. Land Office surveyors established mineral monuments, fixed reference points somewhat arbitrarily positioned, and mineral surveys were made relative to the closest monument. In that way the position of one claim relative to its neighbors could be accurately mapped. Only at a later date, after surveys were run from the valleys into the mountains, could claims be positioned relative to the marker at the southeast corner of Temple Square in Salt Lake City, the point of reference for all surveys in northern Utah. The Land Office took no responsibility in the resolving of conflicts of claims; that was a matter that was usually resolved in the courts. And this was one reason so many mines were involved in time-consuming and costly litigations.

The Prince of Wales was one of several mines that drew the attention of the Walker brothers at a time they were still heavily involved in the Emma Mine. Around the end of 1870 they took steps to buy the mine, but they were astute enough to realize that the Antelope and Wandering Boy were conflicting claims. To avoid future litigation, they bought those mines as well and began developing the Prince of Wales. The works were located high on the east slopes of Silver Fork. A trail traversed the bowl to the pass between Silver Fork and Grizzly Gulch, then dropped down to Grizzly Flat, where it joined the wagon road to Alta. The Antelope had a tunnel going into the northwest end of the Honeycomb Cliffs to intersect the nearly vertical Prince of Wales shaft. Log houses, a bunkhouse, and mine buildings were built at the various shafts and tunnels.[10]

In spite of the Walkers' attempt to avoid litigation, their company was taken to court because of a conflict with the Highland Chief Claim. After two years in the courts, the Walkers won the case, since the Prince of Wales Claim predated the Highland Chief's by one month. But to avoid similar problems in the future, the brothers assumed control and management of the Highland Chief and Wellington, presumably by purchasing them.[11] With that matter resolved, they began to expand their operations. A road was built to replace the trail from the mine to Grizzly Flat to facilitate shipping ore in that direction. New machinery was ordered, including steam engines to drive pumps and hoists, both at the Prince of Wales shaft and the Wellington Tunnel below. To provide water for the engines, a claim was made on three springs high in Grizzly Gulch, just a short distance below Twin Lakes Pass. The springs were developed and a steam engine and pump were installed in a newly constructed pump house. A pipeline was laid up the slope to the north, then for more than a mile along a contour line to the pass into Silver Fork, and over to the mine. There a two-story building was built, twenty by fifty feet in size and suitably protected against snowslides. The upper floor was used for sleeping quarters, while the lower floor housed the machinery and covered the entrance to the shaft. Another pipe carried water down to the Wellington Mine to supply the steam engine there.[12] Since the road between the mine and the pass into Grizzly Gulch was especially dangerous in wintertime, when snowslides often swept down over it, a rail tramway was installed and was covered by a shed for its entire 1,800-foot length. Ore hoisted to the top of the shaft was placed in tramway cars and hauled by mule to the pass, where it was transferred to wagons for the trip down the road to Alta.[13]

As the workings went deeper, the company needed a tunnel to tap the workings at a much greater depth than the Antelope Tunnel. To this end a claim on the Annie Tunnel was filed.[14] Located in Honeycomb Fork, the Annie Tunnel was almost directly below the Prince of Wales workings, but the Honeycomb Cliffs blocked direct surface communication between the two sites. Accordingly, the road to the Antelope Tunnel was extended, winding down the slopes between Silver and Honeycomb Forks and finally dropping to the new tunnel.

The Prince of Wales was one of the more successful mines in the Wasatch Mountains. It was a steady employer of men and consistent shipper of ore, its success reflecting the astute business sense of its owners. After the many improvements were made in 1875 a "tribute system" was introduced, wherein a miner received a percentage of the ore he extracted. This worked well for all parties, since the miners earned more money than they would otherwise and also had the incentive to become more productive. The system was used for several years.[15] The company itself worked the mine until the late 1880s, when it began to allow lessees to extract ore. The mine continued to be operated by lessees until the middle of the 1930s and remained in the hands of the Walker brothers' estate through that time.[16]

Today the Prince of Wales Mine is one of the popular destinations for Wasatch hikers. Many people have visited the shaft, which is at an elevation of over 10,000 feet, and examined and wondered about the boiler and the forty-horsepower steam engine and winch still resting where they were installed in 1875. The more observant also see the remnants of the headworks, with at least one of the sheaves over which the cables ran to hoist the car up the shaft. And below the mine, at the Wellington Tunnel in Silver Fork, another boiler and twenty-horsepower steam engine may be seen, each year being covered by more detritus that slides down the slope. The road that ran around to the Antelope Mine can still be followed, and the continuation road can be followed until it makes its final descent into Honeycomb Fork, where Nature has completely reclaimed it. Up in Grizzly Gulch, a short distance below Twin Lakes Pass, hikers notice an upright boiler and remains of a pump. This was the site of the Prince of Wales pumphouse, where the fifteen-horsepower engine drove a pump that forced the water through the two-inch pipe running to the mine. Although the pipe is buried, its route can be seen up the slope to the north of the pump. At an elevation slightly higher than Twin Lakes Pass, the pipeline's trail still exists along the contour line to the pass into Silver Fork, and the pipe may be seen at several places where it is exposed.

*

Farther down Silver Fork, at the end of today's road, is the tunnel of the Alta Tunnel and Transportation Company. This was a relatively recent venture, having been started by Frederick V. Bodfish in 1912. Bodfish had come to Utah from Colorado and was operating the American Flag Mine in Park City when he became interested in Silver Fork properties. He found a prospector by the name of Fred Schrott, also known as the Lucky Dutchman, who was driving a tunnel into his Lucky Dutchman group of claims. Bodfish believed that if Schrott's tunnel were driven far enough it would intersect the rich Prince of Wales lode at a greater depth than it had ever been worked and could continue farther to tap the mines in Grizzly Gulch, also at great depth. With two associates from Colorado he formed the Alta Tunnel and Transportation Company, bought out Schrott, and in 1913 began developing his new venture. He had a power line built from the main canyon and constructed a compressor house and bunkhouse against the cliff next to the tunnel portal. He also built the road that runs from Giles Flat, location of the present Solitude ski area, into Silver Fork.[17] Up to that time the road that had been built in 1870 was still being used, but the new one provided an easy automobile grade all the way to the mouth of the tunnel. During this time Bodfish made the transition from mine operator to mine promoter. He took an office in the newly completed Newhouse Building and moved into the prestigious Eagle Gate Apartments in Salt Lake City. He managed to make his company well known, with weekly news releases to the newspapers and frequent articles in mining journals such as the *Salt Lake Mining Review* and *Engineering and Mining Journal.*

In the summer of 1915 the tunnel had cut the Lucky Dutchman vein at 1,050 feet. While the company continued driving beyond that point, it gave a lease on the vein to none other than Fred Schrott, the Lucky Dutchman himself, who formed the Mines Development Company and began working the vein from the tunnel.[18] This worked out well for Bodfish, for he could continue driving his tunnel while drawing some income and lots of publicity from Schrott's shipments of ore. When the tunnel encountered water and the flows increased as the tunnel went deeper, the water seriously interfered with the mining operations. Bodfish put a positive spin on the matter, predicting the company could use the water to develop power, and he actually attempted to sell the water to Salt Lake City.[19] But behind the scenes all was not as rosy as it seemed. The development funds were rapidly dwindling. As Bodfish used all his savings for the venture he found it necessary to move his family from the fancy downtown apartment into a house in Holladay. He levied an assessment on the company's stock and drew on the savings of his siblings and in-laws. Bad air in the tunnel forced him to install compressors to ventilate the mine. Each time a known vein from higher elevations was cut, it either proved to be barren or yielded but small quantities of ore. In 1919, at a depth of 3,800 feet, the tunnel entered Tintic quartzite. All agreed that nothing of value lay ahead, and driving was discontinued. Several long drifts were started to get under the Highland Chief and Prince of Wales Mines, both of which had yielded well in years past. Working on a shoe-string budget, Bodfish refused to give up. Bolstered by leases, small ore shipments, and rosy publicity releases, he continued through the 1920s. But the Depression put his company in such serious trouble that his operations ceased, and the tunnel was worked only by lessees. He made another attempt to revive operations in 1938 and reorganized the company in 1940, after which it was reported that the machinery and equipment at the tunnel was sold for scrap. In 1946 another company, Altamina, made a brief and abortive attempt to revive the operation. Bodfish, meanwhile had moved to Oregon, where he worked a mine with his brother. He died there on 3 September 1946.[20]

Today the Alta Tunnel is a favorite hiking destination. Much of the huge dump has been removed for use as road fill, but water still flows from the portal and across the dump to the Silver Fork stream, much as it did in 1920.[21] A gate has been placed across the portal and a dam built inside to divert the water into a pipe that carries it to the Solitude Resort for culinary use. In this respect, the tunnel is now yielding its most valuable product.

*

On the other side of the west ridge of Silver Fork lies a drainage known as Days Fork. While there was much mining activity there over the years, the most prominent and best-known remnant from those days is the remains of the Eclipse Mine. Located at the end of the road, which has become little more than a trail since vehicular traffic was banned, the mine is marked by two rusting boilers that are slowly sinking into the caved-in shaft. The Eclipse dates back to September 1877, when one Bernard Quigg filed a notice for it, the only claim on which his name appears.[22] It is likely that little surface work was done at the mine; instead, its ground was explored through the fifth level of the Flagstaff Mine, just over the ridge to the south. Within two years of its discovery, the Eclipse was shipping about ten tons of ore to the market each day, all of it through the Flagstaff workings. In 1880 the company began to sink a shaft in Days Fork, primarily to provide ventilation and to lower timber, lumber, and other supplies for both the Flagstaff and Eclipse tunnels many hundreds of feet below. Substantial hoisting works and machinery were installed, and boardinghouses and workshops were built, all with shed roofs built against the side hill to protect them from winter snowslides. Indeed, when Alta suffered its devastation in January 1881 the miners at the Eclipse

Mine remained unscathed. To haul the materials to the mine, a good road was built up the fork from the Big Cottonwood road. The dugway about halfway up the canyon was known as the Hirschman Dugway, named after Moses Hirschman, the superintendent of the Eclipse Mine.[23] There is, however, little evidence that Eclipse ore was shipped down Big Cottonwood Canyon; the product of the mine was taken out through the Little Cottonwood access and down that canyon. The machinery was improved and compressors and pumps were installed in the years that followed.[24]

The mine continued to be worked until May 1888 when the works caught fire and all structures were burned. There were conflicting reports as to the extent of the damage, but it appears everything except the boilers and machinery was destroyed.[25] After that time there were no reports of activity from the Eclipse Mine, and from all appearances the works were abandoned, allowing everything that survived the fire to remain on the ground. In the early 1960s one could still see very large timbers, about 12 x 12 or 14 x 14 inches in cross-section, on the leveled ground north of the shaft and a very large stack of cord wood on the hillside above the mine. Since that time Nature has taken care of the timbers, leaving little more than their rotted remains, and the cord wood was consumed in a fire of unknown cause. The two boilers and associated machinery still remain, although one of the boilers is slowly disappearing down the caved-in shaft. One artifact that is admired by everyone who visits the site is a large cast gear with beautifully curved spokes that rests on the ground next to the shaft. Today, the road, albeit now only a faint trail, continues past the mine and into the terminal bowl of Days Fork. It is possible to follow its path first toward the northwest, then turning and traversing the bowl until it intercepts the ridge on the south side, from which point it once descended into Alta. Another route to the mine from Alta came over the divide and traversed the west bowl of Silver Fork, then dropped down through the woods until it intersected the Days Fork road about a tenth of a mile below the Eclipse Mine.[26]

*

The slopes above the Eclipse Mine are full of old mine dumps and prospects, many of them inspired by the early success of the Reed and Benson Mine on the other side of the ridge between Days Fork and the South (Cardiff) Fork. It is interesting that the ridge on which that mine is located was named after the mine, thereby perpetuating the names of two of the locators, while the name of the man who guided the mine's development and operation, as well as that of several other mines in the area, has been all but forgotten. If he had received his just recognition, the Reed and Benson Ridge might well be known today as the Goodspeed Ridge.

In 1870 a group of men including William B. Benson, brothers Major Henry C. Goodspeed and F. Goodspeed, Emit C. Payne, Franklin Reed, and Robert E. Ricker posted four major claims along the ridge between the South Fork and Days Fork in the Big Cottonwood Mining District. The Ophir, Reed and Benson, and Excelsior Lodes were above the South Fork, while the Brilliant Lode was situated at the head of Days Fork.[27] There were other individuals involved in one or more of the claims, including six other members of the Payne family, but the men named above were involved in all four claims.

There can be little doubt that Henry C. Goodspeed was the moving force behind this group. A native of Massachusetts, he had an early inclination toward literature and writing. He came to Utah in 1870 as a representative of the *Chicago InterOcean,* the *New York Graphic,* and the *Boston Traveler,* but almost immediately he became involved in mining. It is interesting that the very first claims posted by Goodspeed and his associates proved to contain very rich ore deposits. In spite of their being in a most difficult location to access, on very steep slopes above nearly vertical cliffs, yet below the crest of the ridge, development began immediately. A trail was built from Central City to the mine, the upper part traversing the precipitous slopes above the South Fork as it approached the mine. It must have been completed in 1870, for early the following spring it was known as the Goodspeed Trail and was shown on a map published about that same time. The following year the first road was built up the South Fork from the Big Cottonwood Canyon road to the Reed and Benson Mine.[28] The final ascent to the mine went up a long and incredibly steep switchback, then along a ledge blasted out of the limestone cliffs. Although the mine was worked in 1871, it suffered legal difficulties and went into receivership. In September the courts ordered the mine and its equipment to be sold at public sale. By that time some 5,400 sacks of Reed and Benson ore, well over 200 tons, were at the Buel and Bateman furnace at the mouth of Little Cottonwood Canyon.[29] Details of the sale are lacking, but it is evident that H. C. Goodspeed and Franklin Reed emerged as the new owners.

Of the original principals, Payne and Ricker disappeared almost immediately. At the time the claims were filed Goodspeed and Emit C. Payne went into business as ore brokers and in October and November 1870 they were shipping ore out of Salt Lake City, although there is no evidence that the ore came from their mine.[30] In December 1870 Payne sold his interest in the mines to one James Jones for $12,000, and nothing further appears of him in the records. By June 1871 the mining brokerage firm had become Majors, Chapman & Goodspeed.[31] William B. Benson continued to work with Goodspeed through the summer of 1871, as evidenced by at least sixteen claims bearing his name, most of them in company with Goodspeed, before he too vanished from the scene. Franklin Reed most likely provided the finances for the venture. He was from Bath, Maine, and probably never set foot in the Wasatch Mountains. His name appears on at least twenty-five claims, all of them in company with Goodspeed, and two of them as late as 1875 and 1876 in conjunction with some of Goodspeed's later ventures.

With legal matters resolved, Goodspeed took control of the mine and began working it in earnest. In the summer of 1872 it was said that the mine could produce fifteen to twenty-five tons of ore per day, but it is not likely it ever shipped at that rate. The difficulties of moving ore down the trail into Little Cottonwood or down the steep road into the South Fork restricted the shipments. So severe was the transport problem that in 1873 the company undertook to construct a tramway to span the cliffs separating the mine from the bottom of the South Fork. It was decided not to use an aerial tram; instead, a trestle was built down over the cliffs. This very steep trestle, sloping down at an angle of forty-five degrees for part of the distance, was equipped with iron rails. Two cars were used, connected together with a cable that wrapped around a drum at the top. The loaded downward bound car pulled the other car up. A single track was used, but a short section of double track placed at the midpoint allowed the cars to pass. The tramway was 1,500 feet long and had its lower terminus in a large ore house. While building this structure, workmen often had to be suspended by ropes to allow them to do necessary blasting. The tramway was completed and running smoothly by September. Recognizing that its operation would be crippled by the winter snows, the company elected to cover the tramway with a snowshed for its entire length. With extensive buildings covering the dump and providing living and working quarters for the men, the mine was able to operate throughout the winter.[32]

This tramway must have been one of the more spectacular man-made achievements in the Wasatch. It was often mentioned in written accounts of the mine, and it was the site of several exciting adventures. After it began operating, but before the snowsheds were completed, the workers began to use it to ride between the mine and the ore house below. The superintendent, George W. Norton, warned the men against this practice on account of the worn condition of the cable, which was to be replaced as soon as a new one arrived. But on Sunday morning, 9 November 1873, in spite of the warning, the mine's blacksmith, Charles Lundgren, climbed aboard one of the tramway cars to ride down to the ore house. Just as the cars passed one another at the midpoint passing track, the frayed cable parted, allowing both cars to accelerate down the steep track, completely unrestrained. Lundgren jumped for his life, but his body went bounding down the steep slope and crashed into some boulders, killing him instantly. The car raced down the track, crashed through the ore house, out the other side, and ended in a ravine some two hundred yards below. Lundgren's shattered remains were taken to the city for burial, while the tramway remained inoperative for another week before the replacement cable arrived. The new one, it was said, was about twice the thickness of the old one.[33]

Being located on the very steep slopes, the mine and its trails were subject to snow avalanches with every heavy snowfall. Many miners suffered close calls or actual involvement with snowslides on their way to or from the mine. One such instance happened after heavy snowfalls in January 1875 when three men en route to the mine were enveloped by a snowslide. Two were found and dug out, but the third, a Mr. Baxter, was carried down under the tramway, where he became lodged against one of the posts. Although he was buried, the timbers protected him and provided a cavity with sufficient air, enabling him to survive. He remained there two days and one night before he was found and rescued, weak but alive.[34]

Three years later the tramway and ore house were all but destroyed by a snowslide. Although there were forty men employed and living at the mine at the time, none was injured. The tramway and ore house were subsequently rebuilt and continued to operate when the mine was being worked sporadically, at least until 1882.[35] By that time photography was becoming more common, but no picture of the Reed and Benson tramway has come to light.

Although the mine produced very rich ore, it was worked with great difficulty. In following the ore fissure, it wound down through the depths, creating what one miner described as a deep rat hole, one that required multiple hoists to raise the ore to the surface. Much of the waste rock was left in the mine, packed into cavities and crevices, causing the mine to be cluttered and more difficult to work. To solve some of these problems and to tap the vein at a much deeper level, Goodspeed and Franklin Reed

filed a claim for the Reed and Goodspeed Tunnel, located on the South
Fork road about a quarter mile north of the Reed and Benson ore house,
and began driving the tunnel.[36] By August 1876 its depth was between
300 and 400 feet. This project became Goodspeed's obsession; its completion was a goal he kept pursuing for the rest of his life. It was an impressive venture: the tunnel was five feet wide and seven feet high, and
was driven in a perfectly straight line. In 1879, when it was 1,300 feet
deep, it was said that from the inside end one could see a man standing in
the portal. By this time the tunnel had struck the Excelsior vein, but at
this depth it was barren. Almost ten years later the Reed and Benson vein
was crossed, but again little ore was found. Still the work continued. In
1891 Goodspeed installed a small hydroelectric plant to provide power
for electric drills and ventilation fans. Finally, in February 1892, the Ophir
vein was cut at a depth of nearly 3,000 feet, and news releases announced
a big strike.[37] However, work continued for another two years with only
small shipments of ore. Then, in August 1894, Goodspeed did a surprising thing: he filed an attachment suit against the Reed and Goodspeed
Mining Company—his own company—for nearly $100,000 owed to him.
This may well have been a legal maneuver to gain sole control of the
company, since Reed had died in 1887. The court ordered the company's
assets sold at public sale to satisfy the attachment. Goodspeed himself
bought the property at the sale for $30,000 and announced the forma-
tion of a new company.[38] It was reminiscent of the receiver's sale of the
Reed and Benson Mine in 1870, when Reed and Goodspeed became sole
owners of the mine. By this time Goodspeed was a sick man and was not
destined to pursue his goal any farther. On 20 January 1895 he died at the
age of fifty-nine.[39] His mine remained idle for several years before being
worked again, this time on lease. In 1902 the lessees formed the Kennebec
Mining Company and bought the property of the Reed and Benson and
Reed and Goodspeed companies.[40] Manager W. J. Craig decided to drive
another tunnel midway between the mine and the lower tunnel, hoping
to tap the two veins that had yielded so much ore in the early days. This
tunnel was begun in 1905 and was developed for over ten years until the
Cardiff Mining Company interests took over the Kennebec in 1917. Craig
continued to work his tunnel as a lessee for several more years, but it
never yielded the riches he thought were there.

Little remains to be seen of the Goodspeed mines today. On the ridge
north of Flagstaff Mountain there are low stone walls, remnants of some of
the cabins that once housed miners. The Goodspeed Trail may still be seen
crossing a flat spot on the ridge west of Flagstaff Mountain, but Nature

has reclaimed most of its length down toward Alta. The trail can be followed north from the ridge for some distance, but it has almost completely vanished on the very steep slopes as it approaches the mine. Nothing remains of the tramway terminal; but from the dump of the Reed and Benson Mine the original road down into South Fork can be followed around the cliffs and onto the steep switchback going down the grassy slope at the extreme southeast corner of the fork. Below the mine the original South Fork road has all but vanished, but it can be seen farther down the canyon. About a thousand feet above the point where the road going up Cardiff Fork crosses to the east side of the stream, the old road goes up the slope on the east and continues up to the site of the Reed and Goodspeed Tunnel, labeled Kennebec Tunnel on contemporary maps.[41] A great deal of twentieth-century erosion created a very deep gully next to the tunnel's dump, but if one crosses to the other side, remnants of the old South Fork road continuing up the fork can be found.

*

On the other side of the South Fork, at the north end of the ridge is a summit called Kessler Peak. Known in 1870 as Kesler's Peak, it was the site of a great deal of early mining activity. A single claim, the Homeward Bound, filed at the end of October 1870, opened the gates for a flood of claims filed the following spring. They included the Silver Bilk, Provo, Little Giant, Infant, McDougald, and Sailor Jack Lodes, among others. All these were on or near the ridge south of Kessler Peak, along the contact zone between the limestone and quartzite formations. As word of these discoveries spread, the peak became overrun with prospectors and many more claims were made. The one to achieve instant recognition was the Provo when the owners, Messrs. Hawkins and Meeks, took a large lump of very rich ore and placed it on display at Wells, Fargo & Company offices in Salt Lake City. The mine was in the news again in November 1871 when the owners brought down a huge boulder of pure galena weighing slightly more than one thousand pounds. By that time the mountain was covered with as much as seven feet of snow, so the trail that had been built the previous July was of little use. The boulder was placed on a skid, or sled, but workmen still had to shovel snow all the way from the mine to the South Fork to bring it down. It was sent to New York as weighty proof of the Provo Mine's value.[42]

After much publicity a few of the mines began producing, although the difficult access kept their shipments small. After 1874 the Provo, once highly touted, disappeared from the news. Three years later, a "dead-broke miner" went to the abandoned Provo and relocated it, renaming it the

Carbonate. While he was clearing a spot to build a cabin he uncovered a large body of galena. He soon had enough sacked to make his first shipment, which attracted no little attention, and in little time he had a buyer for the property. John W. Kerr, of the Hilliard Flume and Lumber Company, and John Lawson bought the claim for $15,000. The "dead-broke miner" was dead-broke no more.[43]

Kerr and Lawson, as the Carbonate Mining Company, immediately began to work the claim, extracting and shipping ore. With typical optimism, they claimed to have 2,000 tons in sight and put ten teams to work hauling ore down the canyon. Thirty animals were used at first to bring ore down from the mine, but within a short time they had seventy-one animals so employed. The ore was sacked and placed in boats, or drags, which the animals could pull down the trail. The difficult haul, they claimed, was not dragging the ore down, but taking the empty boats back up to the mine. When early snows retarded their shipments, they planned and installed a 3,500-foot-long tramway. An improvised affair for use only during the winter, it had four separate sections, each having two boats or sleds attached to opposite ends of a rope twice the length of the section. The rope passed through a stationary pulley at the upper end of the section, so one boat would descend while the other ascended. At the junction of the sections, the boats were transferred to the rope for the next section. It is not clear how effective this tramway was, but the following January the company was reported to be shipping about 1,200 sacks of ore daily.[44]

The adjacent Sailor Jack Mine had entered bankruptcy and was now being operated by lessees. Recognizing that both the Carbonate and Sailor Jack claims covered the same ore body, the owners of the former purchased the latter, bringing in another person for the purchase. But in doing so they created a problem, for there now was a quarrel about who owned how much of each mine. The matter went to the courts, where a judge resolved it, but he also placed the Carbonate Mining Company under the control of a receiver. Operations at the mine ceased for several months until a receiver could be appointed, but then they resumed with a vengeance. It was reported that the ore in the breast was soft, much like gravel, so that two miners could break down in a few minutes what would take the rest of the day to place in sacks.[45] It was reminiscent of the heady days of the old Emma.

One of the first actions of the receiver was to forbid the sale of liquor in the vicinity of the mine. A *Salt Lake Tribune* correspondent remarked,

"Rough on the boys, but the correct thing nevertheless."[46] During this same period a new road was constructed to the mine, this one coming up on the west side, starting at Argenta in Big Cottonwood Canyon. It used the old logging slides from Mill A that ran up the south slopes of the canyon to the ridge above Mineral Fork. From there a road was built traversing a steep slope in a southerly direction until it reached the gully directly below the mines; it then doubled back and followed two sets of switchbacks to gain altitude before reaching the dump of the Little Giant Tunnel. It then followed another long switchback south of the mines to reach the pass at the top of the ridge. The operators may have used wagons on this trail between the mine and the pass above Argenta, but for the rest of the distance the sacked ore was wrapped in cowhides and dragged down. This trail was in use in July 1878, when it was reported that some twenty-five "snakers" were kept on the trail between the mine and Argenta.[47]

In November 1878 the owners negotiated a sale of the Carbonate Mine to New York interests; the reported price was $165,000. The new group organized the Kessler Mining Company, taking the name from the adjacent peak, but adding an extra "s" to its spelling. The receiver was dismissed by the court, and the new company took over.[48] It shipped a great deal of ore during its first year, but then operations slowed down, due either to lack of interest on the part of the owners or because of a general decline in prices paid for ore. By the end of 1880 the mine had been placed under lease. It continued to be operated somewhat sporadically by lessees for the next twenty-five years. Then, in 1908, a lease was taken by Alma Nielsen and some associates. They immediately started the installation of a gravity tramway running from the Homeward Bound Tunnel on the east side of the divide down to the road in the South Fork. With a total length of 4,000 feet, the tram had eight towers, and even though it had only two buckets, or cars, it was claimed to have a capacity to ship one ton of ore each three and a half minutes.[49] While this installation should have cut expenses for hauling ore down to the road, it didn't appear to help the lessees, for two years later the mine was reported to have been lying idle for years.

In the spring of 1910 Ernest R. Woolley, backed by New York capital, purchased the property, formed a new Carbonate Mining Company, and began to push development. While miners blocked out ore in the mine the tramway was overhauled, and by August of that year ore was coming down to the loading station and was being hauled down the canyon to the smelters. But hardly had operations begun before they were stopped by litigation.[50] There is no evidence that the new Carbonate Mining Company was able to operate again.

Three years later, in 1913, James W. Wade and associates leased the property, repaired the tramway, and began extracting and shipping ore. This operation continued through 1916, at which time the Carbonate Mine saw its last operations.[51]

Today the Carbonate Mine is still very much in evidence. The trail from the South Fork still exists and is easily followed all the way to the ridge at Carbonate Pass. The cables for the tramway are still draped up the slope, although the towers have long disappeared, victims of Nature's ravages. The large, fixed cables, one inch in diameter, supported the cars, while the smaller three-eighth-inch cable pulled the cars up and kept them under control on the way down. The site of the lower terminal can be seen below the present-day road several hundred feet up canyon from the trailhead of the Carbonate Pass Trail, but the cables no longer reach that point. About 1916, when the Cardiff company constructed the present road to eliminate the very steep grade at Greens Hill, the cables were dragged down canyon until they were out of the way of the new construction. They can be found in the brush on the steep slopes above the road. At the upper station the cables went around a large wooden drum. Lessee James Wade later recalled that the drum was made with wood bearings, and when the tramway was operating it could be heard for miles, squeaking and groaning, as the cable practically crushed the drum. He said it always took a little push at the top to get the cars started.[52]

Another twentieth-century miner told a story about early operations at the Carbonate Mine. He noted that the Reed and Benson Mine was visible from the Carbonate, and vice-versa, the two mines being separated by "something more than a mile of fresh air." When miners were in short supply the superintendent of one mine would go to the edge of the dump and raise one arm to indicate he would pay one dollar a day more than the miners were getting, or two arms for two dollars more. Miners would spend most of a day going over to the other mine to work there at the higher rate, often only to have the same thing happen in the opposite direction. Finally the superintendents realized they were losing a lot of labor with the miners walking back and forth, so they divided the available men and refused to hire men from the other mine. It is an amusing story, but it probably has not an ounce of truth to it. This same miner also told of sending ore down to Argenta in rawhides in the winter; when the iced trail was in good condition it would take just three minutes for a 450-

pound rawhide load to slide from the mine to the bottom of the trail at Argenta.[53] This also is little more than miner's lore; even though rawhide trails have a characteristic U-shaped cross-section, there are too many places along the way where a rapidly moving ore-filled rawhide would leave the trail and never reach its intended destination.

The Argenta Trail can still be followed for its entire length, although the switchback directly under the mine has suffered greatly from erosion. Also, the trail becomes more and more obscure each year on the section going over the steep traverse above Mineral Fork. There were many other trails in the vicinity, but most have been reclaimed by Nature. One of them, traversing from the Carbonate Tunnel around the west side of the peak to the ridge on the north, could be followed as late as the 1970s, but only several faint sections of it can be seen today. Another, descending the northwest slopes of the peak to a mine, possibly the Little Cora, is used today as part of the popular Kessler Peak North Trail.[54] Remnants of several log cabins as well as mines and prospects are also found along this trail. The remains of a beautifully built stone cabin can be found on the very steep west slope of Kessler Peak, on the north side of the couloir that originates at the summit. It was used to work a mine on the opposite side of the gully, both being accessible at one time by a trail from one of the switchbacks of the Argenta Trail, but it too has returned to Nature. The cabin had a large window on its west wall that framed a spectacular view of Dromedary, Sunrise, and Twin Peaks. Stone walls from cabins may also be seen in the vicinity of Carbonate Pass, directly above the Carbonate Mine.

*

Another important mining operation in the South Fork gave that drainage the name by which it is known today: Cardiff Fork. At the turn of the century an Englishman by the name of Fred W. Price was prospecting in the South Fork. He filed the General Lafayette and Mountain Yueen Claims, both at the head of the fork, in 1901 and 1902. Concurrent with the second, an old-time prospector by the name of Theodore A. Reamer staked the Mountain Cheaf [sic] claim, adjoining Price's two claims. The two men then joined forces to file the Mountain Chief Extension No. 1 the following year, and in 1904 Price filed at least two other claims in the same vicinity.[55] Price and Reamer had become acquainted in the 1890s when they worked for the Denver and Rio Grande Railroad, Price as a painter and Reamer as a carpenter. Reamer had prospected in the Alta area as early as 1871 and had previous association and experience with the ore deposits in the South Fork vicinity. In 1892 he was one of the incorporators in the Jones Mining Company, formed to develop the Little Dora Lode, a claim recorded by Reamer and Frank A. Stoyer, another principal in the new organization. The name of the company came from Thomas B. Jones, a seventy-two-year-old miner who, with the backing of Salt Lake businessman Thomas Miller, worked the mine with some degree of success. In 1899 it was reported to have been producing ore from the 200-foot level in its shaft, and a tunnel was started to drain the property. Work continued another two years before the company ceased operations and the mine was abandoned. Reamer's Mountain Cheaf Claim, filed the following year, was a relocation that encompassed the old Miller mine site.[56] Years later, after the Cardiff Mine had achieved remarkable success, Jones's heirs brought suit against the Cardiff company, claiming Reamer's relocation was a conspiracy to defraud the Jones company, but the court ruled in favor of Cardiff, stating that the rights of the Jones Mining Company had lapsed through inaction over a period of years.[57]

Price and Reamer began working the Mountain Cheaf Mine, presumably through the shaft abandoned by the Jones Mining Company, and at the end of 1905 were reported to have shipped some good ore to the smelters. During this period both men had jobs in Salt Lake City, but the following year, in December 1906, they were joined by Hugh G. McMillin and four minority stockholders to form the Cardiff Mining and Milling Company, the name taken from the city where Price was born: Cardiff, Wales.[58] A total of fifteen claims were transferred to the new company. Continuing work in the deep shaft, they soon encountered water and moved their efforts to the old Jones Tunnel. Before the tunnel could reach the flooded shaft, they received an infusion of capital from former Salt Lake City mayor Ezra Thompson and his business associate, James D. Murdoch, both of whom had achieved considerable success and riches with Park City mines.[59] Thompson immediately took charge and was the driving force behind the company for many years that followed.

Some ore was found, and the tunnel was driven for another year and a half, when a new and important strike was made. The company built a new road to connect the mine to the existing South Fork road and began regular shipments of ore.[60] The ore body grew in size as it was followed deeper, but the depths also tapped large sources of water, seriously impeding operations. Almost immediately it was announced that a new tunnel was planned, starting more than 300 feet lower than the present workings. It was expected to be about 2,500 feet long before it encountered the ore body. The new tunnel was begun at the end of August 1911 and work continued year-round for the next several years.[61] While it generated considerable excitement at first, interest soon waned and it seemed this would be yet another deep tunnel that would achieve nothing. Then, in mid-October 1914, with the tunnel 2,575 feet deep and with little to show for the effort other than a large dump of waste rock, a full face of silver-lead ore was encountered. As excavation continued the ore body grew in size.[62]

From this point forward, things got better and better for the Cardiff Mining and Milling Company. A new road was built to the portal of the tunnel and ore bins were erected. To handle increased ore shipments a new road was built down the South Fork, located somewhat higher and having a more constant grade than the old one. New buildings were con-structed, including a large and commodious bunkhouse on a flat above the mine. A power line was installed by Utah Power and Light Company to allow electrical equipment to be used in the mine. The wires were strung from the Utah Coalitions Mine at Lake Solitude, farther up Big Cotton-wood Canyon. That mine had received electrical service in 1910 when power lines were brought over the divide from Park City. The Cardiff Mine was provided further utility services when a telephone line was constructed over the divide from Alta.[63]

All this activity did not go unnoticed in the world of finance. Before September 1914 the Cardiff company was not even listed on the local stock exchange. A few days before the ore discovery it was listed and was quoted at seven cents a share. Ten days later, about a week after the strike, it was up to twenty-four cents. By November it had increased to fifty cents. The following June its stock price was between $3.00 and $3.50 a share. In September 1915 the company paid its first dividend—twenty-five cents a share—making its stock even more desirable. It followed with periodic dividends for the next year and a half. By the end of March 1916 company shares were selling for over seven dollars, and by the end of May that year the price was between $9.40 and $9.60. Many fortunes were made in Cardiff stock.

All the new activity was not without its problems. To haul the large quantities of ore to the smelters in the valley, many teams and wagons were employed. Almost immediately concerns were raised about the Big Cottonwood Canyon road, concerns that all the heavy ore traffic would interfere with tourist travel to the lakes at the head of the canyon.[64] The Salt Lake County Commission addressed the matter and took steps to have sprinkling wagons keep the road moist to cut down the amount of dust. Officials stepped in and raised questions about the purity of the city's water supply. And well they might, for the end of the season saw as many as seventy four-horse teams going up and down the canyon daily. As many as fifty additional teams were employed bringing ore from the mine down the South Fork to the main canyon, where other teamsters took it to the valley. When snow closed the road in the winter, horse-drawn sleighs were used to bring the ore down as far as possible, where it was dumped and held until wagons could take it the rest of the way to the smelter. Stockpile locations included the mouth of the South Fork, Burnt Flat below Argenta, and Mill B Flat. In May 1917, for instance, it was reported there were about 1,500 tons of ore stacked along the canyon road.[65]

The 1915 season made it painfully clear to those concerned about water purity that something had to be done about all the teams in the canyon. Cardiff president Ezra Thompson proposed a plan to use tractor vehicles to haul the ore, but only if the county would widen and grade the road to make the plan feasible. After much discussion, it was agreed that a $30,000 road improvement be undertaken, the cost to be distributed between Salt Lake City, Salt Lake County, the Cardiff company, and the Boston Development Company, which was working the Maxfield Mine and was also shipping ore down the canyon road. The improvements, to be made between the mouth of the canyon and the South Fork, would include grading and widening the road, building a number of new bridges, and making a significant change at Jones' Hump, where intolerably steep grades and sharp curves were found.[66] The latter was created in 1895 when the Big Cottonwood Power Company's dam and reservoir covered the existing canyon road and a new road was built up a dugway and over a ridge that extended to the edge of the reservoir. Jones' Hump, with its very steep grades and a sharp turn through a narrow cut at the top of the ridge, had long been a nemesis for the teamsters who frequented the canyon road, and it had continued to be such for motor vehicles when they came into use.

The new road improvements were made between April and July 1916. Meanwhile the Cardiff company built huge ore bins at the mouth of South Fork and arranged to have track-laying vehicles and trailers haul the ore to the valley smelters. The Mine Transportation Company, organized to undertake this task, purchased five tractors and sixty trailers to make up tractor-trains with as many as twelve trailers. Each trailer had a six-ton capacity, as much ore as was hauled by a six-horse team. As soon as the road was ready, the motorized carriers were put into operation. On their first day on the job three tractor-trains arrived at the smelter having nine, ten, and eleven trailers—a total of 180 tons of ore.

The Cardiff company earlier had reduced its workforce because it was unable to handle the quantity of ore the men were able to produce. With this successful demonstration the company put its men back to work and prepared to increase its production from 100 tons to 300 tons daily.[67] But the elation was short-lived; it was soon found to be too difficult to handle the long trains on the steep grades in the canyon, even at their slow speed of about three miles per hour. Within three weeks the Cardiff company announced it was going back to horse-drawn transportation, as the tractor-trains had proven unsuccessful, even with only six trailers. This brought a cry of protest from the city commissioners, who felt this latest move was in violation of the agreement. It also brought a cry of glee from the several truck agencies in the city, and in the weeks that followed a number of different trucks were claimed to be able to handle the challenge, including Packard, Jeffery, and Knox. Four-wheel Packard trucks were put to use almost immediately, each truck hauling five tons without trailers and making four trips per day. The company later had these trucks towing a five-ton trailer as well. The Knox tractor, with a five-ton capacity and the power to tow two five-ton trailers, was used extensively, although the county road commission later complained that the tractor's wide steel wheels with diagonal cleats were doing excessive damage to the road. A four-wheel Jeffery truck, with three-and-one-half-ton capacity and hauling one five-ton trailer, also was used. The Cardiff company operated the Packard and Jeffery trucks, while the new United States Transportation Company used the Knox tractors.[68]

Although the city water department wanted to prohibit the use of horses for all ore hauling in the canyon, it relented and allowed teams to be used on the steep grades between the mine and the ore bins at the mouth of South Fork. In 1917 the White Motor Company demonstrated its five-ton narrow-tread ore truck on this road. It was able to follow the wagon ruts and had enough power to negotiate the 23 percent grade on that high mountain road. But it must not have been entirely successful, for it was not until 1921 that the company's trucks regularly made the climb to the mine ore bin. By that time the Cardiff company had rebuilt the section of the road at Greens Hill, near the Tar Baby Mine, eliminating the steepest grade. After this time horses were no longer used for heavy ore hauling except during the winter, when they hauled ore sleighs as far down the canyon as the Stairs Power Station, where motor vehicles took over.[69]

The end of 1917 was the high point of Cardiff operations. The last twenty-five-cent dividend paid by the company was in August of that year. In November it was announced that shipments were reduced by 50 percent due to the low market for metals.[70] The company continued shipping, but it no longer was enjoying the high prosperity of the years following the big ore discovery. It continued to pay dividends sporadically, but they were of decreasing amounts. The last one, in 1924, was for ten cents a share. By that time the company had paid a total of $1,025,000 in dividends. Also by that time it had a new president and general manager, Lynn H. Thompson, who filled the vacancy caused by the death of his father. His brother was elected vice president and director, while Ezra Thompson's business partner James D. Murdoch continued as secretary. Founders T. A. Reamer and F. W. Price remained on the board.[71] The company continued to operate another four years until it closed in 1928. After that time a series of lessees worked the mine, but they were hampered by water in the deeper workings and shipped only modest amounts of ore. In 1955 the Wasatch Drain Tunnel, whose portal was in Little Cottonwood Canyon, was extended until it encountered the Cardiff ore body below the old workings, and ore was extracted well into the 1960s.[72]

Today the Cardiff surface workings are visited by many hikers in Cardiff Fork. The ore bin still stands but has suffered much deterioration. As recently as the late 1960s the trestle between the mine dump and the top of the ore bin was still in place, but heavy floods in the early 1970s brought it down. A few of the trestle's trusses can be seen half-buried in the streambed. A road extended to the ore bin and circled around it so trucks could be loaded from the ore chutes and gates along the sides of the bin. Underneath the bin a boiler may be seen. This was used to provide some warmth during winters to keep the ore from freezing. The same thing was done at the large bins built at the mouth of Cardiff Fork. One miner related a story of using dynamite to break up frozen ore in an ore bin, but the Cardiff company used boilers to keep the ore from freezing.[73]

The foundation of the boardinghouse can still be seen on the flat above the mine. The portal of the Beefsteak raise, now sealed, can be seen at the northeast corner of the boardinghouse. It gave miners covered access to the mine, a definite convenience in the wintertime. The portal to the tunnel below also has been sealed to protect the water supply for the Snowbird resort, since it uses water from the Wasatch Drain Tunnel, which extends under the Cardiff workings.[74] Portions of the ore bin at the mouth of Cardiff Fork remained into the 1960s but have since been removed. It was on the east side of the Cardiff Fork road, south of Big Cottonwood Creek.

It might be noted that Fred W. Price, who had labored many years to bring the Cardiff Mine the success it achieved, could not then rest on his laurels. In 1915 he secured control of many claims on Montreal Hill, only a short distance northwest of the Cardiff Mine, and established the Price Mining Company to work the property. A tunnel located above the Cardiff Mine road was started and a two-story boardinghouse was constructed nearby. Work was pushed for several years, with the tunnel being some 500 feet long by late summer of 1918. But after that time the company was no longer newsworthy. By the time Fred Price died, on 2 April 1931, his company had ceased operations.[75] Today the Price Mine dump towers over the Cardiff Fork road, while water flowing from the tunnel runs down one side of the dump and floods a short section of the road.

*

One final Big Cottonwood mine that should be mentioned is the Maxfield. It was located in the main canyon at Mill A Gulch, very close to the Big Cottonwood Lumber Company's Mill A site. When the mining boom began and mining frenzy swept the populace, many of the prospectors were woodsmen who had been cutting and hauling timber to the sawmills or mill workers or teamsters who were hauling lumber away from the mills. Among this group were the Maxfield brothers, who had been working in Big Cottonwood Canyon since the first days of the Big Cottonwood Lumber Company's activities there. They took contracts to provide logs for the mills; and while much of the Mill A supply came from the slopes on the south side of the canyon, some also came from the gulch on the north side, Mill A Gulch. After the lumber company was dissolved, Richard and Robert Maxfield bought Mill A and were operating it when the mining boom took off. When the brothers began prospecting it was perhaps inevitable that three of the eight claims they filed in 1870 were in Mill A Gulch. One of the three was the Maxfield Lode.[76] It is likely the brothers knew little about working a mine, for nothing of consequence was done with it. But it must have held some promise, for it was purchased by Feramorz Little, former principal of the Big Cottonwood Lumber Company and at that time proprietor of Mill D. While it is not known exactly when this transaction took place, in 1873 it was reported that the mine was being worked by the owners, who included Little's son, James T., and the Maxfield brothers.[77] Feramorz Little had taken part in the Maxfields' mining endeavors before that time, for in the spring of 1871 the three of them, joined by five others, including Charles Bagley from Mill D, recorded the Gen. Thomas Lode at the mouth of Mill A Gulch. In June of the following year Feramorz and James Little joined the Maxfields and Charles F. Decker, another Mill D associate, in filing the Mill A Tunnel Claim at Mill A Gulch, and Little filed the Maxfield Tunnel

Claim "for the owners, to work the Maxfield Lode."[78] While the Maxfield Mine was not often reported in the news, it was well known, for many mining claims filed in its vicinity used it as a location reference.

At this time all the Maxfield Mine workings were located in Mill A Gulch. While they were yielding high-quality ore, it was decided in 1877 to drive a tunnel from the main canyon, only a short distance above the canyon road, to develop the mine and to alleviate water problems in the upper workings. The tunnel claim as filed was named the Little Tunnel, although both Maxfield brothers were included in the notice. Then in March 1879 the Maxfield Mining Company was incorporated, with Feramorz and James Little holding over 93 percent of its 10,000 shares of stock. Although Richard Maxfield was one of the incorporators, he was subscribed for only one share. By this time Robert Maxfield was no longer active in the canyon and his name did not appear in the new organization.[79]

As the tunnel was driven farther and farther into the mountainside it began to tap bodies of rich ore. Large shipments were made and the mine was proclaimed to be one of the most promising in the district. Almost immediately it was announced that the mine was being sold, and it left local hands to become the property of Massachusetts investors. William F. James moved in as manager and continued in that role until the end of the century. Under James the mine was expanded and developed, with considerable improvement of its facilities. The mine complex soon dwarfed the town of Argenta a short distance up canyon. By 1890 there was an engine house, four ore bins, a powder magazine, boardinghouse, bunkhouse, and a stable. The mine operated profitably during the entire tenure of manager James, paying a total of $118,000 in dividends.[80] But the Maxfield was not spared the problems of groundwater, common to most Wasatch mines. Waters began to flow as the ore bodies were followed down below the main tunnel level. Much of the surface machinery was devoted to pumping water out of the mine, and frequently operations halted, sometimes for as long as four months, when the pumps could not handle the flow. Ever-increasing amounts of water, combined with the mining doldrums around the turn of the century, forced an end to the company's operations.

In the early twentieth century the mine was worked sporadically by the company and a series of lessees. In 1911 the Maxfield name was changed to Salt Lake Power and Mining Company, presumably to market electrical power from its small hydroelectric plant that had been installed by William James in the mid-1890s. In 1912 the name was changed again, this time to the Maxfield Power and Mining Company.[81] Then at the end of 1914 the mine was taken over by the Boston Development Company. Incorporated in May 1914 to work a group of claims in Tooele County, the company was a corporate embodiment of Fred H. Vahrenkamp, a Nevada mining promoter.[82] It was his belief that up-to-date machinery could conquer the water problem and that the mine could be placed on a producing basis within a short time. He installed new electrically driven pumps, and after several months he was shipping ore. Although the mine was reported to be dry by the following December, it was also said to be practically closed because of power failure. Heavy snows had taken down the commercial power line, which came down canyon from Brighton, and the stream flows were too low to allow the mine's hydroelectric plant to carry the load.

Vahrenkamp then began promoting a drain tunnel to carry the water away from the deepest workings. This was not an original plan, for William James had suggested a drain tunnel as early as 1892, and in 1904 the company's board of directors had considered a mile-long tunnel starting at Mill B. But none of those proposals approached Vahrenkamp's grand plan, for his tunnel was to start near the mouth of Mill Creek Canyon, presumably in Olympus Cove, and run southeast for a distance of three miles, reaching the Maxfield mine about 2,000 feet below the surface. He formed the Salt Lake Power & Water Company to undertake this project, traveling east several times to promote financing; but the tunnel was delayed while litigation problems besieged the company and finally brought it to its end.[83] However, the Boston Development Company did produce a considerable amount of ore and in 1916 contributed $5,000 toward the joint effort with the Cardiff company, Salt Lake City, and Salt Lake County in improving the Big Cottonwood road. Although the Cardiff Mine shipped much more ore than did the Boston Development Company, the latter also took part in the transition from teams to trucks.

Again the mine was operated sporadically by various lessees until the mid-1920s, when the Bullion Chief Mining Company took over the operations. It enjoyed a number of successful years before the national Depression brought the company's operations to an end. As late as 1951 a new Maxfield Mining Company was incorporated to take over the claims of the original company. Its capitalization of $1,814.40 was the cash value of the claims.[84]

The Maxfield Mine, located as it was on the main canyon road, was a well-known landmark among canyon travelers prior to the middle of the twentieth century. As the road approached the mine it was sandwiched between the mine buildings and the mine dump. The first large building on the left had the turbine waterwheel in the basement level and the compressors, generators, motors, and switches on the ground floor. The boardinghouse, complete with bathrooms and toilets, was on the next floor, while the upper level provided sleeping accommodations for thirty men. A sign on the gable end read, "Maxfield Mine." A short distance farther up on the right were the ore bins, and beyond them the narrow, one-lane road passed under the covered trestle that carried mine cars from the tunnel to the mine dump and ore bins. Under the trestle on the north side of the road were the blacksmith and machine shops, and a short distance away a cottage provided sleeping quarters for the night shift. Today very little remains of the Maxfield Mine. When the canyon highway was rebuilt in 1962 it was raised to the level of the tunnel. The large mine dump was used for fill, although some of it can be seen on Burnt Flat below the mine. A few stone walls remain on the north side of the road, and an old pipe juts out from the closed tunnel, pouring water from the depths of the mine into the light of day, a reminder of the flooding that plagued miners more than a century ago.

CHAPTER 10

Recreation: Brighton Leads the Way

WHILE ALTA WAS THE CENTER OF MINING ACTIVITY in the Wasatch Mountains, Brighton became the focal point of recreation. But recreation is difficult to define. What one person would consider recreation another may consider sheer drudgery. It could be said that some people's daily work is their recreation, for there are fortunate individuals who so enjoy their jobs that they cannot wait to get started each day. As a result, it is difficult to track recreation as an organized or established activity. But, in every life, however mundane it may be, there is some recreation. In numerous cases, pioneers' journals contain comments of climbing to a high point overlooking camp with no apparent reason for doing so other than recreation.[1] Two days after Brigham Young arrived at the newly established Salt Lake City, he and seven others, while exploring the area north of the city, left their horses and climbed to the summit of the mountain they named Ensign Peak.[2] Was that little diversion part of their exploration or can it be classed as recreation? The climb of Twin Peaks less than a month after the arrival of the pioneers has already been described, although it is likely this venture was not purely recreational. In December 1848 Hosea Stout wrote in his journal, "Went alone to explore the mountain ridge immediately South of Emmegration [sic] Canon and followed the ridge almost to the summit. . . ." When the snow became too deep for traveling he came down, arriving at home about sunset.[3] Was this recreation? Most likely. The summit he almost reached is known today as Perkins Peak.

An interesting record of recreation was written by Lorenzo Brown on 4 February 1855. At that time he was working on the construction of Mill A in Big Cottonwood Canyon. The fourth of February was a Sunday,

26

so work on the mill was suspended for the day, and some of the workers sought other activities to fill their time. Brown wrote:

> After breakfast I inquired who would volunteer to accompany me to a high rocky peak which from its peculiar form has been very properly named the Sugar Loaf. Bros Johnson & Bagley announced themselves as being of the party & we three started forthwith & in about 2 hours found ourselves at the foot of the Sugar Loaf.[4] Some 12 or 15 feet above us was a large hole some 8 feet in diameter which our curiosity determined us to explore. We began clambering single file & soon gained a position directly under the mouth it being 5 or 6 feet nearly perpendicular without a foot or hand hold. We overcame the difficulty by placing ourselves myself at the foot sustain Johnson & he in turn supported Bagley who in his stockings was with some peril assisted to ascend to the mouth. When once fairly in he had no difficulty in standing on a nearly level floor in cave about 12 feet deep & as many in width. A part of the sides & bottom were covered with a pitchy resinous substance some pieces we broughs [brought?]. After descending to the foot our next object was to get to the top of the peak which proved no easy task. I started around to the left while the other two started from where they were in a zigzag way & were soon out of each others sight. They followed a back bone which by dint hard climbing and in one place were obliged the hindmost to boost the foremost up a steep rock & he in turn was obliged to pull the other up after. After this they met a large tree that stood directly in their path & could deviate neither to right or left to pass his mgesty [majesty] but were compelled to climb the tree some time & then step from that to the rocks & so on to the top. Not so easy my task. I followed round the base of the mighty rock & finally started up the best place I could find. After climbing near 100 feet the rock grew steeper. I looked up than down to the right then to the left. It was the only chance to go up. To attempt to descend Ioas [was?] death. I was in danger but perfectly cool. I selected with cool deliberation the best course & pushed on. Now I heard my comrades shouting. They had gained the summit. I was not half way up and the worst was to come. I answered their shout. They said I must go back for I could not possibly get up. I asked them to not speak again as I was not on sure footing. Through over exertion my legs began to tremble till they fairly shook but up I went and after 10 or 15 minutes more of the hardest climbing that

I ever done, I found a place where I could sit & rest. I was safe but if during that time I had made one misstep a foot or had slipped I should have been thrown down at least 200 feet before I could possibly have recovered. From where I now sat I dislodged a huge fragment of rock which cleared the steep at a bound & thence went rolling & bounding down the mountain at a fearful rate for at least mile clearing sometimes 20 rods at jump. I soon joined my companions on the summit & laid myself down to rest for a short time whilst they amused themselves by rolling rocks down the mountain side & watched their progress through the thicket of cedars below tearing one up by the roots breaking another in two each taking its own road & each meeting with fresh obstacles. Got back to camp about 2 PM hungry & tired but considering myself well paid for my labor & toil.[5]

After reading this account one can only conclude that Brown was fortunate to have survived to write it. It also shows that the ill-advised tendency to roll rocks down a mountainside was known even then. The "high rocky peak" he called Sugar Loaf can be seen today on the north side of the canyon about a half mile west of Mill A Gulch. It is a prominent limestone monolith that for a short time was known as Peter Cooper Butte, after a mining claim nearby, but was more generally known as "the well-known and prominent limestone butte." The cave Bagley entered is shallow, but relatively inaccessible, as the account suggests, at the base of the south side of the butte. Their explorations failed to reveal to them another cave high on the east side that in later years served as a shelter for miners and prospectors.

It is likely the workers on the Big Cottonwood sawmills made many such pleasure excursions, but hardly any of them left records. One of Brown's fellow workers, Peter Sinclair, left a diary in which he recorded a hike to the eastern head of the canyon in May 1857, when he was working on the construction of Mill E:

in company with the Crowd visited the higest peak of mountains at the head of Kanyon, had a very extensive view of the country all around for hundereds of miles, yes hundereds, the atmosphere is such here in this Latitude and altidude that the eye can penetrate an incredible distance, incredditable to those not experienced, the day being clear gave full scope to our visions. . . . Cut my name and date on some trees enjoyed a while in meditation, thanksgiving, and prayer previous to decending and truely felt well, snow drifts in several places I suppose bordering on one hundered feet deep, possably more or less, a tremendous quantity of snow around, while Provo, Weber, bear and other valley presents a Luxeriant green apearance and no where near in sight is there so much snow or rough and high mountains as where we stand and close around us.[6]

In 1866 Nelson Wheeler Whipple moved his family up to Mill E where he sawed shingles for the Salt Lake Tabernacle. Although he and his boys worked hard cutting timber, sawing it, and packing the shingles to be hauled into town, he still found time to enjoy the surrounding countryside. He later wrote in his autobiography:

> During the summer we thougherly explord the mountains about the head of Big & Little Cottonwoods and the summit between this Valy and Provo or timpanogas Valy in thoes mountains thier are some curious places. These Mountain at thir hight are almost wholy of a ragular granit and numerous lakes at the head of the Stream which come from stupendious snow banks which are deposited in the tops of those mountains.
>
> On our rambles we Discoverd some 7 or 8 Lakes some of which were vury curious one that I will discribe is situated right in the almost solid granit Rock it is about 60 rods long and maybe 40 wide and 40 ft. deep. in it is an Island on a rock rising to the higth of 25 ft on which is several pines and some other trees. here we found in this lone retreet 2 Beavers they were quit tame not having bin used to seeing men. I shot & kild one of them but before we could get to it with a raft it had sunk in 36 ft of water so we lost it at last after a little reflection I felt vary sorrow that I had left the other alone.[7]

The lake where he shot the beaver may have been Lake Martha, since it has an island as he describes, but at that time Lake Mary, being much smaller than the reservoir that is there today, also had an island with trees and was surrounded by granite boulders and cliffs. Whichever lake it might have been, he grossly overestimated its size; until the dam was built to create the Lake Mary reservoir, none of the lakes was as large as sixty by forty rods (990 by 660 feet).

By this time the pleasures of the canyon had been enjoyed by many; hundreds of people visited the lake at the head of Big Cottonwood Canyon for the Pic-Nic party on 24 July 1856, a celebration that was repeated in 1857 and again in 1860. In the 1860s the mining boom broke the domination the lumbermen had over the canyons and brought many new people into the mountains. More than a few miners supplemented their meager mining income by offering food and lodging services to those who worked nearby or to travelers who passed by. One of these men was William S. Brighton, although little did he suspect that his small place at the head of the canyon would grow into one of the Wasatch's most preferred recreation sites.

Early in 1855 William Stewart and Catherine Bow Brighton emigrated to the United States from their native Scotland with their two daughters and William's sister. The younger daughter, Mary, about two years old, died en route and was buried at sea. The family continued to Missouri, where their first son, Robert, was born in June 1855. In 1857, with their six-year-old daughter, two-year-old son, and Catherine expecting another child, they started for Utah in the sixth handcart company, arriving in Salt Lake City on 11 September. Only two months later their second son, William H., was born. This was at the time of the Utah War, so they went south with the rest of the populace during the evacuation of the northern communities. Upon their return, Brighton worked at various jobs, one of them being for Daniel H. Wells, driving teams, harvesting, thrashing, and doing general labor. When the railroad came to Utah he worked on the grade in Echo Canyon. The Brightons had two more sons, Thomas B. and Daniel H., in 1860 and 1864, and during that decade they settled into a home on a large plot on the north side of First South between Eighth and Ninth East Streets. The house would later take the address of 849 East First South. Brighton became the owner of this property in 1873 when he received a deed from Mayor D. H. Wells.[8] The sister who emigrated with them had married Robert Thornley and moved to northern Utah.[9]

One of the jobs Brighton did for Daniel H. Wells in 1864 involved several months work on Mill F, which Wells was building at the time. This was Brighton's introduction to Big Cottonwood Canyon. It is likely the work would have taken him up to Mill E, which was providing most of the lumber used in the new mill, so he would have seen the area that would later take his name. From 1870 to 1872 he was again working for Wells as a lumber hauler. He also may have worked in the canyon between the two periods, but surviving records fail to show that. But he certainly became aware of the mining activity in the Wasatch, because he contracted the prospecting fever himself. In 1870 he and Joseph Elder went on a prospecting expedition, sponsored by David Day, a grocer and merchant with a shop on East Temple (Main) Street between First and Second South Streets. Little is known about Elder or how he and Brighton happened to get together. Elder was prospecting in Little Cottonwood Canyon as early as October 1868, when his name appeared on three claims, teaming with such personages as the Emma Mine's J. F. Woodman and the Flagstaff's John Snyder.[10] It has been claimed that when Brighton and Elder went prospecting together they agreed that Brighton would work the Big Cottonwood side of the divide, while Elder would work on the Little Cottonwood side.[11] Extant records belie this assertion, for the two men filed only four claims together, three on the Big Cottonwood side, one in Little Cottonwood—all in June and July of 1870.[12] After that time Brighton did stay in Big Cottonwood, but Elder worked both sides of the divide. Elder vanished from the scene in 1872 after having placed his name on a total of eighteen claims. David Day's role in financing the prospecting was apparent by the appearance of his name and those of many of his family members on each of the four Brighton-Elder claims. However, it is likely that he had grubstaked only Brighton, for, after the two men parted ways, Day's name never again appeared with Elder's but continued to be associated with Brighton's for another two years.

Apparently Brighton's first cabin was built by his son Robert near his Mountain Lake Lode, one of the four he had claimed with Joseph Elder. It was located well above the site of his later hotels, for the claim was at the very head of the canyon, on the slopes above today's Lake Catherine. By this time Brighton had recorded claims to the Brighton, Day, and Catherine Lodes in the same area, encompassing the uppermost lake.[13] He also filed claims down near the Big Cottonwood Lake, where the Pioneer Day celebrations had taken place years before, and he built another cabin there. It was this cabin that became a popular resting place for miners traveling between Park City and Alta. Word soon got out that Catherine Brighton could serve an excellent meal. Her husband probably saw the potential for more than mining, for he soon applied to the Land Office to buy the land around the lake. He received a Cash Entry patent for eighty acres on 1 November 1875, but the records do not reveal how much he had to pay for it.[14] This was an unbelievable coup, for the lake was recognized as one of the most beautiful in the mountains, almost to the point of being a legend, and now it was his. By that time his location had gained such popularity that he had built a two-story hotel and a number of one- and two-room cabins. The lumber used in this construction came from Mill E, located only a short distance below his cabin.

Brighton probably started building the hotel in 1873, for early that year he placed a mortgage on his home in the city for the amount of $200, due eight months later.[15] This was a trend that continued in the years that followed, for he often borrowed money on his city property early in the year to pay for the costs of opening the hotel and starting the season. The hotel was a distinctive building, for it had vertical siding boards with covering battens. It had seven bedrooms, a dining/sitting room, and a lean-to kitchen. The interior walls were covered with white muslin.

The hotel gained immediate popularity; as early as July 1874 it was reported that Brighton had his hands full with a large party from Salt Lake, as well as guests from Argenta and Silver Springs, all of whom enjoyed fishing, hunting, and horse racing.[16] A few people began to go to his hotel to avoid the summer heat of the city; but it was not yet a resort for the masses. One could not go for a single day's outing because it took the best part of a day to travel up the crude and rough canyon road, and it took another to return. One traveler who rode a wagon up the canyon reported, "the trip is one of the hardest day's labor a man can perform, just to sit on his seat and hold his bones in their proper places."[17] When the tramway from Wasatch to Alta opened in 1875, it provided an easier mode of travel, for it was then possible to go by rail all the way from the city to Alta, and then continue by foot or horseback across the divide and down to Silver Lake. But even then, if a person could not devote three or four days to the visit, Brighton's resort was out of the question. It was a destination for the affluent.

Among those who discovered and frequented Brighton's hotel and cabins were the Walker brothers, who were often in the Alta area attending to their mining interests. In fact, Brighton's granddaughter claimed that it was Joseph R. (Rob) Walker who suggested that the first hotel be built.[18] One visitor in August 1877 reported that Rob Walker and fourteen members of his own and his brother's families were at the hotel.[19] The Walkers found the place so attractive that Rob prevailed upon Brighton to sell a piece of his land, on which the brothers built a summer residence in 1878.[20] Situated on ten acres south of Brighton's hotel and named Silver Lake Villa, it immediately became the focal point of summer activities for the Walker family and friends. Charlotte E. Gilchrist, a frequent visitor at the villa, kept a journal that provides a good insight to activities there and around Brighton's resort.[21] Although she referred to each of the Walker brothers as "uncle," she probably was the wife of Charles K. Gilchrist, an attorney who handled many of the Walkers' legal matters and was a frequent visitor at the villa. Silver Lake Villa soon had a neighbor when Dr. W. F. Anderson built a summer residence on land he purchased from Brighton.[22]

In the next three or four years at least five other cottages were built; however, they either were not on Brighton's land or, if the land was purchased from Brighton, the new owners failed to record the transaction. However, most of those who frequented the resort stayed either in Brighton's hotel, in one of his cabins, or in tents they brought with them. Activities included fishing, boating on the lake, horseback riding, and hiking. Favorite destinations were Lake Mary, then a small lake cradled amongst granite cliffs, Twin Lakes, and Scott Hill. Less-venturesome guests would stroll a short distance down to Mill E or up the dugway on the opposite side of Silver Lake, where they could find a spot to sit, reflect, and enjoy the view. In the evenings bonfire parties in the meadow were enjoyed by hotel guests and campers alike.

Robert A. Brighton, eldest son of William and Catherine, seeing potential in the growing popularity of the resort, applied for and in 1885 received a Cash Entry patent on eighty acres of land adjoining his father's tract.[23] He immediately began selling plots for cottages, some of them for cottages that had been built years earlier, and in 1890 he even subdivided and platted the Silver Lake Summer Resort.[24] This plat took in about half of his eighty acres, centered about the lower end of the Brighton Loop, where one enters the community today. It introduced some streets that remain to this day in fact if not in name: Pine Street, Prospect Avenue, Grand Avenue, Brighton Alley, and Lake Alley.

In 1886 the logging road that already existed around the north side of Silver Lake and up to the Twin Lakes was extended over Twin Lakes Pass and down to the end of the road at the Prince of Wales pumphouse high in Grizzly Gulch.[25] This allowed visitors to travel from Alta to the resort by wagon or carriage rather than on foot or horseback. In 1887 the telephone line from Park City to Alta was built through the Silver Lake resort. A telephone was installed in the hotel, giving it instant communication with the outside world. Then in the spring of 1890 the new Utah Central Railway began narrow gauge rail service from Salt Lake City to Park City. Thus another route to Silver Lake was opened—by rail to Park City and by stage for seven miles over the mountain to the resort. In fact, the railway began promoting a Scenic Circle Route. For a fare of $6.50 the tour went to Silver Lake via Park City, thence to Alta, down the tramway to Wasatch, and then by train to Sandy and Salt Lake City.[26]

By 1892 the resort had grown until it had twenty-five or thirty cottages. That year the county improved the Big Cottonwood road and daily stage service was offered from Salt Lake City directly to Silver Lake, which was now also known as Brighton's Lake or simply Brighton. While the trip by way of Alta was "most picturesque for robust people and lovers of adventure," the stage ride was advertised as such that "any invalid can go." The trip took eight hours.[27]

After the 1892 season William S. Brighton, urged by friends and business associates, decided to build a larger hotel to accommodate more guests. In previous years he and his wife usually borrowed from $500 to $800 to start the season, placing a six-month to one-year mortgage on their home in the city; but in the spring of 1893 they took out a four-year mortgage for $2,000, presumably to help start the new construction.[28] That was followed by another mortgage on the seventy acres at Silver Lake plus an additional ten acres at the northwest corner of Robert A. Brighton's land.[29] In May it was announced that Brighton had let a contract to the firm of Taylor, Romney and Armstrong to build the new hotel. It was a large structure, thirty by one hundred feet and three stories high. The upper floor was not finished; it was used as quarters for the help. In addition to a large dining room, the hotel had fifty sleeping rooms. By the time it was completed that summer, it was reported to have cost $10,000.[30] Brighton's granddaughter later claimed that Bishop George Romney, of the contracting firm, encouraged him to build the hotel. During the construction Romney asked Brighton if he didn't think the building would look better with dormers; Brighton agreed, but later said he didn't realize they would cost him $2,000 more.[31] Incidentally, the Armstrong in the firm of Taylor, Romney and Armstrong was Francis Armstrong, who had run Mill D with Charles Bagley years before.

The 1894 season promised to be better than ever. The big new hotel became the focal point of the resort. The guests were pleased, for no longer did a person have to reserve a room months in advance. Then on Thursday evening, 19 July, tragedy struck. Catherine Brighton, long known as the kindly old lady of Brighton's resort, on entering the dining room, fell to the floor and died. To those who frequented the resort, she was more a symbol of Silver Lake than was her husband. She was known for her ability to go out on the lake in a boat, armed with a fishing pole and worms, and bring back any number of trout for the dinner table. Years earlier, Charlotte Gilchrist had written of a visit to Mrs. Brighton's place, where they were met with friendly greetings and the request to make themselves comfortable. At tea, she reported, "we were treated most generously to right royal mountain luxuries: fried trout; cream biscuits; raspberry shortcake and delicious cream." Now this venerable woman was gone.[32] Yet, as popular as she may have been, her passing was like a single leaf falling from one of the many aspen on the slopes around the hotel. Life went on and the resort enjoyed record crowds through the rest of the season, both at the hotel and in the cabins and campgrounds surrounding it.[33]

The following spring Brighton again borrowed money to start the season, using his home in the city as security. Almost immediately thereafter he acquired a felon, a painful inflammation of the joints of the fingers. The infection spread rapidly, causing blood poisoning. On 28 April it was reported he was lying very ill at his residence. He died that same day and two days later was buried at his wife's side in City Cemetery. Those who mourned him claimed he was highly esteemed among all classes of citizens for his many sterling traits of character.[34] But the old prospector-turned-hotelier had received other, less favorable publicity in his day. He was once accused of being a rabid, fanatical Mormon who pretended not to want gentile trade but who never failed to lighten the ungodly gentile's purse with exorbitant charges. Some said he sold liquor without a license even though he advertised his resort as being run only on temperance principles.[35] At the time of the 1890 elections, it was reported that when the gentile Liberal party was making a strong challenge to the long-time Mormon rule, early telephone communications indicated the Mormons were well ahead. "Brother Brighton was very, very affable then, and his generous heart overflowed its banks, and took in all erring Gentile humanity." Yet, when later news indicated the Liberals were ahead, "the light of joy faded from Brother Brighton's eyes, a dim lusterless, lackadaisical expression succeeded, his inferior maxillary dropped anon, while the rattle of his alveolar processes scared away every burro within half a mile." There were those who said he had stampeded the horses belonging to gentile campers the night before the election, but the non-Mormons got into town anyway and voted.[36] Of course, that was the gentile newspaper's point of view.

During his years in Big Cottonwood Brighton had prospected, recorded, and worked at least twenty-four claims, many of them with his sons. As his hotel business demanded more of his time, he had less time to play the role of miner. After 1885 his name appeared on only three claims, and those may well have been the result of his sons' efforts. They supported him in the hotel trade as well and did much of the labor around the resort. Now that he was gone the sons took over the running of the hotel and kept the resort going for at least two more years. But all was not well, for much of the cost of the construction of the hotel was still owed to the contractor, who foreclosed and took over the property. And the mortgage on the city home caused that property to be sold as well.[37] This was disastrous for two of the sons, Robert and Daniel, for they had built their homes on their father's property. Son Thomas had done the same, but his father had deeded a portion of the property to him.[38] Unfortunately, he had never done the same for the other two; as a result, their homes were lost with their father's.

For the 1898 season the hotel was leased to R. Jay Lambert and William H. Lett, who ran it for two years with apparent success.[39] They managed to keep the hotel busy and scheduled a variety of activities to keep guests entertained. There were daily picnic parties, excursions, tramps to nearby peaks, and nightly bonfires. The latter seemed to be especially favored. On one occasion, to celebrate a thirty-fifth wedding anniversary, thirty-four small bonfires were prepared in a semicircle in front of the couple's cottage, and an enormous stack of logs, thirty feet in circumference and containing thirty-five logs wired together to form a huge vertical cone, was placed in the center. All thirty-five bonfires were ignited to illuminate the party, the large center fire burning long after the last guest had departed. On another occasion a large bonfire built on a raft burned in the center of Silver Lake while fireworks, in the form of Giant powder, were ignited along the shore and from the cliffs of Mount Evergreen above the lake. Indoor activities were enhanced in 1899 with the installation of a piano in the hotel.[40]

Music always played an important role in the resort activities, usually in the form of vocal offerings but sometimes with instruments that might be available. In one case a guitar was the only instrument to be found, but it served to keep the revelers dancing well into the night. On another occasion guests danced happily to the music of one fiddle; a week later they had the six-piece Silurian String Orchestra of Park City furnishing the music. On one occasion guests received a special treat when Willard Weihe, Utah's virtuoso violinist, was visiting the resort and gave an impromptu recital in the hotel dining room.[41]

Weihe, incidentally, was only one of many Utah artists who frequented the resort. Alfred Lambourne, the noted artist and poet, was seeking inspiration in the Wasatch as early as 1871, when he painted the lake at the head of Cottonwood Canyon.[42] He was often accompanied by his friend and fellow artist Henry L. A. Culmer. The two men explored the Wasatch, found their favorite places, and gave many features names that are still used today. Culmer later remembered the two of them having taken a tramp to the divide between Little Cottonwood and American Fork Canyons, when clouds began to gather and the rumble of thunder was heard. The two men turned and fled in an attempt to return to camp before the storm broke. They did manage to reach their camp at Silver Lake before the clouds "culminated in a storm of thunder and lightning more magnificent and terrible than any that had ever been witnessed by those who were in its midst. Our peaceful camp was broken up in the crashing of thunder and incessant lightning flashes which seemed to be literally in and around us."[43] When the storm finally passed and the skies cleared at sunset, a calm fell upon Silver Lake that was so peaceful and inspirational that Lambourne returned to his studio and painted two scenes, the first titled *Storm Clouds* and the second *Sunset after a Storm, Silver Lake*. The latter was a large canvas, twenty-six by forty-one inches, and so well done that it received considerable acclaim. It was soon purchased by W. F. Kirk of Chicago, long a patron of the young artist. Several years later Lambourne painted the sunset scene again on an even larger canvas.[44]

Culmer also was an artist of great ability, and the two men left many paintings of Wasatch scenes. In his later life Culmer had a studio on South Temple Street in Salt Lake City. He had a darkened room in which a featured painting would be hung under special lights, creating a spell that inspired many viewers. Culmer's friend, the violinist Weihe, would take his violin into the special room and play improvised music to go with the spirit of the featured picture.[45] Other artists who were seen at Silver Lake over the years included George M. Ottinger, John Tullidge, Dan Weggeland, Harry Squires, and John Hafen.

*

But we've digressed. Although the hotel had two good years while operating under the lessees, the new century saw the Taylor, Romney and Armstrong Company fencing the property and allowing the big building to stand idle, its darkened windows staring blankly at Silver Lake and the many campers often surrounding it. Then it received a new benefactor, Salt Lake attorney James H. Moyle. He had visited Silver Lake as early as 1887 and was captured by its beauty. In 1890 he bought an acre of land from Robert A. Brighton, although it would be five years before he started building a home on it.[46] During that time he and his family were frequent visitors at the hotel. When William S. Brighton died, Moyle acted as the administrator for his estate and gave the administrator's deeds to the new owners of both the hotel and the city properties. The following year, 1896, Moyle finished his home at Silver Lake, almost in the shadow of the big hotel.[47] It must have bothered him to see the hotel empty and unused, for in August 1900 it was announced that he was taking charge of the building and was making some improvements.[48] Although he promised to open the hotel the following season, that did not happen. Instead, it was announced that he had purchased the hotel and land from the Taylor, Romney and Armstrong Company for $5,000, considerably less than the indebtedness of $10,300 that had placed the property in that company's hands. While the hotel did not open that summer, Moyle did make its dining hall available for dances and rented some of the hotel cottages.[49] In 1902 the hotel's doors were thrown open to visitors again, and it continued to operate for the next quarter century. In 1903 it received a new coat of light green paint, and the following year three bowling alleys were installed. Golf links, tennis courts, and croquet grounds also were installed for the pleasure of the guests.[50] The establishment had a number of different operators over the years, including Hyrum Nielson of Holladay; Carlo Von Puelle, a former Salt Lake City bartender; and Gilbert D. Moyle, eighteen-year-old son of the owner, who with fellow student Victor L. Hall operated the hotel for one year in 1916.[51]

During this period important changes were taking place throughout Utah and the nation. The population of Salt Lake City was growing rapidly, increasing pressures on the adjoining mountainous regions, which, in retrospect, appear to have been very much like the problems faced today. Of greater importance and significance was the introduction of the automobile into daily activities. The county had taken over the maintenance of the canyon roads and, although they were not even close to the quality of highways we enjoy today, the roads were better than they had ever been. As early as 1912 it was reported that "arrangements have been made whereby business men can leave this city Saturday afternoon, spend that night, Sunday and Sunday night at Brighton and return early Monday morning in Salt Lake in an automobile." And the Salt Lake Auto Stage Line advertised that its Brighton stage had new automobile equipment handled by expert drivers.[52] With the growing popularity of the automobile, Sunday newspapers began carrying an Automobile section, in which various out-of-town trips were described. The local canyons were especially popular because they challenged both the driver and the car, and they stimulated competition. In 1916 the Motor Sales Corporation advertised that its "King 8," an eight-cylinder, sixty-horsepower automobile, was driven from the old paper mill at the mouth of Big Cottonwood Canyon to Brighton in fifty-five minutes, all in high gear. The following year, a "Hal Twelve" was driven from the City and County Building in downtown Salt Lake City to Brighton in fifty-eight minutes.[53]

With greater pressures upon Brighton's recreation attractions, the number of area private cabins rapidly increased and several stores were established. It was inevitable that the big Brighton Hotel would get some competition; it came from one of the storekeepers, Thomas C. Davis. As early as 1910 Davis was operating as a merchant in Brighton, and soon thereafter he began building a large log structure at the north end of the resort, where today the canyon road enters the loop road. In 1912 it was entertaining guests and was known as the Log Cabin Hotel, a name that was changed to Balsam Hotel in 1914 and Balsam Inn the following year.[54] It was a large three-story building with a steep-pitched roof, and was capable of housing forty guests. It had a large covered porch at the second level, with a wide staircase leading to it. The hotel gained immediate popularity, although it did not appear to hurt business at the older hotel. When T. C. Davis died, his wife and son, Paul, took over the operation. By the late 1920s the hotel had become a favorite place for outdoor enthusiasts; it even offered quarters for winter skiing and snowshoe trips by such organizations as the Wasatch Mountain Club, East High School Hiking Club, and others.[55] By that time the Brighton Hotel had closed and stood vacant and forlorn to the west of the Balsam Inn. The Davis hotel continued to operate until 1936. On the afternoon of Sunday, 21 March 1937, the building caught fire and was totally destroyed in a spectacular blaze lasting about two hours. Although the resort community had about 100 cabins at that time, no other buildings were damaged.[56]

Paul Davis built a new, smaller log structure on the site of the destroyed hotel, where he ran a store and refreshment stand. It too was destroyed by fire, in July 1941.[57] Undaunted, Davis rebuilt again, the new structure known to skiers as the Davis Club House. In 1955 the building was renovated by Boyd Summerhays, who operated it under the old name of the Balsam Inn. On 7 October 1959 that building took fire and was gutted, leaving only the porch and partial walls standing.[58] This time the structure was not rebuilt, and the site of the many generations of Balsam Inns remains vacant to this day. Incidentally, after the last fire Summerhays opened another business in the city; named the Balsam Embers, it was one of Salt Lake City's finest restaurants for many years.

Another inn that was constructed in Brighton about the same time as Davis's Log Cabin Hotel was the Girls' Friendly Holiday House, sponsored by the Girls' Friendly Society, an affiliation of the Episcopal Church. It was a fairly large log structure located northwest and below the center of the resort, very close to the site of the old sawmill, Mill E. It had a large dining hall, sleeping rooms, and a large porch, and was surrounded by numerous tenting sites. It served as a base for many girls and young women who spent their vacations there, hiking, rafting on the lake, horseback riding, and partaking of the many other pleasures of the mountains. The building was later replaced and was joined by others to make up the present Camp Tuttle.[59]

Two other old-time buildings at the Brighton resort are the Young Ladies' Mutual Improvement Association (YLMIA) girls' lodge and the Wasatch Mountain Club lodge, dating from 1922 and 1928, respectively. In 1921 construction of a summer home for the YLMIA was proposed. The first site considered was the Wasatch Resort at the mouth of Little Cottonwood Canyon, then owned by the Mormon Church. That plan was discarded when the YLMIA acquired a site in Brighton. In 1922 construction began on the log building, which was forty by one hundred feet in size and three stories high. The building was dedicated by President Heber J. Grant of the church the following July and was used by this church organization for many years. Then, on Saturday, 19 January 1963, the building caught fire and burned to the ground, a common occurrence for structures at the remote resort. The building was rebuilt and the MIA lodge remains active to this day.[60]

The Wasatch Mountain Club was formed by a small group of outdoor enthusiasts in 1920. While it conducted outings in all parts of the Wasatch, many of its winter activities took place at Brighton, where it used the facilities of Paul Davis's Balsam Inn. In 1928, however, the club began the construction of its own lodge, located not far from the YLMIA lodge. It was a two-story log building with a fireplace and a large stone chimney. The building is unique among large structures in Brighton in that it has never burned down. It stands today, somewhat remodeled, still serving the needs of club members.

Although it is was not of the same vintage as those already mentioned, one other Brighton structure should be mentioned, if only for its link to the past. In the late 1930s Alf C. Launer had a private residence at the foot of the hill that later became the Majestic ski run. This was at a time when the ski industry was in its infancy but becoming a recognizable force. Launer converted his home into the Launer Clubhouse and housed skiers during their stay at Brighton. By 1940 he had given up his home in the city and moved to the mountains as a full-time resident. The next year the building was renamed the Alpine Rose and was run by Bertha Howard, the woman who had been the cook at the first Balsam Inn when it burned down in 1937.[61] This establishment drifted through the war years of the early 1940s.

During all this time the old three-story Brighton Hotel still stood, empty and deteriorating, a symbol of the former resort and a monument to the man who started it all. But in 1945, after the hotel building had been vacant for over fifteen years, owner James H. Moyle decided it had become a fire hazard and chose to have it razed. It was a sad decision, for the old building housed many memories for former visitors. In an editorial, the *Salt Lake Tribune* recalled how it provided a cool retreat from the city's summer heat and how it had been a gem in a setting of unusual charm, the scene of gaiety and evening strolls and singing in the gloaming. There were more than a few who felt they had lost an old friend when the big hotel was gone.[62] But it was not a total loss, for much of the material salvaged from the hotel was taken to the other end of the resort and used to expand the Alpine Rose, now under the management of Henry S. Florence. The enlarged lodge was capable of accommodating about fifty people in its thirty-three guest rooms and dormitory space. It also had a lobby, dining room, and large sun deck. In 1955 it was under the control of Dr. Guy Wight, former president of the Brighton Recreation Association, who further expanded the structure. In 1963, with the Brighton Investment Company in control, the lodge was briefly seized by the county for delinquent taxes and not having a business license. The matter was quickly resolved and the lodge was opened once again, only to suffer the fate of many of its predecessors at Brighton. It caught fire on 4 June 1965 and burned to the ground.[63] The site on which it stood remained empty for at least fifteen years until the present concrete building known as the Alpine Rose was built.

CHAPTER 11

Water Projects in Big Cottonwood Canyon

W<small>ATER FROM THE CANYONS</small> has been important to Utah residents since the first day the Mormon pioneers arrived, when they diverted City Creek water to soften the dry, sun-baked soil so they could plow. Ever since that time the canyon streams have been used for culinary and irrigation purposes, as well as for power for the many mills established along their banks. Individually, these first efforts were all relatively small endeavors that left few artifacts. There were, however, some projects of major significance, many of them centered in Big Cottonwood Canyon. The first involved the generation of electricity.

When electricity was introduced in Salt Lake City it was recognized that the mountain streams that had for so long powered saw- and gristmills could also provide power for the generation of electricity. Unfortunately, all early electrical systems used direct current, and the transmission of that force over the long distances from the canyons to the city posed a problem that had not yet been solved in the mid-1880s. As a result, the first electrical generation plants were built in town, close to the users, and were powered by steam engines. It was the end of the 1880s before the generation and transmission of alternating current over long distances was demonstrated.[1]

The man who finally harnessed the potential of the Big Cottonwood Canyon stream was civil engineer Robert M. Jones. He came to Utah from Laramie, Wyoming, where he had built one or more hydroelectric plants. His first job in Utah Territory, the installation of a steam-powered electric generation plant in Park City, was pronounced a signal success when electric lights were turned on there for the first time on 22 March 1889.[2] Jones was then hired to construct Salt Lake City's new electric street railway, replacing the old horse-drawn streetcars that had been used for many years. Again, its generators were driven by steam and were located close to the city; but Jones knew of the tremendous potential of the canyon streams and, in fact, had been engaged by Major Henry C. Goodspeed to install a hydroelectric power plant for the Reed and Goodspeed mine in the South Fork of Big Cottonwood Canyon. This relatively small installation was located at "the forks of the stream," the confluence of the South Fork stream and Big Cottonwood Creek, and was intended to power the mine machinery as well as electrical rotary drills of Jones's own design. The plant was constructed during the winter of 1891–92 and was operating by the end of May 1893.[3] It probably did not run for more than a few years, however; Goodspeed died in January 1895 and at the end of the decade his mine was lying idle.[4]

Jones's travels to the Goodspeed construction site certainly gave him opportunities to conduct his own explorations and investigations. On 26 October 1891 he posted a notice for the Stairs water right. Located in Big Cottonwood Canyon, it started about 1,000 feet above the head of the area known as the Stairs, and extended 850 feet below the foot of the Stairs, at which place he posted a mill site location notice. It was the usual practice to file canyon water claims with the local mining district recorder, but in this case the claim never reached the books of the Big Cottonwood Mining District. It is not known whether Jones actually filed the claim or merely posted it on the site, but when he recorded the notice with the Salt Lake County Recorder in May 1893 he added a comment saying that his notice had been posted three times and was each time removed by parties unknown.[5] It is interesting to note that at the end of March 1892 the mining district recorder, James T. Monk, filed a water and power claim, appropriating nearly all the water flowing in Big Cottonwood Canyon from its head down to Argenta, near Mill A Gulch, together with the right to construct conduit, pipeline, or wire to transmit or convey electrical or water power.[6] It is not likely that Monk had sufficient knowledge or ability to carry out a power project such as envisioned in his claim, and it may well have been a reaction to Jones's efforts for Goodspeed in the South Fork, or for himself farther down the canyon. At any rate, nothing ever came of Monk's claim, while Jones pursued his vision relentlessly.

The site of Jones's water right and mill site claim is now well known. The Stairs Power Station, noticed by anyone who has ever traveled the canyon, rests on the mill site. In 1891 the canyon road ran along the stream and climbed steeply up the gorge above the mill site until it came out on a large flat. Except for the gap allowing the road and stream to exit, the flat was surrounded on three sides by rocky ridges, making it a perfect place to build a dam for a reservoir. At that time the flat was the site of the Youngs Peak Lodge, a summer resort constructed in 1888 by two young men, H. L. Hall and Asahel Woodruff, the latter being the son of Mormon Church president Wilford Woodruff. It featured a lumber cookhouse, a bowery, and tents for dining and sleeping.[7] It was named after Youngs Peak, better known today as Storm Mountain, a prominent and imposing peak that dominates the scene to the southwest. The flat was also known as Mill C Flat, for it was the site of the short-lived sawmill known as Mill C more than thirty years earlier.

Jones planned to build an earthen dam across the gap on the south side of the flat to create a reservoir covering some eight acres with depths up to thirty-five feet. To carry the water to the power plant, some 380 feet below at the foot of the Stairs, he chose to tunnel through the quartzite ridge at the west, or down-canyon, end of the reservoir rather than follow the longer path of the streambed. To that end, the task of driving a tunnel 430 feet long was begun in the autumn of 1892. Although Jones expressed his desire to have the tunnel finished by the following spring, difficulties in arranging financing delayed its completion until late summer of 1894. By that time he had been joined by a group of Salt Lake businessmen to form the Big Cottonwood Power Company, organized on 8 December 1893. While Jones was not one of the company's officers, he became its managing director and principal stockholder.[8] With this additional support he was able to push the work during the 1894 season, finishing the tunnel, building the dam, and relocating the canyon road. The latter was necessary because the existing road would be covered by the dam and reservoir. The narrow canyon with its very steep walls posed a difficult problem, but Jones solved it by building a dugway that climbed steeply toward a high rocky ridge that extended out to the very edge of the dam and reservoir. This ridge was so high that he had to blast a cut through it to carry the new road to the other side. There it descended rapidly to the level of the canyon stream, which it crossed on a newly built bridge at the upper end of the reservoir.

The heavy work required for the cut in the new road delayed its completion until well into the 1895 season. The construction of the dam was held up to allow the existing canyon road to be used until the new road was ready. However, the teamsters going up and down the canyon watched the new road as it was being built and were not happy with what they saw. Numerous complaints were filed with the county court, but that governing body had more important matters to attend to, and little was done until the power company closed the old road in October or early November 1895. The road, the court then contended, was a county road and could not be closed without the court's approval. When county selectmen conferred with company officers and visited the site to inspect the road, the company submitted a formal petition asking the county to accept the new road and abandon the old one. The county court decided to accept the petition on condition that the company maintain the bridge it built across the stream for a period of two years.[9]

The teamsters still were not happy with the new road. They contended that the grade was so steep that they either would have to lighten their loads or ruin their stock. In the past they were able to stop and rest their animals on the flat after climbing the steep grade up the Stairs. The new road added nearly a quarter of a mile to the steep climb. The grade, as approved by the county surveyor and the court, was 12 percent on the west side and 10 percent on the east. While the teamsters probably had more descriptive names for the new road, it became generally known as Jones' Hump until it was rebuilt to remove the excessive grades in 1916. The cut atop the ridge was popularly known as Devil's Gate, and it was a favored place to take photographs, especially when automobiles started using the canyon road.

At the end of May 1895 Jones and two other company directors traveled east to float company bonds in New York City. Jones remained to order machinery and materials and consult with other electrical engineers. Much of the material, some thirty-five carloads in all, was on its way to Utah before he returned in August. After this time work on the power plant proceeded at a rapid pace.

By this time R. M. Jones was no longer alone in trying to use Big Cottonwood Creek for the generation of electricity. Although he did not view them as such, he now had two competitors. The first was Frank K. Gillespie, a sometime mining operator and opportunist who was attempting to establish a power plant below the Jones mill site. In 1893, possibly stimulated by Jones's efforts in the canyon, Alonzo B. Richardson, a cashier in the Park City bank, came over to Big Cottonwood Canyon and filed four water appropriation and water right claims with the mining recorder. The appropriations were for water flowing between Mill B and Jones's reservoir, and between the tail race of Jones's power plant and the Butler sawmill at the mouth of the canyon. The water rights were more important; they gave Richardson control of the water flowing between Jones's power plant and the first irrigation ditch dam, about a half mile downstream, as well as another section of Big Cottonwood Creek below the mouth of the canyon.[10] He also had the two water rights recorded in the County Recorder's office.[11] That seems to be the extent of Richardson's activities in Big Cottonwood Canyon, for in the following January Frank Gillespie bought Richardson's rights, intending to use them in conjunction with his own claims.[12]

In the summer of 1894 Frank Gillespie started working in the canyon below the Jones mill site, digging a ditch or flume to carry water to his own mill site. He attempted to convince the street railway people to use his hydrogenerated power, claiming it would be less expensive than electricity from their steam plant. In this way he became associated with Orson P. Arnold, for many years the superintendent of the Salt Lake City Street Railroad. As a result, it was generally supposed that the street railway company was behind Gillespie's efforts.[13] But that was not the case, as was made apparent the following spring when Jones's second competitor surfaced. What Gillespie had managed to do was to convince Arnold and the streetcar company that hydroelectric plants were the way to go. Indeed, in the summer of 1894 a number of tents had been erected and a house was built at the mouth of the canyon to serve as headquarters for engineers and contractors who were running surveys in the canyon. It was assumed that they were quietly pushing an electric railroad enterprise, although many of the engineer's surveying pennants were on prominent crags and peaks high on the sides of the canyon.[14] It later became apparent that they had been surveying for the flume for a new but yet unborn power company.

The Utah Power Company was formed on 30 April 1895 by a number of prominent and wealthy mining men from Salt Lake City and Park City to provide electricity for heating, lighting, and the propulsion of railways. They were joined by Salt Lake City ex-mayor Francis Armstrong, the same man who with Charles S. Bagley once ran Mill D in Big Cottonwood Canyon and who now became president and manager of the new company. Orson P. Arnold became a minority stockholder.[15] The company planned to build a power station at the mouth of the canyon, take its water at the tailrace of Jones's power plant, the same water source that Gillespie was developing, and convey it to its power plant by a lengthy flume.

Undaunted, Gillespie and his men continued working on the south side of the stream, while workers for the new company built their flume on the north side of the canyon, the two groups often working close enough to "exchange sarcastic courtesies."[16] Gillespie believed that the water right was his, while the Utah Power Company claimed it had relocated the right. An unnamed official of the Big Cottonwood Power Company said his company respected Gillespie's location but that it seemed to have been jumped or relocated by "a Mr. Arnold" in the interests of the other power company.[17] The tense situation came to a head on 29 January 1896 when Gillespie's men installed a flume to carry water from the dam below Jones's plant. Armstrong and another man, both armed with double-barreled shotguns, asked Gillespie to remove the flume. When he refused, Armstrong ordered his own men to tear it out and destroy it, which they did. Outnumbered, Gillespie and his men offered no resistance.[18]

The next day, Gillespie went into town and had another flume made to replace the one destroyed. This time he took the water from the stream above the dam Armstrong claimed as his. Gillespie had a dynamo weighing 6,000 pounds delivered to his powerhouse site during the night of Sunday, 2 February. He already had installed the turbine wheel and penstock from the old Granite Paper Mill, which had been destroyed by fire on 1 April 1893. The next day, his plant was started and a number of arc lamps that had been erected in the vicinity were lighted. He also delivered electric current to Salt Lake City over the Big Cottonwood Power Company's newly erected transmission lines, but the wires were not connected to anything in the city, so the power served no useful purpose. Still, Gillespie did become the first to deliver hydroelectric power to Salt Lake City. Unfortunately, a few hours later a mud slide damaged much of the machinery at his plant.[19] He repaired the damage, but his operations were short-lived. In April a visitor to the canyon reported that the Gillespie "plant" was a small tent with an arc lamp outside, and that the plant furnished power by which the Big Cottonwood Power Company was able to light its operations at night, thereby hastening the completion of its own plant.[20] When that company finally started generating electricity, there was no further need for Gillespie's small operation; its turbine stopped turning and its lights went out. Gillespie moved over to Little Cottonwood Canyon, where he formed the Cottonwood Water, Power and Electric Company, Limited, and was granted a franchise to build transmission lines along county roads to deliver the power he could generate. He also attempted to introduce electricity into Alta, but his efforts came to naught.[21]

During the winter of 1895–96 the Big Cottonwood Power Company installed 2,200 feet of forty-nine- and fifty-inch steel pipe to carry water from the reservoir to its turbines, constructed its powerhouse, and installed four Pelton waterwheels and four 500-kw General Electric generators, as well as other associated equipment. The generating equipment was run for the first time in a successful test on 18 May 1896. The equipment then remained idle until the completion of the distribution system in the city. Finally, on 15 June 1896, the streets of Salt Lake City were lighted by electricity generated in and transmitted from Big Cottonwood Canyon.[22] At that time it was claimed that there were only two larger power plants in the United States: one at Niagara Falls, New York, and one at Sacramento, California.[23]

*

As if chastened by the unfavorable publicity received in its encounter with Gillespie and his men, the Utah Power Company maintained a low profile, rarely appearing in the news. In April 1896 a correspondent going up the canyon to the Big Cottonwood Power Company mentioned seeing the wonderful piece of fluming put in by the Armstrong syndicate. And indeed it was. It was an open flume about five feet wide and three feet deep, constructed of two-inch planks. Its upper end was at the dam, directly below the Stairs Power Plant. It passed under the canyon road, then followed a winding path along a contour line on the steep and rugged cliffs on the north side of the canyon, dropping about one foot per thousand feet to allow the water to flow. At some places trestles were constructed to carry it across deep gulches, and at other places it seemed to cling to the side of the cliffs with no apparent means of support. It ran 8,342 feet, over a mile and a half, until it reached a point some four hundred feet above the site of the power plant, just above the mouth of the canyon. There it emptied its content into a four-foot-diameter steel pipe, a penstock that delivered the water at high pressure to the Pelton waterwheels in the plant. For over one hundred years visitors to the canyon have marveled at this engineering achievement, their view sometimes favored with the sight of an inspector walking a walkway on top of the flume. In the wintertime its leaks would cause great sheets of ice to flow over the cliffs above the canyon road, and occasionally a catastrophic break would cause a flood that would carry tons of mud onto the road below. Yet the flume was maintained for more than a century until 1998 when it was replaced by an underground pipe running close to the highway.

As if to distance itself from the Big Cottonwood Power Company, the Utah Power Company contracted with Westinghouse for its two generators and electrical apparatus. The Westinghouse company sent a specialist, P. H. Knight, to Utah to supervise the installation, after which time the Utah Power Company convinced him to stay and take charge of its operation. A transmission line was run along the east side of the valley to Fort Douglas, then down South Temple Street to the street-railway steam power plant on Second East Street, where the high-voltage alternating current was converted to direct current for the streetcars. The company had its generators running in February 1897 and thereafter provided all the electrical power for the street railway, much to the delight of city residents who had objected to the inky black smoke that previously poured from the steam plant's chimneys.[24]

The power plants of both companies are still landmarks in the canyon today, especially the Big Cottonwood Power Company's plant, a beautiful old brick building occupying such a strategic place in the canyon bottom that the highway has to wind around it. On the wall facing the road is a huge masonry and plaster plaque that carries the name "The Big Cottonwood Power Co., Stairs Station, 1895." The building and its machinery have received numerous renovations since 1895, but it still generates electricity as it did over a century ago. Although its original name appears on the building, the company operated only one year before being absorbed into the Union Light & Power Company, organized in August 1897 to consolidate a number of small electric companies. R. M. Jones soon left Salt Lake City and distinguished himself in the construction of numerous hydroelectric plants in Colorado and New Mexico. He died at Carlsbad, New Mexico, on 2 March 1916.[25]

Today the Utah Power Company's plant is somewhat obscured by the water treatment plant that is located between it and the highway. Its stately brick building has a large semicircular plaque over its door that proclaims, "Utah Power Company, Erected 1896." In 1901 the Consolidated Railway & Power Company gained controlling interest in the Utah Power Company, but the latter maintained its corporate identity until 1935, when it deeded its system to the Utah Light and Traction Company. Over the years, the installation became known as the Granite Plant. Today, after a century of mergers and consolidations, both plants are part of Utah Power, A Division of PacifiCorp.[26]

*

Over the years the reservoir for the Stairs Power Plant has filled with debris brought down by the stream and today doesn't look like a reservoir at all. On the flat below the dam the Forest Service built the Storm Mountain picnic ground, and the highway going up the canyon makes many sharp turns that were not there in R. M. Jones's day. At that time the road, after passing the power plant, stayed close to the stream as it went into the gorge, crossed a bridge to the south side of the stream, and climbed steeply to the flat, crossing another bridge over the Stairs Gulch stream on the way. It was above the flat where Jones built the dugway that was named after him. The road remained that way until 1916 when Salt Lake City and Salt Lake County negotiated with the Cardiff Mining and Milling Company and the Boston Development Company to stop using teams to haul their ore down the canyon. One of the terms of the agreement was that the county widen and grade the road from the mouth of the canyon to the Cardiff ore bins at the South Fork, build a number of new bridges, and improve the road at Jones' Hump. The problem there, the very steep grade, could be nearly eliminated by tunneling through the ridge that formed the Hump. So formidable was this job that it was expected to cost nearly one-third of the total road improvement project. The tunnel was abandoned, however, when bids were received, for they showed it would be less expensive to make an open cut. A contract was awarded and the blasting began. When the work was completed the road went through a cut on the alignment that still exists today, except that the cut was then much deeper and very narrow.

As a postscript to this story it should be mentioned that the Big Cottonwood Power Company also had plans to build another plant farther up the canyon. In an intricate series of maneuvers, the company acquired the Mill B water rights, which at that time were owned by Alva Butler and David B. Brinton. First, in three separate quitclaim deeds, James Maxfield (representing earlier Mill B owner Richard Maxfield), Alva Butler, and D. B. Brinton sold their claims to the property to H. A. Bagley on 19 January 1894. Several weeks later, on 3 February, Bagley sold the property back to Alva Butler, who, on the same day, sold it to William H. Rowe, Joseph W. Summerhays, Robert M. Jones, and George M. Cannon, all principals of the electric company.[27] A few months later, on 14 June 1894, the same men filed a water right appropriation and claim as well as a mill site.[28] The latter was a 280-by-725-foot plot at the mouth of "Borck Fork" (Broads Fork) to be used for powerhouses and other buildings. Water was to be taken from Big Cottonwood Creek above the dam for Mill B and carried some 3,300 feet to the mill site. Although work was started for this project, it never went to completion. The only remnant to be seen today is the right-of-way for the flume or conduit to carry the water. In spite of a century of erosion, it can still be followed along a contour line on the south side of the canyon from the stream crossing at the S-turn to the Broads Fork stream.

*

Another major water project in the canyon started in 1905 when the Brown and Sanford Irrigation Company made application to appropriate sixty second feet of water, to be stored in suitable reservoirs in Mill B

South Fork. The company had been operating since 1874, when it completed its ditch to irrigate lands south and west of Big Cottonwood Canyon. It took water from Big Cottonwood Creek near the old tollgate, close to today's water treatment plant, and in 1884, when water rights were being formalized, laid claim to one-fifth of the usual flow of water in the creek.[29] The company had a problem common to all irrigation projects: in the spring when stream flows were high the water needs were small, and later in the summer, when irrigation was most necessary, stream flows were small. Its solution was to use the Three Sister Lakes—Blanche, Lillian, and Florence—to store water until it was needed, then release it to flow down the natural watercourses to the diversion gate.

Before construction of the dams could begin, the company was advised that the reservoir sites were part of the U.S. Forest Reserve and it would be illegal to do any work without proper consent. Application was made to the newly created U.S. Forest Service in 1905, but it bogged down in bureaucratic procedures until the spring of 1908. Construction finally began on 9 July 1908 and continued for two years, hampered both by the need to carry all materials up the steep trail on pack animals and by the shortness of the season, before it was completed in September 1910. The first water was released from the reservoirs in July of that year.[30] Almost immediately it was recognized that a greater storage capacity was needed. Application for additional capacity was made in 1912 and construction started in the following year to raise the heights of all three dams. This was a larger job than the original construction, for the height of the Lake Blanche dam was nearly doubled. Work proceeded slowly. The company claimed it was storing water as early as 1914, but that was only by using the original dams. Repeatedly the company went back to the state engineer to request extensions of time to complete the project. When that official had granted extensions to his legal limit, the company had to apply through the Third District Court. Twice they went to the court, receiving four-year extensions each time. Finally, the construction was completed in October 1934.

Today when hikers approach Lake Blanche they are confronted with a stone dike that has to be climbed before the lake is visible. An interesting fact about the Lake Blanche dam is that the original one had no spillway: when the reservoir filled, the water would leave the reservoir where the dike now stands and flow down over the cliffs to join the natural stream channel in the fork below. It was Nature's own spillway, but it had to be blocked when the dam was raised. Accordingly, the dike was built and the dam was equipped with its own spillway, allowing all the water flowing from Lake Blanche to go into Lake Lillian, then into Lake Florence, and finally down the natural watercourse. All three dams were given a cement cap as they were completed. At the south end of the Lake Lillian dam someone placed an inscription in the wet cement: "Lake Minnie J. A K." Whoever JAK may have been, he or she had the wrong name for the lake. Minnie was once the name of the lake in Albion Basin but never of this one. When the cap was placed on the lower dam someone inscribed the name, Lake Florence, and four people, three of them Maxfields, wrote their names and a now-illegible date in the cement.[31]

Salt Lake City gradually purchased Big Cottonwood water rights in the years that followed. Eventually the irrigation reservoirs were no longer needed and fell into disuse. The dams were breached in 1972.[32]

If not for the Brown and Sanford Irrigation Company's difficulties and slow progress with the dams at the three lakes in Mill B South Fork, we might see the remains of another dam today. In July 1913 Charles W. Hardy, who was serving as Brown and Sanford's engineer, was near the head of Broads Fork surveying for a reservoir.[33] His plan showed a dam 325 feet wide, which would form a lake holding 360 acre-feet, about half the size of the Lake Mary-Phoebe reservoir above Brighton but four times as large as the Lake Blanche reservoir in Mill B South Fork. Nothing ever came of this plan, probably due to the pressures of completing the improvements under way at the Three Sister Lakes reservoirs. Also, within a year of the Broads Fork survey Hardy suddenly died, causing a considerable perturbation in Brown and Sanford's operations. Whatever the reasons, the head of Broads Fork remains undisturbed by human improvements to this very day.

*

The next big water project was undertaken by Salt Lake City at the head of the canyon. As the city grew it had to reach farther and farther to provide an adequate water supply. During the 1880s the city population more than doubled, so officials negotiated to purchase Parleys Canyon water, providing the former users with water brought from Utah Lake through the Salt Lake and Jordan Canal. A conduit was constructed from the mouth of the canyon to the city, providing sufficient additional capacity to satisfy demands almost to the turn of the century. Then, in 1898, when an extended period of drought made it apparent that additional water supplies had to be developed, City Engineer Louis C. Kelsey undertook an extensive survey of Mill Creek and Big and Little Cottonwood Canyons. His first report, containing sixty pages of typewritten text, maps, and photographs pertaining only to Mill Creek Canyon, was presented to the city council at the end of November 1899. The following month he presented additional reports for the two Cottonwood canyons.[34] Meanwhile, to protect the purity of its water supply, the city began to purchase watershed lands to the north and east, much of it from the Union Pacific Railroad Company, who held it by virtue of its transcontinental railroad land grants. In 1900 and 1901, at the city's request, the Land Office withdrew from public entry about 200,000 acres of vacant and unappropriated lands comprising much of the watersheds of the Wasatch Mountain canyons draining into the Salt Lake Valley.[35]

In a special election on 3 January 1905, Salt Lake City voters approved a bond issue to expand the city's water system. It included funds to purchase Big Cottonwood water rights and to construct a conduit to carry water from the mouth of the canyon to the small reservoir at the mouth of Parleys Canyon. The Parleys conduit then carried the water to the city.[36] The diversion dam for the new pipeline was near the location of the present Big Cottonwood water treatment plant. While the purchase of water rights continued over a period of many years, the conduit was in operation in about a year's time.[37] The new water supply did not solve all the city's problems, however, for during the periods of high water demands, in the hot late-summer months, the stream flows were low and sufficient water had to be released at the diversion dam to satisfy those farther downstream who still had valid water rights. What was needed was a place to store water until it was needed. This, of course, was not a new concept; Engineer Kelsey's reports in 1898 and 1899 considered such possibilities.

In 1900 Salt Lake attorney J. M. Thomas, possibly inspired by Kelsey's report, had moved faster than the city and appropriated all waters flowing down the gulches below the Twin Lakes and Lake Mary at the head of Big Cottonwood Canyon, proposing to build dams at those lakes to store water. Now that the Big Cottonwood conduit was in operation, he offered to sell his rights to the city. Accordingly, on 15 and 16 September 1906 the city council, guided by Thomas, traveled to Park City by rail, then to Brighton on horseback. There they inspected the Twin Lakes, Dog Lake, and Lakes Mary and Phoebe, while Engineer Kelsey described the possibilities—one large dam for the two Twin Lakes and another for Lakes Phoebe and Mary. Despite a sudden early-season snowstorm that enveloped the group and caused them to lose their bearings, the party eventually arrived at the Brighton Hotel, where they spent the night. The next day, on their way down the canyon, they detoured to visit Lakes Blanche, Lillian, and Florence for consideration as reservoir sites, although at this time the

Brown and Sanford Irrigation Company had already embarked upon such a venture. It was reported that Thomas was asking $35,000 for his water rights. Although there was some question as to whether the six-year-old rights were still valid, since no work had been done to build the dams described in the appropriation, the city eventually bought the rights for $2,000 in order to extinguish all existing claims and remove adverse rights on all lakes in Big Cottonwood Canyon.[38]

In 1912 the city commission appropriated $5,000 for preliminary work on the two dams. The following May another $16,000 was appropriated, with the expectation that all work would be performed by the city water department. However, threatened litigation forced a reconsideration, and the job was put out for bids. By the time awards could be made the season was well advanced and little was done on the Lakes Phoebe and Mary dam that year. In 1914 the contractor, Owen H. Gray & Co., discovered the difficulties of working at such an elevation and so remote a site. Snows did not leave until the end of June, leaving only three months before snow began to fall again. In the interest of economy, materials were taken to Park City by rail, then hauled over the divide to Brighton on wagons. From there, special light wagons with four-horse teams were used when the wagon road to the dam site was dry and pack animals were used when it was wet. As the excavation proceeded it was found that nearly three times the estimated amount of material had to be removed to reach suitable bedrock, causing the schedule to slip drastically. The next year, 1915, nearly all of the remaining work was completed, and the dam was allowed to impound water the following spring while the contractor finished the job. Located at the lower end of Lake Phoebe, the 80-foot-high, 330-foot-long dam formed a reservoir that covered both Lakes Phoebe and Mary and held about 240 million gallons of water. Its contract price was $69,000.[39]

When the Twin Lakes dam project was put out for bid in 1914 only two contractors responded, one of them being Owen Gray; but both bids, being well above the estimate, were rejected. Bids were received again the following April, at which time the job was awarded to James Stewart & Co. This firm certainly benefited from the work done by Owen Gray in establishing supply lines and by the water department in preparing the site. The construction proceeded rapidly, suffered no unforeseen difficulties, and was completed by the end of the 1915 season. The reservoir behind the dam was somewhat larger than was the Lake Phoebe-Mary reservoir, holding more than 300 million gallons. It cost the city about

$72,000.[40] Both contractors and the water department before them built cabins at their construction sites. Only one of them, near the Lake Mary dam, remains today. The dams and reservoirs, of course, may be seen much as they appeared when they were completed, except for periodic repairs and reconditioning over the years. Also seen and enjoyed today is the wonderful trail that was built between the two dam sites during the construction period.

There is an interesting postscript to the story of the Twin Lakes dam. In May of 1917 the city commission received a communication demanding that it remove its Twin Lakes reservoir off the property of the Old Evergreen Mining Company. This property was in the form of the patented George mining claim, part of which now was under the dam and the lower end of the reservoir. The company demanded $10,000 for the surface rights. Negotiations followed, but it was not until two years later, at a 14 July 1919 meeting of the city commission, that Commissioner C. Clarence Neslen stated, "I am offered the surface rights of the George mining claim, which lies partly under the water of the Twin Lakes reservoir, for $183.35." The other four commissioners responded with a resounding, "Sold!" Neslen was authorized to close the deal.[41]

*

The biggest water project ever proposed for Big Cottonwood Canyon never was implemented. Most would consider this fortunate, for it would have changed the face of the canyon forever. When the two dams above Brighton were placed into service it was obvious that they could not satisfy the growing water demands of the city for very long. Indeed, before the end of the decade a proposal was raised to construct new reservoirs in Emigration, Mill Creek, Big Cottonwood, and Little Cottonwood Canyons, as well as build a conduit between the latter two canyons. It was placed before the voters in a bond election on 27 April 1920 but was soundly defeated.[42] The water department continued to acquire water rights through purchase or exchange for Utah Lake water, and it appropriated moneys for additional reservoir studies. But these efforts received little publicity and all seemed quiet on the water front until 1924, when an exceedingly dry summer forced the city to impose watering restrictions and emphasized a need for more water-storage capacity. In November 1926 the *Salt Lake Tribune* published a lengthy article about the water supply problem. It was stated that two reservoir sites were under consideration in Mill Creek Canyon and three in Big Cottonwood. The Mill Creek sites were at the Boy Scout camp about two miles above the mouth of the canyon and at the Elbow. The first could provide a reservoir holding 1,000–2,000 acre-feet, depending upon the height of the dam; the upper reservoir would hold 1,000 acre-feet. The Big Cottonwood sites were at Argenta, Mill D, and Mule Gulch. While the latter was considered an excellent site geologically, its reservoir capacity was small and the cost per acre-foot relatively high. It never was a serious contender.[43] The dam at the Argenta site, located just above Mill A Gulch where the mining camp with the same name had been located, would impound 12,000 acre-feet, more than twelve times the size of the Twin Lakes reservoir. The Mill D location would provide an even larger reservoir, about 13,500 acre-feet.[44]

There were, however, some who felt that small reservoirs in the canyons were at best a temporary solution. As early as 1920, a group of businessmen suggested that the city should inaugurate planning on a broader basis to provide a real water supply, one that would be adequate for many years of growth. This could be done by looking past the local canyons and reaching out to the waters of the Weber, Provo, Strawberry, and even Duchesne Rivers.[45] That may have been a radical proposal, but in retrospect it was a seed that took several years to germinate. A *Tribune* article in November 1926 included maps showing the feasibility of bringing water from the upper reaches of the Weber River, an idea that was further supported in a proposal by H. W. Sheley, a Salt Lake hydraulic engineer.[46] By this time it was no longer a farfetched concept, but it did complicate and confuse the city's planning. Accordingly, Mayor John F. Bowman appointed a water advisory board to study the possibilities and make recommendations as to a future course of action. Meanwhile the mayor filed an application with the state engineer for the use of water from the Duchesne watershed and storage in the upper Provo River canyon in case such a project were recommended by the advisory board. He further stated that he opposed a proposition made by a city commissioner that immediate steps be taken to build a reservoir in Big Cottonwood Canyon.[47]

By the time the advisory board completed its report at the end of March 1929, the Provo River storage concept had coalesced into the Deer Creek project. Accordingly, the board recommended a Deer Creek reservoir for an "outside" supply and an Argenta reservoir for an "inside" supply. Accepting the recommendation, the mayor immediately began campaigning for public support and proposed a $5 million bond issue, $3 million for the Argenta reservoir, the rest to secure water in the Deer Creek project.[48] In spite of the perceived urgency of the matter, the city commission did not act for nearly a year, by which time the Deer Creek project had been set aside. Finally, in February 1930 the city commission agreed to pursue the Argenta dam and reservoir, to be financed by a $3 million bond issue, and unleashed one of the all-time great local public controversies.[49] Everyone got into the act. The *Salt Lake Tribune* published a series of at least seventeen daily articles exploring all aspects of Salt Lake City's water problems—past, present, and future. Editorials supporting or questioning the proposals appeared almost daily. The mayor began a campaign, speaking before many civic organizations and anywhere else he could find someone to listen. Radio stations broadcast discussions of the issues, both for and against. And there were many of the latter. There were those who questioned the design of the dam, the geology of the site, whether the watershed would provide enough water to fill the reservoir, and whether the city had rights to water that was not already being stored or used.

The debate intensified when the city commission at its 1 April meeting passed an ordinance providing for the election to be held on Tuesday, 6 May.[50] Newspaper forums then overflowed with letters from people favoring or opposing the issue. The most crushing blow came when members of the Utah Society of Engineers voted overwhelmingly against the Argenta project. The society had appointed an executive committee to review material pertaining to the project before giving its support. While the committee supported the idea of increasing the water supply, it decided there were too many questions remaining unanswered to support this specific plan. As if responding to the society, the mayor immediately agreed to hire additional experts to make more definite reports.[51] But his action raised more questions in the collective mind of the public, which began to wonder whether it was being told the whole truth. The *Salt Lake Tribune* in editorials urged the city fathers to either satisfy and dissipate criticism and objections before 6 May or defer the election until that could be done.[52] On the penultimate day of April the mayor announced that the city administration had exhausted all its means to obtain expert advice on the project, but he believed that all questions raised by the opponents had been answered satisfactorily.[53] Secure in that belief, he continued to predict that the bonds would receive the approval of the electorate.

On 6 May, the day of the election, Salt Lake City was drenched by a torrential downpour. It was, perhaps, symbolic that so much water should fall on the day that voters were asked to approve water bonds. In spite of the weather the populace turned out in record numbers and defeated the issue by a ratio of more than four to one. It was said that never before in the history of Salt Lake had a bond issue been so utterly swamped under a flood of votes. The administration had been unable to rally any support beyond the numerical strength of the city employees.[54]

In the postmortem nearly everyone agreed that the voters were expressing their lack of confidence in the administration rather than opposing water development. While the defeat of the Argenta proposal may have saved Big Cottonwood Canyon from the intrusion of a very large man-made lake, it did pave the way for more productive water projects for Salt Lake City, including development of artesian wells, acquisition of Little Cottonwood water rights, construction of the Little Cottonwood conduit, and ultimately the realization of the Deer Creek reservoir and aqueduct that still provides much of the culinary water used in the Salt Lake Valley.

The Argenta dam would have been 220 feet high at the center and 800 feet wide. It would have impounded 12,000 acre-feet of water, with its lakebed covering some 143 acres. The reservoir would have been a mile and a quarter long, reaching nearly to the top of the grade below Cardiff Fork. Its waters would have filled the lower reaches of Butler Fork as well as the seldom-visited lower part of the Cardiff Fork stream, which enters Big Cottonwood Creek a short distance below Butler Fork. The canyon road would have been moved onto the north slope, following a contour line above the high-water line. Although it was said that easy curves and light grades would take the road to the level of the top of the dam, examination of the north slopes in the vicinity of Mill A Gulch leaves one to wonder how it could possibly have gained 220 feet of elevation with only light grades and easy curves. Had it been built, however, many of today's visitors, never having seen the canyon without the dam and reservoir, would probably enjoy the beauty of a large body of water in the early summer as well as express disgust with the bare dirt slopes leading down to the low water levels later in the season.

The Argenta project did not completely die with the 1931 bond election. The city had obtained rights to build the dam and reservoir on federal lands, rights that now had to be renewed annually until construction was started or the project officially abandoned. Rather than have this annual imposition, the city commission requested that Utah's congressional delegation draft legislation giving the rights an indefinite duration. In February 1933 Utah Senator William H. King introduced a bill to provide protection for Salt Lake City's municipal water supply. While it covered most lands within the Wasatch National Forest east of the city, Section 3 of the bill specifically set aside the Mill D and Argenta sites as municipal water supply reservoir sites for the use and benefit of Salt Lake City. The bill was passed on 26 May 1934.[55] Years later, when the Central Utah Project was formulated, the Argenta dam was included as one of its objectives, and as recently as 1964 R. E. Marsell, a water consultant for the Utah Water and Power Board, predicted, "Some day the Argenta Dam will be constructed, but at a cost of several times as much as it would have cost had it been built in 1931."[56]

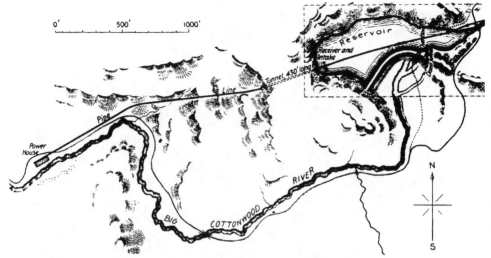

Figure 63. Map of Big Cottonwood Power Company project. Youngs Peak Lodge was located in what became the left center of the reservoir. Note that the original road ran across the flat where today's Storm Mountain picnic area is located, crossed the stream, and continued up canyon. Jones's new road climbed up a dugway, crossing the hogsback through a deep cut, then dropped to cross the creek over a new bridge above the upper end of the reservoir. It was commonly known as "Jones' Hump." (*Engineering News*, 1 October 1896)

Figure 64. The cut at the top of Jones' Hump, sometimes called Devils Gate. (*Improvement Era*, June 1915)

Figure 65. The Big Cottonwood Power Company plant under construction. Note the canyon road winding behind the left side of the building, heading for the Stairs. (*Engineering News*, 1 October 1896)

Figure 66. Bridge on the original canyon road in the depths of the Stairs between the power plant and the reservoir. The bridge abutments can still be seen. (SLT, 13 June 1915)

Figure 67. Another view of the Big Cottonwood Power Company plant about the time construction was completed. The dam in the foreground is diverting water into a flume for the Utah Power Company's plant farther down the canyon. (UHS)

Figure 68. The Big Cottonwood Power Company's reservoir was much larger and deeper in 1896 than it is today. (*Engineering News*, 1 October 1896)

Figure 70. Big Cottonwood Power Company plaque mounted on the north wall of the power plant building. (Author's photo)

Figure 69. The Big Cottonwood road at Storm Mountain reservoir after the county removed the grade at Jones' Hump in 1916. The cut looks very much like the one that had been made at the top of Jones' Hump, but the steep grades on each side were eliminated. Today the cut is considerably widened but follows the same alignment. (UHS)

Figure 71. Utah Power Company plant as it appeared in the early 1900s. Its flume and penstock can be seen on the hillside above the building. Today the building is obscured by the water treatment plant near the mouth of the canyon. (UHS)

Figure 72. Flume for the Utah Power Company clinging to the cliffs high on the north side of Big Cottonwood Canyon. The flume was replaced by an underground pipe in 1998. (Author's photo)

Figure 73. Utah Power Company plaque, mounted above the building's large door. (Author's photo)

Figure 74. Artist's drawing of the proposed Argenta dam and reservoir in Big Cottonwood Canyon. (SLT, 23 March 1930)

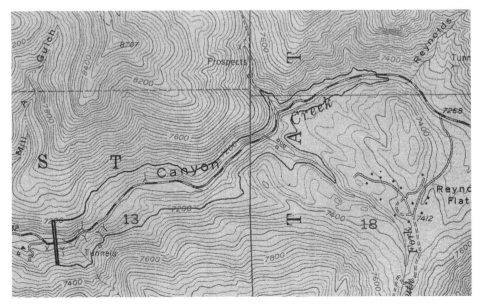

Figure 75. Map showing the extent of the proposed Argenta reservoir.

Figure 76. Power station at the mouth of Mill Creek Canyon. Built by the Mill Creek Power Company in 1910, it operated until 1949. Notice the penstock coming down the hill from the end of the pipeline that ran along today's very popular Pipeline Trail. ("Utah Power & Light Company: History of Origin and Development")

Figure 77. Lime kilns in Mill Creek Canyon were active at the beginning of the twentieth century. (Author's photo)

Figure 78. Quarries on the west face of the mountains between Mill Creek and Parleys Canyons. Parleys Canyon is in the background. (UHS)

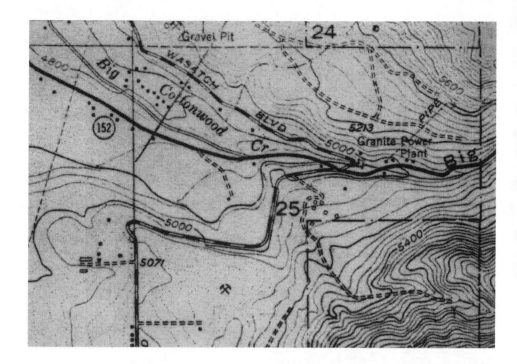

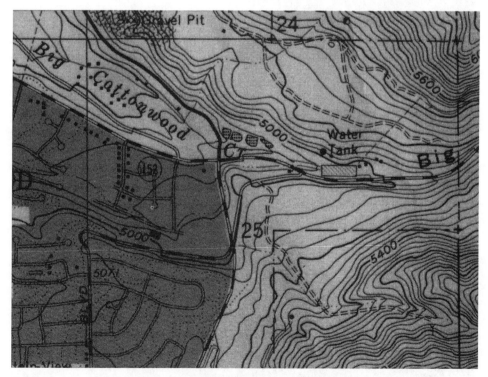

Figure 79. The mouth of Big Cottonwood Canyon. The upper map shows the route of the original Wasatch Boulevard as constructed in 1919 and as it still existed at mid-century. The lower map shows how Wasatch Boulevard was rerouted in the late 1950s before the water treatment plant was constructed. (USGS Draper Quadrangle maps, 1952 and 1963)

Above left and above:
Figure 80. Early views of the narrows of Big Cottonwood Canyon, with Youngs Peak rising overhead and the canyon road pinched between stream and cliff. C. R. Savage photographed the scene (left) and Henry L. A. Culmer sketched it (right). (Photo: UHS; sketch: *The Contributor*, February 1892)

Left:
Figure 81. The narrows of Big Cottonwood Canyon, with Youngs Peak (Storm Mountain) rising overhead, as it appears today. (Author's photograph)

Figure 82. Alfred Lambourne. (Lee Greene Richards, *Portrait of the Poet, Alfred Lambourne* [1920] oil on canvas, Springville Museum of Art; gift of Margaret Cannon, SMA 1987.026)

Figure 83. Lake Blanche and Sundial Peak in Mill B South Fork, Big Cottonwood Canyon. The lake is one of the "Sister Lakes" and the peak is one of the "Pillars of the Wasatch" that Lambourne loved so well. (UHS)

Figure 84. Sam McNutt in the doorway of his home near Whipple Fork; his visitor is Billy Schaaf. (UHS)

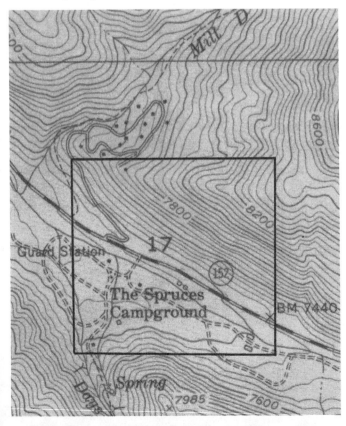

Figure 85. This map shows the extent of land set aside by presidential proclamation in 1906 to be used as a forest nursery. The site was later used for the Community Camp, and it then became the Spruces Campground.

Figure 86. The store at the Community Camp, where food was available "at Salt Lake prices." (UHS)

Figure 87. Buildings at Argenta in the early twentieth century. The road is next to the stream, well below today's highway. (UHS)

Figure 88. Consolidated Jefferson Company shaft house and dump. It was located on the west face of the Wasatch, south of Little Willow Canyon and above today's Wasatch Boulevard. It was one of the last gold mining operations in that area. (SLMR, 15 December 1904)

CHAPTER 12

Mill Creek Canyon—Names and Places

MILL CREEK CANYON has about 21.3 square miles of watershed area, placing it fourth in size among the Wasatch Mountain canyons draining into Salt Lake County. As has been explained in the introduction, the canyon took its name from the stream originating there, and that stream was named by Brigham Young in a special conference on 22 August 1847. Today a good county road extends more than seven miles up the canyon, ending a short distance above the site of Archibald Gardner's upper sawmill, where Soldiers Fork and Big Water Gulch drain into Mill Creek stream. Salt Lake County has provided milepost markers along this road, mile zero being at the intersection of Parkview Drive with 3800 South Street, near the mouth of the canyon (3700 East and 3800 South). Features within the canyon will be located relative to the highway mileage from milepost zero.

*

Mill Creek Power Company. A scant two-tenths of a mile from milepost zero the road enters the narrow V-shaped gash that marks the mouth of the canyon. On the north side a vertical scar can be seen descending from the top of the ridge. This is a faint monument, one of a few reminders of a hydroelectric power generation venture that operated in the canyon for sixty-five years.

In the first years of the twentieth century Francis M. Lyman, Jr., was an engineer, surveyor, and manager for the Shoshone Power Company in Salt Lake City, a company presided over by his father. Lyman saw the possibilities of harnessing the mountain streams for the generation of hydroelectric power. He had applied for water rights and planned three large plants on the Duchesne River on the south slopes of the Uinta Mountains. While those plans did not come to fruition, he also surveyed much

259

of the mountainous area east of Salt Lake City. In 1903 he applied to the state engineer for a permit to appropriate waters of Mill Creek Canyon for power purposes, an application that was granted in 1905, and convinced John P. Cahoon, a Murray businessman, to join him to develop a Mill Creek power plant. One of Cahoon's businesses was the Salt Lake Pressed Brick Company, which could use much of the power that might be generated; and so he and his partner, Melvin M. Miller, joined Lyman to form the Mill Creek Power Company on 24 January 1905. By that time the Salt Lake County Commission had granted Lyman a franchise to construct power lines along county roads from the mouth of the canyon.[1]

The first power plant was built in 1907 below Porter Fork, near the mouth of Burch Hollow. It was a one-story building, about twenty-six by forty feet, that housed a waterwheel and a 300-kw generator. A rubble masonry dam was built at the mouth of Elbow Fork, less than two miles above the plant. Water was carried through a twenty-two-inch wood-stave pipe that was laid in a trench on the north side of the canyon. At its lower end it joined a penstock of eighteen-inch riveted steel pipe, leading down to the plant. The new company had applied for and received a franchise from the county commission to construct a power line from the plant site to the Salt Lake Pressed Brick Company's location.[2] Electricity was delivered over the twelve-mile power line to the brick company's yard, where most of the plant's output was consumed. The remaining power was sold to other users.

Early in 1910 the company got a five-year contract to supply power to the U. S. Smelter Company at Midvale. The existing power plant was inadequate to fulfill the contract, so construction of a second plant was started. It was located at the mouth of Mill Creek Canyon, on the north side of the stream. The water to drive it was taken from a reservoir directly below the tailrace of the upper plant. From there it was taken through another wood-stave pipe that ran along the north slope of the canyon for nearly three and one-half miles to a point above the power plant, where it entered a steel penstock that carried it down to the waterwheels that drove two 560-kw generators. Shortly after construction began the Knight Investment Company absorbed the Mill Creek Power Company and became the Knight Power Company. Construction was completed by the new concern.

The Knight Investment Company was formed by Provo businessman Jesse Knight to hold and develop mining claims in the Tintic district. To provide more electric power for the Tintic smelter, the company built a power plant at Santaquin. It later built another plant at Snake Creek to supply the power needs of Park City. These two power plants were merged with the Mill Creek Power Company to form the new Knight Power Company. That company was replaced by the Knight Consolidated Power Company in January 1912, which later that year was incorporated into the Utah Power and Light Company when it was organized on 6 September 1912. The new company continued to operate in Mill Creek Canyon until April 1970, when the upper plant was torn down.[3] The lower generating station had operated until 7 April 1949, when the pipeline became unsafe for further use. During the year that followed the pipeline was removed.[4] Although the company maintained its water appropriation for a number of years that followed, the lower generating plant never operated again and was later dismantled. Its concrete floors and foundations could be seen at the site as recently as the mid-1970s. Today private homes occupy the site.

After the two Mill Creek power plants were decommissioned the right-of-way for their pipelines became a favorite hiking trail, especially popular because it is nearly level. In recent years the portion between the upper power plant and Church Fork has been extensively reworked and has become a well-used biking trail. The lower pipeline trail, in following a contour line, winds in and out through deep gullies. Where the pipe made a right-angle turn a steel elbow was used, held in position by deadmen and steel rods to resist the large forces created by the water as it changed its direction of flow. After the pipeline was abandoned the exposed portions were dismantled, leaving little more than weatherworn staves discarded on the slopes below and concrete footings from the few trestles spanning gullies. The steel elbows, being heavy and well secured to their restraining deadmen, were left in place. Buried sections of the pipeline were allowed to remain to their fate. Eventually the staves rotted away and the earth above them collapsed, leaving the wire hoops exposed like bare ribs of some ancient reptile.[5] At the head of the penstock, above the mouth of the canyon, a steel-pipe "Y" may be seen. Wooden stave pipe was attached to one arm while the penstock, long since removed, connected to the other end. The third opening points up the ridge where a pipe ran for some distance before terminating with an open end. The purpose of this section of pipe was to minimize the effect of a water hammer that might occur if the flow were abruptly turned off at the power station. In that case the water could surge up the open pipe and run out the end. Indeed, if the north side of that ridge is viewed from the streets below, a rocky gully can be seen starting at the top of the ridge and running straight down to the drainage below. This was caused by water pouring out of the surge pipe and running down the slope. While that may not have happened often, the power plant was in operation for at least sixty years, during which time a lot of water from such actions could have run down the hill.

*

Russell Ditch. A short distance above the mouth of the canyon, about mile 0.3, the faint remains of the Russell Ditch may be seen. Over the years there were many irrigation ditches that drew water from Mill Creek stream, but this was the only one that extended into the canyon itself. It was shown on a 1900 map and may well have been constructed by Daniel Russell, an 1848 pioneer who settled at the mouth of Mill Creek Canyon.[6] It carried water to irrigate lands high on the bench to the north. The ditch can be followed until it leaves the canyon and crosses the west-facing slopes. At that point it has been obliterated by recent real estate development.

*

Limekilns. At mile 0.7 a faint trail, once a dirt road, heads westerly on the north side of the canyon. If the trail is followed, the crumbling remains of a large limekiln soon come into sight. Kilns such as this once were commonplace east and north of the city. Lime was an important ingredient to the developing city from the first days of its settlement. Less than a week and a half after arriving in the valley, Albert Carrington went into the mountains to look for limestone.[7] Apparently he found none, for the next month the Mormon brethren were instructed to report any discovery of chalk, lime, coal, and various other minerals.[8] However, nearly a year passed before it was reported that Joel Parrish and Charles Chase had discovered limestone in Emigration Canyon. At that time it was generally believed there was no limestone to be found in the mountains. Using a blacksmith's forge, the men were soon able to burn 150 bushels of lime a day, the first record of burning lime in the Salt Lake Valley.[9] In retrospect, one is led to wonder if the men seeking limestone were able to recognize the rock if they saw it, for limestone is very abundant in the Wasatch Mountains. The entire ridge north of Mill Creek Canyon is composed of a thick bed of limestone, one that reappears repeatedly to the north, alternating with beds of sandstone. Once the material was recognized, lime quarries blossomed all along the face of the mountains.

The Mill Creek Canyon limekilns hold the distinction of being the southernmost kilns to be found in the Central Wasatch. They drew stone from the slopes directly above. Traces of wooden slides that carried the stone down to the kiln may still be seen. The enterprise had two cooking vats into which limestone was poured and four furnaces where fires were built to cause heated air to rise through the kilns, reducing the stone to quicklime. The remaining structure clearly shows two generations of construction, for the east and west halves were built at different times. Its first half was present as early as June 1894, when it was used as a reference in a claim. At least four other claim notices mentioned the kiln, the last in January 1904, but it is not known if it was still operational at that time.[10] It was not alone in the limestone industry in this vicinity, for just north of the mouth of Mill Creek Canyon, on the upper Bonneville Bench level, are a number of scars marking sites of limestone quarries. They were being worked at the turn of the century by the Union Lime and Stone Company, an enterprise involving Simon Bamberger, a man well known both as the builder of the interurban rail line along the Wasatch Front and as one of Utah's governors.[11] By 1904 the quarries were being operated by Perry Kalbaugh, who installed tracks on the shoreline bench from the quarries to the edge of Parleys Canyon. The rock was loaded onto cars, moved to the edge of the canyon by mule power, and then transferred to railroad cars with the use of a gravity tramway. Kalbaugh claimed to be able to provide as much as 500 tons of rock per day, most of it going to smelters for use in their operations.[12] Except for the very noticeable scars of the quarries and some of the tramway's concrete foundations at the edge of Parleys Canyon very little remains to be seen from this operation.

*

Rattlesnake Gulch. A wide spot in the road at mile 1.1 offers a parking area for those who care to venture up the popular trail in Rattlesnake Gulch. The trail also offers access to the pipeline trail some 500 feet above. The name suggests that the gulch offers a haven for the venomous reptiles, which may have been the case, for the place was known as Rattlesnake Hollow in 1891 when W. H. Stout and H. K. North located the Rattlesnake Lode there.[13] Those two men apparently worked their claim for several years, because they were still in the area three years later when they filed another mining claim and a water claim for a mill.[14] A patent was issued for the lode in 1896, but by that time North had a new partner, William H. Dodge.[15] The place continued to be known as Rattlesnake Hollow at least through 1901; but sometime during the years following the name mutated to Rattlesnake Gulch.

*

Camp Tracy. Miles 1.1 to 2.8. The presence of this Boy Scout camp in the canyon can hardly be missed, for it extends over a mile and a half along the road and has numerous structures and other facilities distributed along its length. Curiously, this camp has its roots in Mill Creek Canyon's mining activity. Although Mill Creek never was as important to the mining scene as were the canyons to the south, there was a small amount of activity over the years, mostly in claims that were filed and then forgotten. But in 1902 there was a sudden flurry of activity in which at least eighteen placer claims were filed. Placer claims were used for surface mining operations, such as quarrying stone or sand. They were seldom found in the mining districts to the south, where the desired minerals were found below the surface. In the Hot Springs Mining District to the north, however, where alternating beds of limestone and sandstone offered many quarrying possibilities, placer claims were quite common. Placer claims could encompass an area as large as 160 acres, a quarter section, whereas claims on mineral lodes were limited to an area of 600 feet by 1,500 feet, a mere 20.66 acres. If a person were filing claims with the intent of amassing land area, placer claims were more effective than mineral claims.

The first Mill Creek Canyon placer claims in the year 1902 were filed by members of the Taylor family, local residents and landholders, in conjunction with a John H. Powers of Wisconsin.[16] Later that year a series of twelve placer claims were filed by a consortium headed by Lewis W. Pitcher and Charles L. Furey of Chicago. These claims, named after states, took in much of the canyon bottom from Rattlesnake Gulch to Porter Fork, for a total of 1,920 acres, or three square miles of land. In 1903 all of the claims, including those held by the Taylor family, were transferred to the Western Development and Construction Company, incorporated in Utah to do general contracting, railway construction, and mining business, and to operate pleasure resorts. The incorporators included Pitcher and Furey of Chicago and Powers of Wisconson.[17] In addition, in 1903 the Taylors and Powers filed three more placer claims and transferred them to the company.

In January 1904 a group of local residents got into the act by filing eleven placer claims encompassing 1,320 acres, many of them overlapping earlier claims, although two were as far up the canyon as the Elbow. Whether they were attempting to relocate earlier claims that had not been worked or were acting as agents for the Western Development and Construction Company is not known, but within four days of their filing the claims were transferred to that company. That same month the company began to transfer many of its holdings to the Salt Lake Southern Railway Company. This was a Delaware corporation, originally named the Salt Lake and Suburban Railway Company, whose bylaws were filed in Utah in July 1902. Its purpose was to deal in railways, electrical power, mining, quarrying, and building materials. It had only three subscribers for its capital stock: Lewis W. Pitcher and Charles L. Furey of Chicago and Alvin V. Taylor of Salt Lake City. Taylor, an attorney, was the company's Utah agent and a member of the family involved in the original placer claims.[18] There is no evidence that the company attempted to work any of its claims in the canyon, but it did issue $2 million worth of bonds through the Knickerbocker Trust Company of New York, using its claims as collateral. It also worked quietly to secure patents on its land holdings, both from the United States and the State of Utah. The last patent was received in 1908, after which time the Salt Lake Southern Railway Company slid into oblivion.[19] In 1915 the company's charter to do business in Utah was revoked for nonpayment of license tax.

Meanwhile, Taylor put the company's holdings to good use by allowing the flats between Rattlesnake Gulch and Church Fork to be used as a camp for local boys' groups, such as the YMCA and Boy Scouts of America. The site soon became known as Taylor's Flat. As its popularity grew, Taylor moved to make the place a permanent camp for boys. On 21 May 1919 he, as the sole surviving director and trustee for stockholders of the Salt Lake Southern Railway Company, presented to the Boy Scouts of America a deed for most, if not all of the company's remaining holdings in Mill Creek Canyon, amounting to some 1,320 acres of land.[20] This magnanimous gesture was not without limitations, for much of the property was encumbered by various tax liens and the trust deed with the Knickerbocker Trust Company of New York, and some of it had already been sold for back taxes. It took the Boy Scout organization's lawyers five years to untangle the encumbrances, the final move being to wrest titles from the defunct company through a suit in Third District Court.[21] In the interim, the scouts honored their benefactor by establishing Camp Alvin V. Taylor on their new property. The christening ceremonies were held on 14 May 1920. Although Troop 51 had already built a cabin, the first official improvement was the installation of a forty-foot flagstaff, a pine pole brought from Mount Timpanogos.[22]

Although it had been planned to build a commodious weekend camp with logs from the timber on the tract, this improvement had to wait another three years. In 1923 Salt Lake banker and philanthropist Russel L. Tracy gave the scouts a gift of a two-story lodge measuring thirty-five by seventy feet. The building was constructed at a cost of $10,000. Tracy had lost both of his sons: one died in infancy and the other, at age twenty-one, had died following an operation only two years earlier. The lodge was a memorial to his sons. At the dedication service Tracy called the new building a "wigwam," explaining that the name came from an Indian word meaning an assembling place. Thereafter the lodge was known as the Tracy Wigwam, located in Camp Alvin V. Taylor. The latter name continued to be used throughout the 1920s, but gradually the Tracy Wigwam name was applied to the entire camp, while Taylor was all but forgotten.[23]

Tracy Wigwam served the scouts well for over sixty years before it was razed to make way for a newer, more modern structure. When the new lodge was dedicated on 10 October 1987 the Wigwam name was dropped, and the camp is now known as Camp Tracy. Near the main entrance stands a stone marker with a bronze plaque, the sole remaining reminder of the original benefactor, Alvin V. Taylor.

*

Green Canyon. Mile 2. This drainage, on the south side of Mill Creek Canyon, is now completely encompassed by Camp Tracy. It once was the site of one of Alva Alexander's sawmills or shingle mills after he received permission from the county court to build a logging slide in the canyon to bring down timber for his mill.[24] The source of the Green name is not known, but it was in use as early as 1891, at which time the canyon was called Green Hollow.[25]

*

Church Fork. Mile 2.8. As already noted in an earlier chapter, this fork gained its name in 1854 when Abraham O. Smoot had the task of constructing a building on Big Kanyon Creek to house sugar-making machinery, the building eventually known as the Sugar House. To get timbers needed for the structure, Smoot petitioned the county court for the right to make a road and control the timber in this fork. Since he was doing this work at the direction of Brigham Young, his petition was made in behalf of the trustee in trust of the church.[26] As a result, this tributary canyon became known as the Church Fork, and it continues to be known by that name to this day.

The lower end of Church Fork has a number of picnic sites, with a paved road that climbs several hundred feet to the highest ones. From that point a dirt road continues but quickly dwindles to a trail, the very popular Grandeur Peak Trail. A short distance from the end of the road the trail crosses a small flat, where it intersects the pipeline trail. When the Mill Creek electric power plants were in operation the wooden-stave pipe ran across this flat on its way to the lower power plant. The company, having rights to use water from the Church Fork stream, had installed a wooden dam and intake above this point to allow water to be diverted and added to that already in the pipeline.[27]

*

Thayne Canyon. Mile 3.0. This tributary on the south side of Mill Creek Canyon was known by several names before gaining the one that has survived. In 1859, when Enoch Reese was operating the sawmill on the flat at the mouth of this fork, it was called Little Mill Creek Canyon.[28] The canyon became known as Ellsworth Canyon after Edmund Ellsworth ran the mill.[29] Then when John Thayn took over the mill and ran it for many years the fork took his name, although spelled Thayne, and that name continues to be used to this day. Thayne Flat, at the base of this canyon, has trailheads for two trails. One goes up the drainage to the ridge between Thayne and Neffs Canyons and then drops into the latter to meet that canyon's trail. The other is the popular Desolation Trail, which climbs to the Salt Lake Overlook on the west ridge of Thayne Canyon. This once was part of a trail leading into Green Canyon, but it was used as the Mill Creek end of Desolation Trail, constructed by the U. S. Forest Service in 1967 and 1968. It goes all the way to Lake Desolation in Big Cottonwood Canyon, for a total distance of some seventeen miles. At the time it was built it was the policy of the Forest Service to keep grades gentle, so it has an interminable number of switchbacks, discouraging all but the most determined hikers from going farther than the overlook. For many years the trail was heavily used by motorcycle traffic, until such vehicles were banned because of the environmental damage they were causing. Mountain bike traffic was also banned within the Mount Olympus Wilderness after that area was designated, so the higher portions of this trail now see but little use.

*

Log Haven. Mile 3.75. This stone and log building has housed a popular restaurant for nearly forty years, but it was originally built as a private residence. The land on which it stands once was part of the large holdings of the Salt Lake Southern Railway Company and was part of the large transfer made to the Boy Scouts of America by Alvin V. Taylor in 1919. However, by that time Salt Lake County had attached the land for back taxes and then sold it to E. S. Hallock.[30] In 1923 the land changed hands again when Hallock sold it to L. F. Rains, who built a summer home on the site. Rains imported logs from Oregon and had them hauled up the canyon by horse-drawn wagons. It took two years to build the home, which was to be an anniversary gift to his wife. While that may sound like the plot for a romantic folktale, in this case it was fact, for on 23 July 1925 he gave his wife, Ruth B. Rains, title to the property, "in consideration of the love and affection which he bears towards his wife."[31]

In 1937 the building and property was sold to Zane W. Miller, who remodeled it into a year-round home. It remained in the Miller family until 1964, when it was sold to Stanley W. Sprouse.[32] It was the latter owner who converted it into a restaurant and ran it for many years. It has since changed hands several times but still remains an attractive year-round dinner spot in the heart of Mill Creek Canyon.

*

Porter Fork. Mile 4.0. The name of this fork, on the south side of the canyon, comes from Chauncey W. Porter, who was operating a sawmill in the main canyon at the mouth of the fork late in 1850. At that time the drainage was known as the South Fork of Mill Creek Canyon. But in June 1853 Porter petitioned the county court for permission to build a road up the South Fork and control the sawing timber found there. After his request was granted and the road was built the fork assumed his name.[33] Porter Fork was the scene of a modest amount of mining activity. There was a legend of a man by the name of Indian Pete who operated a mine in Porter Fork and was killed in a snow avalanche. There actually was a published report that the body of Indian Pete was found on Decoration Day 1889, head down with feet sticking up through the melting snow. While nothing is known about Indian Pete, his name was perpetuated ten years later when the Indian Pete Mining & Milling Company was organized to operate a series of claims in the southwest corner of Porter Fork and extending into the upper reaches of Thaynes and Neffs Canyons. In 1915 the company filed an Indian Pete Claim, followed over a period of two years by the Indian Pete Numbers 1 through 6 Claims. The last reference found to this mine or mining company was in 1929 when members of the Wasatch Mountain Club took a midnight hike to the Indian Pete Mine. When they returned, no one, with the possible exception of the leader, knew for sure where the mine was, except that it was somewhere in Mill Creek Canyon.[34]

*

Bowman Fork. About 0.4 miles up Porter Fork a drainage enters from the east. This fork was named after Isaac Bowman, who received a grant from the county court on 21 September 1853, allowing him to build a road into the fork and charge a toll for timber or poles hauled out.[35] His road, several times improved, still can be seen. In fact, the Bowman Fork Trail goes up the road until the trail reverses itself and climbs out of the bottom of the drainage. At that point the road can be seen continuing up the bottom of the canyon. In the other direction, it continues down through the cabin area to the Porter Fork road. While Bowman had a limited tenure in the canyon, his name has survived for nearly a century and a half.

*

Burch Hollow. Mile 4.2. This was the site of the upper Mill Creek power station and the diversion dam for the pipeline to the lower station. The dam still exists, together with remains of the gates for the pipe. Nothing remains of the power station, which was dismantled in 1970. It had been a one-story building measuring about twenty-four by thirty-eight feet, and had a peaked roof. The road up Burch Hollow provided access to the pipeline; it doubles back to reach the pipeline trail about where the wood-stave pipe met the steel penstock.[36]

*

Elbow Fork. Mile 5.8. The Elbow name was typically used when a canyon made a right-angle turn, as Mill Creek does at this point. The name was in use as early as 1856 when the county court mentioned Gardner's grant "above a point known as the Elbow."[37] Here are found the trailheads for the Mount Aire, Lambs Canyon Pass, Terraces, and Pipeline Trails. The Elbow was the site of the dam and reservoir feeding the pipeline for the upper power plant. Remnants of the dam—blocks of concrete, timbers, bolts, steel rods and a few pieces of the headgate—may still be seen along the creek directly across the road from the small parking area. The dam remained for well over a decade after the power plant was decommissioned, until it was destroyed by crews clearing the streambed of debris during the heavy runoff years in the mid-1980s. It is interesting to note that when the dam was built the location of both it and the road were reversed from what we see today. The canyon road crossed to the south side of the stream just below the Elbow, then continued past the dam and reservoir, staying west of the stream. Another branch road went up Elbow Fork on its way to the Mount Aire Pass and Parleys Canyon.[38] At some later time the dam was rebuilt and the road relocated to where we see it today.

The observant hiker heading up the Mount Aire Trail may notice a series of five or six flat areas separated by low stone walls on the right side of the trail. In 1936 and 1937 crews of men employed through the Employment Recovery Act were working in Mill Creek Canyon, building trails and campgrounds, and making other improvements. The supervisor of the Wasatch National Forest, in explaining some of that work, said, "Among other camps with special attractions are Maple Grove, . . . and Elbow, where an extensive trail system begins to Mount Aire."[39] It is known that the Civilian Conservation Corps and other government agencies that were active at various places throughout the Wasatch often set up temporary living quarters in the areas where they were working. While no other documentary evidence has been found to explain the flat sites, they may well have been used as tent sites by the workers making the improvements the supervisor described. The site was later used as a picnic ground, as shown on the United States Geological Survey's Mount Aire map of 1955.

*

Alexander Basin. Mile 7.4. It has been generally assumed that this fork and basin were named after Alva Alexander, who was active in the canyon during the early logging years. Alexander and his son ran at least two mills in Mill Creek Canyon, his shingle mill being in the vicinity of this fork. The location, however, was in Archibald Gardner's grant, so the mill either was run with Gardner's blessings or was actually Gardner's mill and was being run for him by Alexander. As for the fork and basin, it is likely that Alexander was cutting timber for Gardner's mill and put in a trail and slides for that purpose, thereby leaving his name behind. Today's Alexander Basin Trail goes up one of the old logging slides for some distance, and the slide can be seen continuing up the slope after the trail leaves it.

*

Wilson Fork. Mile 8.4. This fork, heading south-southwest from the main canyon, is little used today. While little more than a faint trail beckons the traveler, a keen eye can find suggestions of logging trails and slides. There can be little doubt that this fork provided some of the timber for nearby Gardner's mill. It is very likely that the name of the fork comes from the man who opened a road into the fork to remove the timber. Directly across the canyon from Wilson Fork is another shallow drainage with a road that extends a scant quarter mile from the main canyon. The mountainside above the end of the road has several faint logging slides, a mute testimony to the fact that all the slopes in the vicinity of the mill were logged.

*

Soldier Fork. Mile 8.8. This fork heads south-southwest from the lower parking lot near the end of the Mill Creek road. A trail goes up the fork, over the dividing ridge between Mill Creek and Big Cottonwood Canyons, then drops down into Butler Fork. There was a legend that the fork was named after a soldier who was found dead there, but nothing has been found to substantiate this story. No other origin for the name has been found, but in November 1865 a claim was located near the head of Big Water in Mill Creek Canyon that was said to be "above Lt. Whitney's & Co. US Wood Ranche."[40] This was several years after the arrival of Col. Patrick E. Connor's troops and the establishment of Camp Douglas. By this time the camp's voracious appetite for wood, both for construction and fuel, had led wood-acquisition parties to go farther and farther afield to find timber. The reference to Lt. Whitney's wood ranch suggests that the soldiers were taking timber from the upper part of Mill Creek Canyon, in the vicinity of the drainage that is known as Soldier Fork. While this may be a logical supposition, nothing further has been found to verify it. The first use of the name that has been found was in 1903 when two claims were filed on the divide between Butler Gulch and Soldier Gulch in Mill Creek Canyon.[41]

*

Big Water Gulch, Little Water Gulch. Mile 8.9. Big Water Gulch drains into Mill Creek from the south at the end of the paved road. Little Water Gulch drains from the south about a quarter mile farther up the canyon. Trails, as well as flowing water, are found in both drainages. Their names likely originate in the pioneers' propensity to repeat a name, then differentiate between the two sites by labeling them "big" and "little." This has resulted in such combinations as Big and Little Sandy, Big and Little Cottonwood Canyons, and Big and Little Mountains, to name a few. The earliest use of the names that has been found was in 1865 for Big Water and 1899 for Little Water, but both names were used frequently after the latter date.[42]

*

Neffs Canyon, Norths Fork, Thomas Fork. Neffs Canyon is found less than a mile south of the mouth of Mill Creek Canyon. As explained in an earlier chapter, its name came from Franklin Neff, who, with Orrin Porter Rockwell, held a county court grant to build a road into the canyon and extract timber from the slopes and tolls from the road. Their road ran diagonally down the western slope to the mouth of Mill Creek Canyon. The first use of the Neffs Canyon name that has been found was in 1856, when E. H. Hiatt petitioned the county court for exclusive use of the timber in the fork that would become known as Thomas Fork.[43] The latter fork's name came from Charles W. Thomas, who built a road or trail into it in 1860.[44] The other major drainage in Neffs Canyon is Norths Fork, located just above the mouth of the canyon. The origin of its name

is not known. It is possible it came from Levi North, an early settler in
the area who was known to have been active in the canyons. The first use
found for the name of either fork was in 1898.[45]

The Neffs Canyon Trail extends into Peterson Basin, "a greatly subdued
cirque" at the head of the canyon, where some modest mining activity
took place.[46] One branch goes over the ridge into Thayne Canyon, while
another leads to the pass over the divide into Big Cottonwood Canyon.
Neffs Canyon and both tributary forks, hold remains of dams, pipelines,
and flumes. These were built after the water was claimed to irrigate lands
below the mouth of the canyon, lands that became the Mount Olympus
Fruit and Livestock Farm.[47]

*

Mount Olympus. On the south side of Neffs Canyon rises the imposing north face of Mount Olympus. Because its summit is nearly two thousand feet lower than those of the higher peaks to the south—Twin Peaks
and Lone Peak—the mountain does not stand out when viewed from the
center of Salt Lake City. This probably explains why it did not receive
special attention when the early settlers were bestowing names upon surrounding features. When people settled away from the city, especially along
Mill Creek and in the area between Mill Creek and Big Cottonwood, the
mountain became more significant. Yet, despite its daily presence, towering overhead and offering views that could not be ignored, the mountain
failed to receive a generally accepted name. It is likely that local residents
had favored names for the peak but failed to record them. The only one
that has come to light was from Archibald Gardner, who wrote that he,
his brother Robert, and Thomas Broderic scaled the Twin Peaks between
Mill Creek and Big Cottonwood Canyons. They ascended from the Big
Cottonwood side, but found the route so extremely difficult that they
came down on the Mill Creek side.[48] They must have come down into
Neffs Canyon, but the event happened about 1850, at a time when Neffs
Canyon had not yet received its name. While we think of Twin Peaks as
being the summit several miles farther south, the view of Mount Olympus from the west side shows the two summits very clearly and distinctly,
and certainly gives reason to think of it as Twin Peaks.

The first use of the Mount Olympus name that has been found was in
1899 when the Mount Olympus Fruit and Livestock Farm Company was
incorporated.[49] This company operated on a large plot of land held by
the incorporators, the Taylor and Stout families. It was located south of
the mouth of Mill Creek Canyon and extended nearly to the mouth of
Neffs Canyon. While the origin of the Mount Olympus name is not

known, it may well be that the Taylors and Stouts, living under the impressive north face of the mountain, were inspired to use the name of the mythical abode of the Greek gods for their company and that the name carried over to the mountain as well. One of the members of the Taylor family, Alvin V., the same man who was involved in the Boy Scout camp in Mill Creek Canyon, also started the Mount Olympus Spring Water Company, which used water from Neffs Canyon as its product. Although both the farm and spring water companies were operated by the same people, they were separate entities until they merged in 1929 under the water company's name. And so this earliest use of the Mount Olympus name is carried forth to this day, although the company has since changed hands and been reorganized.

On the west side of Mount Olympus, close to Wasatch Boulevard, is a large outcrop of quartzite that for many scores of years has been a favorite place for those who want to practice rock climbing. Today it is known as Petes Rock, named after Odell Peterson, an avid rock climber during the 1940s. The rock has always been a prominent feature; as early as 1871 it was described as the large rock butte near the base of the mountain. In 1895 it was described as "the big rock which is about one mile SE of Nielsons store Hollidaysberg." Also that same year, it was described as "what is known as the big rock, the one that has the snakes and lizzards painted on it."[50] The snakes and lizards have disappeared, as have several generations of numbers designating climbing routes, but Petes Rock still enjoys its prominence and popularity.

*

Tolcat Canyon. This drainage is on the west slope of the Wasatch immediately to the south of Mount Olympus. Its north fork drains the area between the north and south peaks. The name's origin is not known, but it was in use as early as 1879.[51] It gave prospectors considerable trouble, for in their attempts to spell the name it came out in many variations: Talcott, Folkjor, Tolkjar, Tolcar, Tolgeer, Tolker, and Talkets, to name a few.

*

Hughes Canyon. This drainage is on the west slope of the Wasatch about a mile and a half north of Big Cottonwood Canyon. Its creek bed crosses Wasatch Boulevard around 6400 South but is obscured by the large development of homes that all but restrict access to the canyon. Extant USGS and Forest Service maps show Hughes spelled as "Heughs." Such strange spellings are commonplace in Wasatch Mountain history, since many of the people who frequented the area were marginally literate and created quaint and amusing names when attempting to match their spelling with the spoken name. However, in this case we must blame some latter-day mapmaker for the curious name. With but few exceptions, area prospectors and miners unanimously called the place Hughes Canyon, the name being used as early as the spring of 1871.[52] The "Heughs" mutation has been found as early as 1908, when it appeared on a Salt Lake County map.[53]

No origin for the name of this canyon has been found. Although there were several men by the name of Hughes active in the mining district, none of them was involved in a claim in this canyon. However, in 1876 an interesting item appeared in the *Salt Lake Tribune.*

> About five years ago a young Gentile named Hughes settled on and pre-empted a quarter section of land lying between the mouths of Big and Little Cottonwood canyons. When he took up his abode there a number of Danites from South Cottonwood waited upon him with their shotguns, and ordered him to leave the land, as it had been granted to one of the brethren by the Legislature years before.
>
> He declined to go, and from that time he has cultivated some forty acres of the land, though not without being harassed by the priests.... Several times of late armed men have gone to his place and ordered him to leave, and a few nights ago he was driven into the house by a number of these ruffian visitors, who discharged their guns at his door and window.
>
> A man from this city happened that way the next morning, to whom these facts were related, and who reports them to us. Hughes, he says, is anxious to sell out if anybody will buy his claim, but he does not intend to be driven off by a mob of Mormons, after he has spent the labor of five years on the place.[54]

CHAPTER 13

Big Cottonwood Canyon—Names and Places

BIG COTTONWOOD CANYON includes about 48.47 square miles of watershed area, placing it second in size among the Wasatch Mountain canyons draining into Salt Lake County. As explained earlier, the canyon took its name from the stream originating there, and that stream was known as "The Cottonwoods" as early as January 1848. It was variously known as Near Cottonwood, Great Cottonwood, Big Cottonwood, and simply as the Cottonwood. Today Utah Highway 190 runs about fourteen miles up the canyon—as far as Brighton. Its green milepost markers are found every mile along the canyon highway, except for brief periods when they might have been destroyed by snowplows or construction crews, and they are used in this chapter to locate places under discussion. Milepost 0 is located on Fort Union Boulevard where it intersects 2300 East Street, somewhat below the mouth of the canyon. The major intersection at Fort Union and Wasatch Boulevards, usually considered to be at the mouth of the canyon, is at mile 1.9.

In the nineteenth century the road into the canyon followed the stream from Knudson's Corner, near the 6200 South interchange on I-215, then followed the stream into the canyon. When Wasatch Boulevard was first built as far south as Big Cottonwood Canyon, in 1919, it followed the contour of the north slope and crossed the stream on a bridge below the Granite Power Plant, about at the lower end of today's water treatment plant. This is also where the Brown and Sanford ditch took its water from the stream as early as 1874, where the Big Cottonwood Lumber Company's toll gate was located, and where Salt Lake City built its diversion dam to direct water into the Big Cottonwood conduit in 1906. In 1952 both the canyon highway and Wasatch Boulevard were realigned, the latter to cross a new concrete bridge where the boulevard crosses the stream today.[1]

Some thirty years later the bridge was replaced by pipes and a large fill during the I-215 construction.

The water treatment plant was built in 1956–57. It was the city's second treatment plant, the first having been built in City Creek Canyon in 1954 following a threatened blacklisting of the city by the U.S. Public Health Service because of marginally pure water. While the City Creek plant solved that immediate problem, the need for further water treatment was recognized and led to the construction of the Big Cottonwood plant. Although it went into operation in December 1957, its dedication was not held until the spring of 1959.[2] The Granite Power Plant, which used to be a prominent feature at the entrance to the canyon, now is hidden behind the treatment plant and goes unnoticed by most canyon travelers.

The large flat above the mouth of the canyon on the south side, now covered with homes, was the site of the Civilian Conservation Corps (CCC) Camp F-38 in the late 1930s. It was built in 1935 and served the CCC until it was closed in July 1942.[3]

About a mile into the canyon, mile 2.8, the road passes through a dark grey slate formation. It was here the miners met in March 1870 to form the Big Cottonwood Mining District. The springs that were then known as Slate Springs are no longer apparent, possibly buried by the road construction that has taken place since then.

At mile 3.2 the highway makes a ninety-degree turn to the right. On the north side, the paved surface of an earlier road can be seen climbing up well above the stream, only to come back down about a quarter mile farther up canyon.[4] It may be difficult to imagine today, but at one time the bottom of the canyon at this point was difficult of access. If one looks at the steep cuts along the highway and can imagine the slope that once was there, one might appreciate why a dugway—known as Smoots Dugway—was built to bypass this part of the streambed. It is not known who Smoot was, but the route may have existed before the days of the Big Cottonwood Lumber Company. The first reference to the name that has been found was made in December 1872, after miners entered the canyon.[5] Smoots Dugway continued to be used well into the twentieth century before the slope was excavated to allow a new road to be placed closer to the stream; but that was long enough for it to receive pavement when the canyon highway was first paved.[6]

At the upper end of Smoots Dugway the highway runs between vertical quartzite cliffs and the stream. At the lower end of this very narrow section of the canyon there once was a small diversion dam for the Butler Ditch, which carried water down the side of the canyon to irrigate lands to the south. The ditch was later replaced by a concrete pipe, which can still be seen on the south banks today. This narrow part of the canyon was a popular spot for artists and photographers alike, with its one-lane dirt road huddling close to the overhanging cliffs and competing with the stream for the limited available space. Towering overhead was the imposing Youngs Peak, probably named after Joseph Young, who received control of the canyon in 1852. Today Youngs Peak is known as Storm Mountain. With the abundance of scenery in the canyon bottom, most travelers fail to look up and notice this equally scenic summit.

*

Stairs Power Station. At mile 3.9 the canyon highway curves around the brick building that is the Stairs Power Station. Today the power plant stands much as it did when it was constructed more than one hundred years ago, although the highway has been improved and widened several times. The original road up the Stairs may be seen if one walks along the north side of the stream from the lower end of the bridge above the power station. Remnants of the paved road can be followed into the depths of the gorge, where it ends at stone abutments that once supported a bridge across the stream. On the other side of the stream the old road disappears under the huge bank of fill for today's highway. Robert M. Jones's dam still hovers above the Storm Mountain picnic ground, and even though the reservoir has filled with a century's load of detritus, the conduit still runs through the ridge to the west and carries water down to the power plant. The reservoir is located on Mill C Flat, where the Big Cottonwood Lumber Company's Mill C once was located, and which was the site of Youngs Peak Lodge in 1888. Today's highway curves through a deep cut just above the picnic ground, the same cut, albeit considerably widened, that was made in 1916 so that mining companies could use motor vehicles instead of teams to transport their ores. Jones' Hump was above the cut, over the top of the spur, but so much rock has been blasted away that the deep notch at the hump no longer remains. The route of the road up to Jones' Hump can be followed through the trees opposite the picnic ground but no longer leads all the way to the top.

*

Mule Hollow. The drainage that flows into the canyon from the north just above the reservoir site, at mile 4.9, is now known as Mule Hollow. Since the flat on which the reservoir was built once housed Mill C and was known as Mill C Flat, it was perhaps inevitable that the drainage to the north would first be known as Mill C North Fork. Around the turn of the century it became known as Mule Cañon, then as Mule Gulch, a name that lasted four or five decades before it got changed to Mule Hollow.[7] Although mid-century residents from the area bemoaned the change, the Mule Hollow name gained permanence through its appearance on the USGS Mount Aire map in 1955.

*

Whipple Fork. At mile 5.3 a seasonal stream flows into Big Cottonwood Creek from the north slope. This is Whipple Fork, named after Nelson Wheeler Whipple, who had his last sawmill in the canyon bottom nearby and drew his logs from the fork. Whipple, who figured heavily in the chapter on the lumber industry in Big Cottonwood Canyon, was a squatter at Whipple Fork; that is to say, he never held title to any land in that vicinity. The men who did hold title were Richard Maxfield and Sam McNutt, both of whom homesteaded their lands some years after Whipple was gone from the area.

The Maxfields were long-time residents of Big Cottonwood and have been mentioned frequently in this narrative. Richard Dunwell Maxfield Jr. had been active in the canyon for many years, having filed claims with his father, Richard D. Maxfield, since 1880, when he was twenty-one years old. Eventually he settled at the upper end of Mill C Flat, where he built a home for his family and did some farming. He claimed water from Whipple Fork and Stairs Gulch as early as August 1895. After a number of years on the site he applied for and received a homestead certificate for eighty acres taking in all of Mill C Flat and the lower end of Mule Hollow. With the construction of a few cottages and campsites, the Maxfield residence became an attractive destination for visitors from the city. Daughters Josie and Lois catered to the guests, and by 1912 the place became known as Maxfield Lodge. When the two girls married, their husbands also took part in the operation, and by 1915 it was advertised as the Maxfield Lodge Resort, featuring furnished cottages and tents as well as trout and chicken dinners. They even had an automobile to meet guests at the end of the streetcar line in Holladay. After Richard D. Maxfield Jr. died in 1931 the two daughters continued running the resort until 1956, when they sold it and retired. During their years at the lodge they gave names to many of the surrounding features, including the renaming of Youngs Peak as Storm Mountain. Whenever the peak became clouded over, they said, it always meant a storm, and all the cloudbursts came out of Storm Mountain Gulch.[8]

The new owner ran a Swiss chalet type of restaurant until, four years later, a fire destroyed the kitchen. Almost immediately it was sold to real estate developers, who tore down the old lodge building and built the large stone and glass building that remains to this day. For a number of years it housed a restaurant, appropriately known as Maxfield Lodge, that was an attractive and pleasant place to dine. After the restaurant closed, a variety of tenants followed. In the process the Maxfield Lodge name was lost, and today very little remains of the original settlers and their enterprise.[9]

Sam McNutt was the kind of person people remember but never know anything about. He came from Ireland and was in the Wasatch as early as 1886, when he filed a mining claim in Mill B South Fork. In 1895 he filed a claim at the mouth of Whipple Fork, and there he found a home. He built a small cabin, part wood and part canvas, where he lived while he puttered about his claims. He would visit the Maxfields, his close neighbors, and play his fiddle in exchange for a few drinks. When Maxfield applied for a homestead, Sam McNutt did the same, and in 1909 he received a homestead certificate for 160 acres of land taking in all of the canyon bottom from Maxfield's land to Mill B North Fork. In those days canyon property did not make a man rich, but his property did provide McNutt the wherewithal to borrow money. He repeatedly placed mortgages on his land and somehow managed to get them released. Sam McNutt died of a heart attack in Salt Lake City on 7 December 1936.[10]

*

Broads Fork. This fork is on the south side of the canyon, with its stream flowing into Big Cottonwood Creek at mile 5.7. Its present trailhead, however, is at the S-turn, about a half mile farther up the canyon. The person who gave his name to this fork is not known, but it probably was someone who opened a logging trail into the fork while working for the Maxfield brothers to supply timber for Mill B. The logging slide and trail still exist and can be found today. On the downhill side of the stream crossing, today's Broads Fork Trail descends straight down the former slide for several hundred feet before making a right-angle turn to follow a more gentle grade. The logging trail, however, continued straight down the slope, paralleling the stream. It can be followed to within about a hundred feet above the canyon bottom, where it turns toward the east and makes a descending traverse toward the stream. Erosion has removed the very lower end of that trail.

The Broads Fork name has seen several variations over the years. In 1877 Nelson Wheeler Whipple recorded in his journal, "I went over the ridge into Broughts fork." In 1893 Frank K. Gillespie called it Brocks Fork when he filed an appropriation on its water. In 1894 the principals of the Big Cottonwood Power Company called the tributary Borck Fork when they bought the Mill B water rights and located the Mill B Electric Mill

Site at the mouth of the fork.[11] In 1908 it was called Broads Gulch in a series of claims, and the following January a miner called it Brodsfork. In 1913, Charles W. Hardy, engineer for the Brown and Sanford Irrigation Company, called it by today's name, Broads Fork, when he was surveying for a proposed dam and reservoir to impound its waters.[12]

*

The S-Turn. Today the S-turn in the Big Cottonwood road, at mile 6.2, is such a prominent and well-known feature that it tends to obscure some of the area's natural and historical interest. Mill B was located at the apex of the lower half of the S-turn. Mill B Flat, used first to stack lumber and later to stockpile ore during winter months, has been excavated for the lower half of the curve and filled for the upper half. The canyon road originally came up onto and across Mill B Flat, then across the stream to continue on the south side. After it passed the Mill B South Fork stream it crossed Big Cottonwood Creek again to climb steep grades on the north side at the stream level. Today remnants of Mill B Flat and the old canyon road may be seen in the center of the arc of the lower half of the S-turn. The old road had been paved before it was abandoned, and it can be seen headed for the stream crossing, where the bridge has been replaced by a fill over large pipes. Bridge abutments for the bridge across the canyon stream above Mill B South Fork can still be seen. The S-turn was built in 1932 as a joint venture between the Forest Service and Salt Lake County.[13] Before that time, the road followed some steep and precipitous grades variously known as the Cat Stairs, Wildcat Stairs, or Upper Stairs. There is evidence of an earlier road having bypassed this area by climbing the slopes on the south side of the canyon. A graveled roadway, with some stone retaining walls, can be seen on the south side of the stream at mile 6.7. It climbs as it heads down canyon until it overlooks Mill B South Fork, but from that point its route is not apparent.

*

Mill B North Fork. This fork surely was used to supply logs for Mill B. Although no documentary evidence has been found to support this conjecture, there are indications of logging trails to be found in the fork. Hidden Falls, a short distance up the Mill B North Fork stream from the upper part of the S-turn, was indeed hidden before the S-turn was constructed. At that time the only indication the traveler had of Mill B North Fork was its stream, which the road crossed then and crosses now at milepost 6. Only those who ventured to hike up the streambed were rewarded with a view of the falls. Josie Reenders and Lois Recore claimed to have named the falls when they were living in the canyon.[14]

*

Mill B South Fork. Mill B South Fork also provided logs for Mill B. The 1855 contract Brigham Young had with the Maxfield brothers to supply logs for Mill B specifically called out the first south fork above Mill B as one of the sources.[15] Today the observant hiker will notice a number of logging trails to one side or the other of the hiking trail. The upper part of Mill B South Fork, the area around and above Lakes Blanche, Lillian, and Florence, was very popular with hikers in the 1880s, as it is today. The lakes were known as the Three Sisters, the drainage around and above them was called Hidden Valley, and the peaks towering over them were the Pillars of the Wasatch. The names were popularized by artists Henry L. A. Culmer and Alfred Lambourne. Lake Blanche was named for Miss Blanche Cutler by Harry Squires, an artist friend of both Culmer and Lambourne who spent much time in the mountains with them. Lambourne named the middle lake Lillian after his daughter, later Mrs. Andrew Walker; and Culmer named Lake Florence after his black-eyed daughter, later Mrs. David J. Varnes. Culmer later suggested that perhaps the lakes had been named before, but that the artist had a great advantage over others because, "in cleaning his palette, he paints in large and enduring letters upon a rock the name with which he has christened the spot."[16] That may be true, but the fact that he and his colleague publicized the names in their writings probably helped more than any name painted on a rock.

Lambourne, especially, was taken by this area. It was he who kept using the term the "Pillars of the Wasatch," and he was responsible for the naming of the ever-popular peak known as the Sundial. He once called it a "vast natural sun-dial," but more often referred to it as a gnomon, one whose moving shadow upon the snow "measures as upon a dial the passing moments of the untold centuries."[17] So much did Lambourne love Hidden Valley that in 1893 or 1894 he took occupancy of a cabin near the shores of Lake Blanche on 21 June and stayed there all summer, roaming and exploring the valley, the surrounding peaks, and the many little glens. His isolation was his joy, despite having his family in the city. He had a similar venture when he spent fourteen months attempting to homestead Gunnison Island in the Great Salt Lake. From his island he could see in the distant haze the heights of the Wasatch, and now, from his summer home in Hidden Valley, he could look out to the northwest and see the lake and its islands. Two remote spots, so vastly different, yet both taking a special place in his heart. On 20 September he packed his meager belongings and wrote, "After today the smoke will arise from my cabin chimney no more." He recorded his pleasures of that summer in his book *A Summer in the Wasatch* so that all of us who follow him may share them too.

Another lake in Hidden Valley, about a mile southeast of Lake Blanche, under the ridge at the head of Mineral Fork and in one of the glens explored by Lambourne, was mentioned as early as August 1872. In February 1874 the miners in Mineral Fork met and decided that the lake should be known as Lake Ada May, and it was shown by that name on a rather distorted map of the mining district published the following month.[18] It is not known who Ada May was, but, while the lake remains, its name failed to survive.

It should be noted that the Three Sister lakes that visitors in the nineteenth century saw were not quite as we see them today, for that was before the dams were built enlarging them. That story was covered in an earlier chapter.

*

Mineral Fork. The lower end of this fork is very narrow and deep, making access particularly difficult. When prospectors began to work there, the fork apparently had no name. It is not likely that loggers had worked this fork, since it was midway between Mill B and Mill A, and good stands of timber were available much closer to both mills. The first mining claims were identified as being in the Second South Fork of Mill B. The earliest such claim was recorded on 6 December 1870, and that name for the tributary continued to be used until 2 August 1872, when the present-day name of Mineral Fork was used by C. J. Johnson for his Bright Point Lode.[19] The actual source of this name is not known, but the name was commonly used where mining or prospecting activities were taking place, probably reflecting the hopes that much wealth would be drawn from the prospect holes that were appearing there. While Mineral Fork did support numerous mining ventures, it was not to provide the yields that came out of Cardiff and Silver Forks to the east or the Little Cottonwood Mining District to the south.

There is evidence of at least three generations of trails and/or roads into Mineral Fork. The first was, of course, only a trail. It began at Big Cottonwood Creek a short distance above its confluence with the Mineral Fork stream. The trail went down canyon, rising above the creek, until it reached the Mineral Fork stream, then turned south and climbed along the left bank of the tributary for several hundred feet. When the slope became too steep, the trail turned toward the southeast and climbed steeply, turning to a southerly direction as it gained elevation. Parts of this trail can be seen today. About 100 feet below the first switchback on the present road the trail can be seen dropping toward the northwest. It can be followed with relative ease until it disappears into the dense willows along Big Cottonwood Creek. In the other direction, the trail went up through the first short switchback of the road, but all traces have been obliterated by the road builders. It can be picked up again at the east end of the short switchback, where it turns and goes directly up the slope, reaching the road again about 135 feet below the upper end of the long switchback. A short segment of the trail can be seen along the road and in the woods farther up the canyon. Many upper parts of this trail show the characteristic U-shaped depression created by pack animals and the dragging of rawhides.

When a wagon road was cut into Mineral Fork, it followed the trail for much of the distance, except where the grades were too steep. This is why the long switchback was graded, although it too had some sections that were too steep when the third-generation, the truck road, was graded. The second-generation road can be seen about midway up the long switchback, where it climbs steeply to the upper road. The second-generation road continued on the east side of the creek, following the present road and the Silver Mountain Mine trail. A short distance, perhaps 200 yards, from the road, the latter trail makes nearly a right-angle turn to the left; at this point the road continued to the south, staying on the east side of the stream. With care, it can be followed almost as far up as the Wasatch Tunnel.

Above the Wasatch Mine, where the canyon bottom rises rapidly to the upper basin, many switchbacks can be seen climbing the steep slope. The first of these was graded about 1899. The first mention of this road was in a mining claim filed on 2 September 1899. The claim was located "on the west side of Mineral Fork at or near what is called the Zigzag Road."[20] Other claims noted the "Zigzag wagon or cart road." Can there be any doubt about what road was described here?

The third-generation road was built in 1936 to provide truck access to the Wasatch Tunnel and the Regulator Mine.[21] Ore bins were located on the south side of Big Cottonwood Creek at the foot of the road. Debris from these bins remains and can be seen between the Mineral Fork road and the creek. While it may seem that the bins are very low, it must be remembered that the main canyon road was then down at the stream level, not as high as it is today. Some remnants of old highway paving can be seen on the north side of the stream. During the latest period of operation ore was brought down to these bins in ten-wheel dump trucks. The usual procedure was to back the truck up or down the first short switch-back and the longer second one, where the stream from the springs now flows over the road. At one time that corner had two bridges built with heavy timbers crossing the Mineral Fork stream so that the road made a great sweeping turn, allowing trucks to make the corner without backing, crossing the stream twice in the process. Part of one of these bridges re-mains, while the rest moved downstream during heavy runoff years in the 1970s and 1980s. A driver from the 1951–52 period insisted he always backed his truck over the second switchback and never turned around on the bridges. The corners of the switchbacks still have high mounds of dirt to protect against trucks backing too far and making an unscheduled trip down into the gorge.

Where the road leaves the gorge and the full expanse of the upper fork comes into view it turns and crosses the creek to continue up on the west side. Just before the stream crossing there is a trail going off to the left, southeast. This is part of the second-generation road and the Silver Moun-tain Mine trail, providing access to the Silver Mountain Mine high on the east slopes of Mineral Fork. This was a fine, if somewhat overgrown, trail until the mid-1970s when, during heavy runoff, the stream left its usual bed and ran down the trail, eroding very deep ditches that are only now, more than twenty-five years later, beginning to show signs of filling in. About a half mile from the road the trail climbs out of the forest and fol-lows a steep traverse on the north side of the gully, finally reaching some cliffs and a high waterfall that has a heavy flow in the spring but only a trickle later in the year. The trail continues above the low cliff but offers no clue as to how pack animals were taken over this minor obstacle. It stays on the north side of the drainage; however, the path is steep the rest of the way to the mine dump.

The Silver Mountain mine was recorded in January 1871. It was called Silver Mountain Mine No. 2 to differentiate it from a Silver Mountain Lode already recorded at the head of Silver Fork. It was worked with in-different results for nearly ten years before January 1880 when the Silver Mountain Mining Company was organized. The British Tunnel Claim was recorded and work began on the tunnel we see today.[22] When the tunnel reached the lode, a big strike of silver-lead ore was reported. In 1883 the company built a road up the crude trail that existed and began shipping ore. The company worked the mine for several years before turn-ing it over to lessees. Its better days were behind it well before the turn of the century.[23] As a point of interest, the tunnel had a track that was made of planks with a metal strap nailed on top. The ties were deeply notched so the planks could fit into the notch and be held there with wedges driven next to them. Unfortunately, none of these artifacts survive today; the ties and rails, being combustible, were used by visitors who felt the need to have a fire.

*

Meridian's Cabin. About a quarter mile above the Mineral Fork road, just past milepost 8, there is the remnant of a stone cabin above the north side of the highway. While it is not readily visible to motorists traveling up or down the canyon, occasionally a person will happen to look up, notice the angularity and odd color of the remaining wall, and wonder what it was all about. The cabin was built by Antonio Alexander Meridian, who, with his wife, Anna, immigrated to Salt Lake City from Russia in the late 1920s. They settled on south Main Street and opened a business called the Monumental and Sculptor Works. In Big Cottonwood Canyon Meridian found a source of stone to use in his business, and in July 1929 he filed a claim for the Meridian Quarry Lode.[24] The stone had a characteristic purple tint, which may be seen in the remnants of the cabin. He modified his claim in 1936 to increase its size, had it surveyed, and applied for a patent.[25] By the time the patent was issued in June 1941 the Meridians had been conducting their Main Street business for more than a dozen years, during which time they had built the cabin on the north side of the canyon high above the road.[26] A winding trail led up to the cabin from the road, which at that time was in the canyon bottom on the north side of the stream. Then, on 1 July 1943, Anna Meridian died, leaving no survivors other than her husband. Except for a brother, Nicholas, who lived in Ogden, and incidentally ran the Intermountain Monumental Company there, Antonio was now alone.[27] While he continued to run the business, renamed the Meridian Monument Company, and maintain his home in the city, he began to spend more and more time at his quarry. Canyon residents from the 1940s remembered him as being somewhat reclusive. In 1949 or 1950 he closed his business and moved from his home to a room at the Hotel New York on Post Office Place in the city. There he died on 29 October 1952 at age seventy-four.[28]

His abandoned cabin stood until the early 1960s, when the new canyon highway was built at its present location. The necessary cut into the hillside undermined the outer walls of the structure, causing them to collapse and fall, leaving only the rear wall standing. On the other side of the stream there is what appears to be a stone retaining wall for a segment of a road, a road that goes nowhere, as it leads to a cliff. This was where Meridian prepared the stone he took into the city. At two places he carved

"Private Property" and "Keep Out" on the quartzite rock, inscriptions that are faint reminders of the man who once lived and worked at this place.

*

Burnt Flat. About a quarter mile above Meridian's cabin was a large flat area in the bottom of the canyon. While a flat can be seen between the road and the stream today, it is not quite the same as it was before the mid-twentieth century. When the present road was graded in the 1960s much of the dump material from the Maxfield Mine was redistributed, changing the appearance of the canyon bottom in that area. However, a century earlier the flat was used as a staging area for lumber cut in the sawmills above that point. Lumber was hauled from the mills and stacked on the flat to await the teamsters who would haul it into the city. One day in 1864 or 1865 a fire burned all the lumber stored there and the place was thereafter known as Burnt Flat.[29] During the mining era the flat was used as one of the staging points for ore. In wintertime, ore would be hauled from the mine to Burnt Flat or to Mill B Flat on sleighs, then dumped to await teamsters with wagons to haul it the rest of the way to the smelters. In one instance, a miner, Frank Trager, who was fatally injured in the shaft of the Silver Key Tunnel, was buried on Burnt Flat.[30] The most recent reference found for the name was in August 1936 when Antonio Meridian, in an amended notice for the Meridian Quarry Lode, used "the foundation of an old house on Burnt Flat" as a reference point.[31]

*

Maxfield Mine. At mile 8.6 a pipe sticks out from the bank on the north side of the highway and pours a considerable stream of water into the roadside ditch. This once was the tunnel for the Maxfield Mine, described in an earlier chapter. The mine itself was up Mill A North Fork, which meets the road about a hundred yards east of the tunnel.

*

Mill A North Fork (Mill A Gulch). The mouth of this fork, at mile 8.7, was considerably excavated during the road reconstruction in the 1960s. Mill A, located near this point, had a millpond on the stream, one that was later used to drive machinery at the Maxfield Mine. When prospectors moved into the canyon they repeatedly referred to a "well known and prominent Limestone butte" at the head of the west branch of Mill A North Fork. Actually the butte stands above the slopes on the north side of Big Cottonwood Canyon, but the easiest access was by way of the west fork of Mill A North Fork, then using a trail that climbed to the south ridge and, on the other side, dropped down to the butte. If one stands at the mouth of Mill A North Fork and looks up the north slopes, down canyon from the fork, there can be no doubt about where this butte is located. It may not be well known today, but it is prominent.

In June 1878 David Maxey and Richard Greenway filed a claim for the Peter Cooper Lode, located near the butte, and for some time thereafter the limestone formation was known as Peter Cooper Butte.[32] As other miners came into the area, the name was lost, and it went back to being simply the well-known and prominent limestone butte. In later days it was known as White Butte, and more recently as Poison Dog Peak.[33] The butte had another claim to fame, for in its upper flanks on the east side is a cave that was well known among the miners.[34] It has two cavities. The first is large enough for a man to stand erect and houses a cast-iron stove with a stovepipe going out a small manmade hole next to the cave entrance. The adjacent cavity has a raised floor, providing a natural sleeping loft for two or three people, and was undoubtedly used for that purpose by more than one miner over the years. Today the stove is in advanced stages of disintegration, but the cave, while no longer well known, is still there. Unfortunately, trails that took the miners to it have all but disappeared, making access quite difficult.

In November 1878 the Afghan Claim was recorded by John McDonald, who happened to be the mining recorder of the Big Cottonwood district at the time.[35] His claim was located high in the west fork of Mill A North Fork, only a short distance below the ridge. Four days later he filed the Afghan Spring and Water Claim, located some 1,500 feet below his Afghan discovery. He claimed "all water furnished by the spring for use of mine and household purposes and concentrating ore," as well as five acres of land.[36] When one climbs up the west fork, into what is now known as Maxfield Basin, there is little indication of any water, but high in the fork, where the bottom becomes wider and travel becomes easier, there are a good number of springs bringing water to the surface. It was one, or perhaps several, of these that John McDonald named Afghan Spring in 1878.

*

Argenta. At mile 8.8 the old canyon road crossed the creek on a bridge whose abutments still remain. This is where the town of Argenta was located. In 1872 the mining activity in the South Fork and on Kesler's Peak was such that it seemed reasonable to have a community closer to the center of activity than was Silver Springs. At that time the site of the future town of Argenta was known as Camp Carr, probably named after

J. I. Carr, who, in company of James F. Wardner and Peter Fischer, was working several claims in the area.[37] On 27 July miners met at Camp Carr and agreed to establish a town site on the spot. A plat was prepared, showing streets to be forty-five feet wide and lots twenty-five feet wide, extending back on the south side to the mountain and on the north side to the creek. Permanent settlers were to receive lots gratis. One of the attractions of the location was the strong flowing spring on the south side of the canyon just above Argenta. It was given the name Diamond Spring at the same time the settlement was named.[38]

Almost immediately a move developed to form a new mining district centered at Argenta. A special meeting was held on 11 August 1872, when the Argenta Mining District was formed, taking in the entire western half of the Big Cottonwood Mining District, from the South Fork to the western face of the mountains.[39] The recorder for the new district was located in Argenta, while the now-fractured Big Cottonwood district kept its headquarters at Silver Springs. On 10 July 1876 the new district was absorbed back into the Big Cottonwood Mining District, returning the focal point for all mining activity to Silver Springs. In 1878, however, when recorder Richard Greenway died suddenly, his successor, John McDonald, moved the recorder's office to Argenta, where it remained for the rest of the century. In 1878 the town could boast a hotel, saloon, store, and post office, but it probably never had a very large population. In March 1880 it had but one resident, James T. Monk, the deputy mining recorder, although later that year a Mrs. Graham took up residence there with five good milk cows and proposed to provide buttermilk and cream as well as pies and cakes for the miners during the season.[40] Monk became the district recorder that same year and held that post for seventeen years, with Argenta as his headquarters. The Argenta post office continued to be designated as a polling place for elections as late as 1904.[41] Today nothing remains to be seen except for Diamond Spring, which still brings forth a heavy flow of water.

*

Butler Fork. This drainage is on the north side of the canyon at mile 10.1. It is named after the Butler brothers, who had a steam sawmill operating in the fork for one season in 1877. They then moved their mill to the head of the canyon but left their name behind for us to know and use.

*

Cardiff Fork. When driving up the Big Cottonwood highway today Cardiff Fork comes into view at mile 10.8, where the "Meeting of the Glaciers" sign stands. From this point one can see up Big Cottonwood

Canyon as well as up Cardiff Fork to the south. This is "The Basin," referred to by Lorenzo Brown in his journal in January 1855 and probably named by Kesler, Little, and Wells during their exploration for the Big Cottonwood Lumber Company the previous year.[42] It is all very open and visible. But things are not as they appear, for the Cardiff Fork drainage goes behind the glacial moraine on the west side of the Basin and flows into Big Cottonwood Creek well downstream. Imagine the plight of the early traveler coming up the canyon: just below Butler Fork, at milepost 10 on today's highway, he found the confluence of two streams, both coming out of equally deep and steep ravines. Each branch leads into a large continuation of the canyon, one towards the east, the other towards the south. What would be more logical than to call the latter the South Fork of Big Cottonwood? That did happen, but not right away. Mill D was located in the main canyon above today's Cardiff Fork road. Surely some of the logs for the mill were brought out of Cardiff Fork, but we don't know what, if anything, that southern tributary was then called. We do know that in the summer of 1867 Nelson Wheeler Whipple negotiated with Feramorz Little, who was running Mill D, for some timber land "at the Basin in Big Cottonwood Kanyon."[43] Had there been an accepted name for the fork heading south, Whipple probably would have used it; he was in the habit of writing names of places and appending a qualifying "so calld" to show that the name wasn't his. He did build his mill there and by late summer of the following year, 1868, was producing lumber and shingles. Except for his not-too-near neighbors at Mill D, he, his sons, and his hired hands were the only ones there. They knew where they were, so they didn't need names to describe the place. As prospectors began roaming the mountains, names became more important. They needed to identify where in these wide mountains their claims were situated. To do so, they used the nearest mill to identify their location.

Just a mile down canyon from the confluence of the two forks of Big Cottonwood was Mill A, so Mill A was used to identify the major fork branching off to the south: Mill A South Fork. The first claim to use this name was the Kimball Brothers Lode, recorded on 27 August 1870. The name was not universally recognized, for when the soon-to-be-acclaimed Reed and Benson claim was recorded several weeks later, on 13 September 1870, the location was noted simply as on the south side of Big Cottonwood Canyon. And when Whipple's sons—Daniel, Samuel, and Edson—tried their hand at prospecting and located a claim a short distance up the canyon on the east slopes of Kesler's Peak, they modestly called the tributary Whipple's Fork. Others must have recognized this as a major fork of the canyon, for as early as November 1870 the Union Pacific Lode was recorded as being in the South Fork of Big Cottonwood Canyon.[44] Both names, Mill A South Fork and South Fork of Big Cottonwood Canyon, continued to be used for several years, but eventually the South Fork freed itself from all claims of Mill A. It continued to be known as the South Fork of Big Cottonwood to the miners, but the U. S. Geological Survey settled with the name Mill D South Fork on its 1907 Cottonwood Special map and continues to use that name until this day. In spite of that, the fork is commonly known as Cardiff Fork.

The Basin is called Reynolds Flat on contemporary USGS maps. But it has been known by a number of names over the years: Mill D Flat and Mill G Flat, to name a few. With Mill D located on the flat, it was inevitable that it would be known at times as Mill D Flat. Mill G was the mill built by Nelson Wheeler Whipple across the Basin from Mill D. Curiously, the names referring to the mills were used by miners around the turn of the century, well after the mills had been abandoned.

The Reynolds Flat name came from Ammon Reynolds, a late nineteenth-century Murray butcher who grazed his cattle in Big Cottonwood Canyon. Possibly to house himself and his herders while they were in the canyon with the herd, he built a cabin on the north side of the Basin. Like many others before him, he succumbed to the lure of precious metals and did a little prospecting. In 1894 he filed his only claim, it being about a fifth of a mile north of Big Cottonwood Creek in what he called Mill G North Fork, not far from his cabin. He may have worked the claim briefly, but by 1903 not only Reynolds but also his cabin were only a memory. That year two claims were filed in the vicinity of Reynolds's old claim, both being "where Reynolds cabin once stood."[45] Although the butcher's activities in the canyon were rather obscure, he was known well enough among local residents for his name to survive to be used on the USGS map. It was also assigned to Reynolds Gulch, where his cabin once stood, and to Reynolds Peak at the head of the gulch.

Another interesting set of events in or near the Basin ended in a sad episode in Big Cottonwood history. In 1870 one of the mining men who came into the Wasatch was Captain Thomas W. Bates, a "well known mining engineer and capitalist."[46] Since he later appeared in the company of men associated with the Emma Mine transaction, it may be assumed that he knew them well, possibly before he came to Utah. He first appeared in the Alta mining scene in August 1870 when he filed a claim in the company of six other men, including one of the Walker brothers.[47] He continued to be active in both Big and Little Cottonwood for three years. In the spring of 1871 he and a group of New York men filed a claim on Kesler's Peak; although filed as the Bates & Selover Mining Company, the mine became known as the General Baxter, after one of the claimants, H. H. Baxter. General Baxter, it will be remembered, was associated with Trenor Park in the purchase of the Emma Mine from its former owners and its sale in England. Another claimant was James M. Selover, a New York financier and Bates's partner in the mining company.[48] To support his operations in the South Fork Bates erected a cabin, known as the Bates House, that was soon accompanied by several others. Although Bates had left the Wasatch by 1874, his cabins remained to be used by others and continued to be known as the Bates House. The location became a local center of activity and in 1878 a general merchandise and miners outfitting store was opened there. The following year the collection of cabins became known as Batesville and the post office at Argenta was given the Batesville name.[49]

Winter weather generally forced those miners with small mines to leave the mountains, so the Batesville cabins were usually empty during the cold seasons. An exception to this generalization was the woodchoppers, those men who were cutting timber to supply sawmills or the demand for logs in the city. They found it more convenient to cut the trees in the wintertime, when the logs could more easily be taken off the slopes and stacked near the road. One such worker was Charles Taggart. Usually the loggers sent their families to lower elevations during the winter, but Taggart chose to keep his wife and five children with him, and all of them took residence in a cabin he built at Batesville. One night in mid-February 1882 an avalanche tumbled down from the slopes and in an instant destroyed Taggart's cabin, taking with it the lives of the seven occupants. The incident was not discovered for several days, but then a rescue effort was begun. Taggart and his wife were found in their bed with their infant child. The other four children were found in another bed. All died in their sleep, apparently unaware of the tragedy that befell them. Their bodies were placed in rawhides and taken down the canyon for burial. The same avalanche also crushed Batesville's five houses and two stables, thereby ending the life of that community as well as the lives of the Taggart family members.[50] With the demise of Batesville, the canyon post office again assumed the name of Argenta.

There are a number of names in Cardiff Fork that should be mentioned. On the west side at the mouth of the fork is Kessler Peak, the name coming from Frederick Kesler, as described in the narrative on the Big Cottonwood Lumber Company. The summit seen from the main canyon road, however, is the lower of two summits, and was known as Reeds Peak. Will C. Higgins, of the *Salt Lake Mining Review*, in describing the Big Cottonwood Consolidated Mining Company in 1911, wrote that Reeds Peak was the lower neighbor of Kessler Peak, writing that the company's properties were "overshadowed by Reed's peak, and higher in the clouds by Kessler's peak."[51]

The Reeds Peak Mining Company, formed in 1907 by a group of men from Park City, was located on the east side of the lower peak. At first a tunnel high above the canyon floor was worked, but then the company moved down and drove a tunnel immediately below the South Fork road to tap their lode at great depth. It did them no good, however, for the tunnel in all its 1,700-foot length encountered no great amount of ore.[52] The peak probably had its name long before this company was formed, named after Franklin Reed, of Reed and Benson Mine fame. This cannot be firmly substantiated, but Reed can be linked to several claims in this vicinity in the 1870s. One of them was the McDougald Lode of 1871. One of the seven men involved in this claim was Henry C. Goodspeed, a friend and associate of Reed's. It is believed that the McDougald Lode is one and the same as the Reeds Peak Mining Company's upper mine. That same year a trail was built to the mine, the same trail used by the Reeds Peak company and the one that remains today as part of the Kessler Peak North Trail. The mine must have had great promise, for Goodspeed himself filed claims for a McDougald Mill site and a Goodspeed Tunnel to work the lode. In 1876 the McDougald Tunnel claim, including Franklin Reed's name, was filed.[53] Goodspeed continued to work the mine through 1880, at which time he was vice president of the McDougald Mining Company.[54] After that time the mine was abandoned and was relocated by other parties several times. One relocation was called the Eagle Nest Lode, the name certainly inspired by the mine's location on the face of a high cliff with a narrow cliff-hanging trail leading to it.[55]

Farther up the South Fork a spur, known as Montreal Hill, juts out into the fork from the west. The name comes from the Montreal Lode, filed by H. C. Goodspeed, among others, in July 1871. Within a year he joined Franklin Reed to file two tunnel claims to work the lode. The spur was called Montreal Hill as early as June 1874.[56] On the north side of Montreal Hill is a flat, several hundred feet above the canyon floor, known as Vina Flat. The name comes from the Vino mining claim filed in 1872. It was followed by two Vino tunnel claims. In 1875, when the second tunnel claim was filed, the flat was known as Montreal Flat, but the following year it was called Vino Flat. Then the Vino name suffered a mutation and became Vina. The new name appeared in a newspaper report in 1878 and in a claim notice in 1881. The Vino name was never used after that, and the flat is known as Vina Flat to this day.[57]

*

Mill D North Fork. At mile 11.2 the Mill D North Fork drainage meets the canyon road. Here was the traditional trailhead for Dog Lake and Lake Desolation until the early 1990s. At that time the Forest Service built the alternate trail that starts about a quarter mile down canyon and traverses the north slopes to enter this fork, thereby bypassing the cottages on the east slope of the fork. It is likely that the 1854 scouting party for the Big Cottonwood Lumber Company came into Big Cottonwood by way of this fork. After Mill D was built, much of its timber came from this drainage. Trees were often cut during the winter, when the snow was deep, leaving tall stumps standing in the summer months after the snow melted. The two lakes in this drainage are sumps; that is, they have no surface drainage. The origin of Lake Desolation's name is not known, but the Dog Lake name certainly came from the presence of salamanders, also known as dog fish.

*

The Spruces. The Spruces campground, at mile 11.5, was a Forest Service operation almost since the day the Wasatch National Forest came into being. The national forests had their origin in an appendage to an act passed by Congress in 1891. It was a common practice then, as it is today, for congressmen to attach to an important act an enabling paragraph for their favorite pork-barrel project to benefit their constituents. Such attachments are not always bad, for in this case it gave the president of the United States the power to declare forest reserves, which ultimately became national forests.[58] This addition was made at the insistence of the secretary of the interior at a last-minute conference and was passed during the closing hours of the congressional session, without having been considered by either the House or Senate.[59] The act, of course, was not universally embraced. When President Grover Cleveland added 21 million acres to the national forest reserves in 1897, doubling the total area, his action was met by a storm of protest. Perhaps that discouraged further use of the act for a number of years, for when Theodore Roosevelt took office in 1901 the total forest reserve area was only 46.4 million acres. Roosevelt, however, used the presidential power of the act to increase forest reserve area more than fourfold during his administration, raising it to a total of 194.5 million acres. So much did the senators from the northwestern states resent both the act and the president's use of it that they attached a rider to the Department of Agriculture appropriations act, withdrawing the president's proclamation powers in the states of Oregon, Washington, Idaho, Montana, Colorado, and Wyoming, reserving to Congress the right to create forest reserves in those states.[60] The president signed this bill into law, but before he did so he issued proclamations setting aside thirty-two additional reserves, containing 17 million acres of forests, in the six states enumerated in the act.

At the beginning of the twentieth century Utah had but two forest reserves: Uintah and Fish Lake.[61] In 1900 the Department of the Interior withdrew from public entry all unappropriated lands in the Wasatch Mountains surrounding Salt Lake City. While there was mention of placing this land into a forest reserve, the action was actually taken on behalf of the city to protect the watershed of the Salt Lake County canyons as well as that of Utah Lake.[62] Then, in 1902, the Department of the Interior withdrew large tracts of land extending from Cache County to Utah's southern border, encompassing nearly 5,000 square miles, until the creation and extension of forest reserves could be thoroughly investigated.[63] As might be expected, this action drew the ire of Utah's governor and its congressional delegation. "Think of it," Governor Heber Wells said, "five thousand square miles of land, three million acres...." Congressman George Sutherland had a long interview with Director Charles D. Wolcott of the U. S. Geological Survey, while Senator Thomas Kearns discussed the matter with the president. But all was to no avail, for later that same year the president issued a proclamation establishing the Sevier Forest Reserve, encompassing the Tushar Mountains and the Markagunt and Paunsaugunt Plateaus.[64] Many other forest reserves followed, including the Salt Lake Forest Reserve, taking in 95,440 acres east of the city, established on 26 May 1904. That same month saw the establishment of the Grantsville reserve, and in August 1906 the Wasatch reserve was created. In 1908 these three were merged to form the Wasatch National Forest.[65]

Forest reserves were initially administered by the Department of the Interior through the General Land Office. That department had little money, few personnel, and limited authority to deal with the timberlands under its control. In 1905, however, the forest lands were transferred to the newly created U. S. Forest Service in the Department of Agriculture, at which time the management of the lands began in earnest. The new service immediately undertook to establish several forest nurseries where trees could be started and then transplanted to slopes that had been denuded by logging, fires, or grazing. One such nursery was placed in the

Wasatch Mountains, in Big Cottonwood Canyon. The chosen site was near Mill D Flat, and on 6 June 1906 President Roosevelt issued a proclamation setting aside 160 acres, a quarter section of land, for the nursery. The plan was to raise trees in a lathhouse for one year, then set them out in nursery rows for two years, after which time they would be transplanted to mountain slopes. Some experimentation was also to take place in an attempt to determine which trees would be best adapted to the region. This probably explains why a number of species of trees not native to the Wasatch are found in the vicinity of the Spruces. The work at the nursery went forward zealously; in the spring of 1910 it was reported that 19,000 trees had been shipped to destinations away from the Wasatch and that about 50,000 trees would be planted in Big Cottonwood and Parleys Canyons. The following year it was reported that the nursery held 5.9 million trees of different ages, from seedlings to three-year-olds, mostly Douglas fir. Some 250,000 were planted in Big Cottonwood Canyon alone that year.[66] The nursery continued to operate at least until 1918 and certainly contributed immensely to the reforestation of the Wasatch.

In this second decade of the twentieth century the Wasatch was receiving increased pressures from those visiting for recreation. The advent of the automobile and the operation of several motor stage lines into the canyons gave the public access it had never enjoyed before. And this produced an impact upon stream sanitation. Ever since Salt Lake City began using Big Cottonwood water for culinary purposes, sanitation had been a primary concern. City and county officials made repeated inspections of the camps within the canyon, but in 1915 they found it necessary to advise residents to boil water used for drinking purposes. The blame for the problem was placed on the mines, horse-drawn ore wagons, and campers who disregarded ordinary rules of sanitation. Shortly thereafter the Forest Service agreed not to allow campers in the canyon forests without special permits from the county health department. If this ruling was implemented it may not have been enforced, for three years later Forest Service officials again threatened to place camping under permit.[67] Instead, it undertook to develop campgrounds with adequate sanitation and convenience facilities. Using the old nursery site at the mouth of Days Fork, the Forest Service built a campground in 1919, installed toilets and camp fireplaces, and erected twenty tents, ten old army tents furnished by the governor and ten government tents in its own possession. It was called the Community Camp and was an immediate success. Its tents were in constant use. The following year, the supervisor of the Wasatch

National Forest approached the Rotary Club, asking it to place a tent in the camp. The response was a declaration that the Rotary Club was not a one-tent organization; it appointed a committee to investigate and act, and within two weeks the club was the sponsor of twenty-five tents.[68]

In 1921 the Utah Outdoor Association was organized, a joint venture of the local Forest Service office and the Commercial Club. Its intent was to promote outdoor recreation and make the Wasatch National Forest a stimulus to the health and enjoyment of the people. To this end it began to run the former Community Camp, now known as the Utah Outdoor Camp. It even provided weekend bus service from the city to the camp and a store where food and fresh milk could be purchased at Salt Lake prices. Activities included hiking, bonfires, orchestra concerts, dancing, and lectures by University of Utah professors. In July 1922 there were between 125 and 150 people camped there, and plans were made to enlarge the camp to accommodate 800 people.[69] In 1927 the camp boasted fifty-six tents, numerous cabins, the store, playgrounds, and an open-air theater. The camp continued to be sponsored by the Utah Outdoor Association at least until 1938. By that time it had grown to include tents pitched over wooden floors, cabins, facilities for baseball, horseshoe pitching, and volleyball, as well as a small swimming tank. In the winter of 1936–37 a ski jump and toboggan slide were added at the mouth of Days Fork, and tennis courts were flooded for skating. A large log shelter, open at both ends and having a large fireplace, tables, benches, and a lunch counter, was built for both winter and summer activities. Most of these improvements were made with CCC labor, and many of them remain to this day.[70] The camp was later named the Spruces and continues to be as popular today as it was in the 1920s.

*

Days Fork. Days Fork, being close to the location of Mill D, was a good source of timber for the mill. Exactly who gave his name to this tributary is not known, but the name was used from the earliest days of the mining boom. On 14 September 1870 the Brilliant Lode was recorded as being situated at the head of Days Fork.[71] During the late 1860s James M. Day was active in mining across the divide in Little Cottonwood Canyon, but there is no evidence that links him to Days Fork. Day was a fairly common name in the Mormon community, and it may well be that one of the contractors providing logs for the Big Cottonwood Lumber Company mills had an employee named Day who opened a trail or logging slide in the canyon.

*

Greens Basin. This beautiful spot is on the south side of the canyon, above the upper end of the Spruces campground, about mile 12.2 on the canyon highway. It most likely was named after Alvin W. Green, who ran a sawmill directly below the basin. Above the basin is a spring that flows year-round and served sheepherders for many years. A faint trail climbs out of Silver Fork to the slopes above Greens Basin, and other trails drop down to this spring. An inscription on a large aspen tree along the trail reads, "David Probst 8/12/41 Lots of Bear," and has an arrow pointing down the trail. Others include "Herding Sheep for David Probst," and "Lynn Epperson 8/19/41 Midway Utah." This is interesting because it places sheep in Big Cottonwood Canyon as late as 1941.

Grazing was taking place in the mountains as long as men were working there. At first it was only the animals used by the workers, but as the loggers moved out the herders moved in, further decimating the vegetation. Also, as if cattle grazing were not bad enough, the sheep herds moved in with a tenfold increase in numbers of animals. It was said that bands of sheep on the mountainsides could be spotted from town by the dust clouds they raised.[72] A nineteenth-century sheepman remembered that when sheepmen used the tops of mountains as a trail in driving sheep, the whole mountaintop became a dust bed.[73] Campers complained that the streams were befouled by sheep all the time. In 1897 it was said that one flock of 2,000 sheep crossed the stream at Mill D three times in a single week.[74] When water purity became a greater concern, an attempt was made to limit grazing. In 1916 the grazing authorization for the Wasatch Forest Reserve was cut by 2,000 head of cattle and 25,000 head of sheep; but that still allowed 12,600 head of cattle or horses and 61,100 head of sheep in the reserve. By 1924 it was stated that grazing of livestock was prohibited on the watersheds of Big Cottonwood, Parleys, and City Creek canyons.[75] In spite of this, as late as 1946 the Forest Service was issuing permits for limited grazing in the watershed canyons. It allowed 1,200 sheep to graze at Silver Fork during the period 26 July through 31 August, for which it charged ten cents per head per month.[76] Above Greens Basin and just over the ridge from Silver Fork evidence of this sheep grazing can still be seen. In 1954 all domestic animals were banned from Parleys, Big Cottonwood, and Little Cottonwood Canyons.[77]

*

Beartrap Fork. This drainage is on the north side of the canyon at mile 12.5. In 1860 when Sir Richard F. Burton visited Salt Lake City, Feramorz

Little, at that time one of the principals in the Big Cottonwood Lumber Company, took Burton on a trip to Mill E. In his writings, Burton tells us the source of this fork's name. "On our left," he wrote, "in a pretty grove of thin pines, stood a beartrap. It was a dwarf hut, with one or two doors, which fall when Cuffy tugs the bait from the figure of 4 in the centre."[78] Burton didn't say who had placed the bear trap, but later in the decade Mag Littleford was trapping bears there. He had a cabin near Beartrap Fork, where his wife, Margaret Bailey Littleford, on 26 April 1868, gave birth to a son, Seth E., who was claimed to have been the first white child born in Big Cottonwood Canyon.[79]

*

Mats Basin. This basin is located between Greens Basin and Silver Fork, also on the south side of the canyon. Mats Basin is not often visited since there is no good trail into it. There are faint trails coming down from the ridge above, indicating it was used in connection with the spring above Greens Basin. Mats Basin is supposed to have received its name from Mat Ellison, who brought logs down from it in the 1880s.[80] There was a Matthew T. Ellison active in the area between 1880 and 1911. He was a party in the filing of at least eight claims between 1880 and 1904, most of them in Days Fork, not far from Mats Basin. While he may have done some logging, there is no documented evidence to indicate that he ever owned or ran a sawmill.

*

Silver Fork. It is unlikely that this fork had a formal name before the mining boom. While Mill F was only a short distance up canyon, about where the Solitude ski area condominiums are today, the higher elevations and more gentle slopes on both sides of the canyon provided a good supply of timber close to the mill. Even if the fork was used for timber, there was no reason to bestow a name like Silver Fork—that is a name a miner would use. Indeed, early recorded claims reflect that propensity: Silver Mountain, Silver Star, Silver Queen, as well as several Silver mining companies. When the prospectors began recording claims in this area the first descriptions used nearby Mill F as a fixed reference, and called it the first south fork below Mill F. However, in July 1870 the name Silver Fork came into use.[81] That year Nelson Wheeler Whipple observed, "During this Summer many Silver mines were discovered in various parts of this [Big Cottonwood] kanyon but more especily in what they call Silver Fork."[82]

The Silver Fork community, once a collection of summer homes, now has many year-round residences. The original community was called Silver Springs, named after the many springs along Big Cottonwood Creek in the area. Today some of the springs may be seen, several directly under the bridge on Honeycomb Road. One of the notable attractions at Silver Fork is the Silver Fork Lodge, where one can get a meal with a view or a room for the night. This building seems to thrive on remodeling, for over the years it has seen many additions and is still seeing them today. The original small building was put up by Frenchy Hudson in 1945, using materials from surplus ammunition boxes. Tom Egbert took over around 1947. He called the place Uncle Tom's Cabin and had a gas pump out front to service thirsty automobiles. In 1954 Ted and Ethel Glines bought the place and added the dining room and second floor, some of the materials coming from the Cardiff Mine's bunkhouse in Cardiff Fork. Jim and Avis Light assumed control in 1968 and ran the lodge for fifteen years before selling it to the present owners.[83] Who that experienced it could ever forget sitting in the large rustic dining room, looking out the windows to watch hummingbirds flitting about in the summer, or clouds moving across the summits on a stormy day? Although the building has only some fifty years of history, it has left its mark in the memories of thousands.

*

Willow Lake. At mile 13.5 there is a trail on the north side of the road with a Willow Lake sign. This is the accepted route to the lake, although the road into the fork comes down to the upper end of Silver Fork, about two-tenths of a mile down canyon. The fork was originally known as Willow Patch Fork, circa 1888, and later as Willow Heights.[84] It was in this fork that Julius Kuck built a cabin and homesteaded 160 acres. The land fell into the hands of Frank E. Bagley, as related in an earlier chapter. The Bagley family later had dairy cows on the land. The cement floor and foundation of their barn can be seen off to the right when going up the trail.

*

Giles Flat. Contemporary maps, such as the USGS Brighton 7.5-minute quadrangle, show Giles Flat and the Redman Forest Camp at the same place. Actually, Giles Flat was farther down canyon, where Mill F once was located and where the Solitude condominiums are located today. It is shown there on older USGS maps.[85] The name comes from Hyrum L. and Minnie J. Giles, who were active in the upper Big Cottonwood mining scene from 1892 to 1915. During that time they filed at least sixty-eight claims and actively worked a number of them.

Hyrum L. Giles was born in Wales in 1850, where his father, Thomas D., had been a miner. However, after a mining accident left him sightless, the elder Giles learned to play the harp and earned considerable fame as the "Blind Harpist." In 1856 Thomas Giles, his wife, and three children emigrated to America and started for Utah in one of the early handcart companies. His wife and daughter died en route, and when Thomas fell ill, his two sons were taken into a different handcart company, one that endured much suffering when it was caught on the trail by the early winter storms. The remnants of the Giles family were reunited in Utah. There Hyrum learned to play the violin; his brother learned piano and organ. The two often accompanied their father in concerts or playing for dances. Hyrum turned his attention to mining and was involved in the Park City district before moving to Big Cottonwood in 1892. There many of his early claims were on Scott Hill, but he moved down to the lower end of Mill F East Fork, which he called Spring Branch, and filed the Little Dollie Claim. This was followed by Little Dollie Numbers 2 through 8 and the Little Dollie Tunnel.[86] The latter ran under today's canyon highway, just above its junction with the upper Solitude road. Another of his claims was the Giles Flat Lode, giving a second name to Mill F Flat. In 1902 he incorporated the Giles Mining and Milling Company to work fifteen of his claims.[87] Within a year he had added sixteen more claims to the company's holdings, taking in all of Giles, or Mill F, Flat, the northeast slopes above the canyon road from Willow Patch Fork up to the Brighton–Park City road, and most of the Mill F East Fork drainage. He worked the claim quite successfully until 1915, when he sold out. The new owners formed the Cottonwood-King Mining Company to work what was generally known as the Giles Group of thirty-one claims, one of the largest acreages in the Big Cottonwood district.[88]

During his tenure in Big Cottonwood, Hyrum Giles was a well-known presence. He never lost his love for music; as late as 1898, while he was driving the Giles Tunnel, he was playing the violin for dances in Park City. Hyrum Giles passed away on 22 February 1924, leaving behind a widow, nine children, and his name for a flat in Big Cottonwood Canyon.[89]

*

Redman Forest Camp. This Forest Service campground is at mile 14.9, between Silver Fork and Brighton. Its name comes from Benjamin F. Redman, who, with Emmett G. Hunt, received a patent on the Monarch No. 1 Lode in 1907.[90] The lode was first recorded in 1894 by J. F. and Blanche Brim, who recorded many claims around the turn of the century.[91] On occasion they also teamed up with Emmett Hunt, completing a full circle back to Redman. It is not known how long Redman and Hunt worked the mine, or if they even worked it at all. B. F. Redman was not a miner but a Salt Lake City businessman. He founded the Redman Moving and Storage Company, based in Sugar House. Many older residents of the city remember the large storehouse topped by the big Redman neon sign on the roof, a red beacon in the night. Redman also was active in civic affairs and organizations. He was the first member of the Rotary Club in Salt Lake City and was active in the chamber of commerce, state fair association, and Utah Automobile Club, among others. He was identified with many progressive ventures and gained the distinction of being the first paying passenger on a commercial airline in the United States. The event happened on 23 May 1926, when he and J. A. Tomlinson climbed aboard a Western Air Express Douglas M-2 mail plane at Salt Lake City, sat on folding seats in the mail compartment, and flew to Los Angeles. Actually, there were four passengers who flew on that first day of commercial passenger service, two on the eastbound flight and two on the westbound flight. In recognition of his efforts to arrange the Salt Lake City terminus for air mail flights and his work as chairman of the Aviation Committee of the Salt Lake City Chamber of Commerce, Redman was given ticket number 1, making him the first commercial passenger.[92] The airline had made its first flight over the route only a month before, on 17 April 1926.[93] In 1913 and 1914 Redman served as grand marshal and president of the wards of the Wizard of the Wasatch, an unusual organization that sponsored the Wizard's Carnival, a giant street fair that began with a parade and continued as a block party for another three nights. In 1914 it was estimated that 60,000 people attended the parade and carnival. Presiding over it all was Hat-Um-Ai, the great Wizard of the Wasatch, and Queen Sirrah of the Oquirrh. Hat-Um-Ai is a transposition of "I Am Utah."[94]

Amid all this activity the twenty acres of land at the Monarch No. 1 Lode was used for thirty years as the Redman family's summer resort, where they had a cabin in a meadow and view of Scott Hill. They called it Mountain Park. In July 1936 Redman donated the land to the U.S. government; the Wasatch National Forest supervisor accepted the land on behalf of the government and explained that it would be used to develop a new camp to be named Redman Forest Camp. Benjamin F. Redman died on 12 January 1945.[95]

*

Brighton–Park City–Midway Road. At mile 15.7, just below Brighton, a road branches off to climb the east slopes. This is the Guardsman Pass road, which crosses the divide to go to Park City or Midway on the east side of the mountains. In the early days of the mining boom a road was built from Park City into Brighton. That road came across Scotts Pass, went down the bottom of Mill F East Fork, then climbed over the ridge northeast of Brighton and dropped down into the community. Much of this road can still be seen. It crossed the ridge between Mill F East Fork and the main canyon about 300 feet below the winter closure gate on the Guardsman Pass road. Its remnants can be seen on both sides of the ridge. It can be followed down the steep west slopes until it meets the present road about 500 feet above its intersection with the main canyon highway. The road served until 1954, when the Utah National Guard needed a training project for its 115th Engineer Group. At that time a road was proposed to link Brighton with Little Cottonwood and American Fork Canyons if the Utah County Commission could secure rights-of-way.[96] Within a month the plans changed after the Wasatch Mountain Recreation Association came forth with a suggestion that the road connect Brighton with Midway. It would also connect to Park City by way of an existing road to Bonanza Flats, on the east side of the divide. This was not a new idea; in 1926 the Good Roads Committee of the Salt Lake Chamber of Commerce and the Kiwanis Club at Park City had contacted the Salt Lake County Commission with a proposal to build a highway to connect Brighton and Park City, thereby making a loop of the Big Cottonwood and Parleys Canyon roads.[97]

The National Guard spent several years building this road, working only during its summer encampment, then left it to the state to be paved.[98] Once it was opened the old Park City road was abandoned. The Guardsman Pass road remained as a dirt road, some years as a very rough dirt road, for several decades before it was paved in bits and pieces. The new road followed the old Park City road out of Brighton, although at a lesser grade, but it then took a different route after climbing the first ridge and crossed the divide to Bonanza Flats. The pass it crossed, known today as Guardsman Pass, is about eight-tenths of a mile south of Scotts Pass.

*

Brighton Area. An entire chapter has dealt with Brighton, its origins and activities. In the mountains surrounding it there are many names, some of which have stories that are begging to be told.

Silver Lake, of course, is the heart of Brighton. From the day the first person laid eyes on it there has been an attraction that drew people and burned its memory into their minds and hearts. Possibly the first recorded name for this lake was "Plesant Lake," used by Frederick Kesler in his journal when he was scouting timber resources for Mill F. When the miners came into the area they called it Cottonwood Lake, Big Cottonwood

Lake, or Big Lake.[99] When it was discovered the lake held an abundance of fish it became known as Fish Lake or Trout Lake. Catherine Brighton is supposed to have given it the Silver Lake name.[100] If so, she did it in 1872, because in September 1871 she and her husband had filed a mining claim "situated a little west of the Big Lake or Fish Lake." A year later, in September 1872, William S. Brighton filed a claim located "a little south of east of Silver Lake," the earliest use of that name that has been found.[101]

Mount Evergreen is the mountain that towers immediately over Silver Lake. It draws its name from the June 1870 Evergreen mining claim, located on the flanks of the peak.[102] The mine was worked with a vengeance, being supplemented by extension claims in all directions, taking it to the other side of the peak into Mill F South Fork. It became so well known to miners in the area that one newspaper correspondent used the name Evergreen Gulch for Mill F South Fork.[103] In 1874 the Evergreen Consolidated Mining and Tunnel Company was incorporated to work the mine and several surrounding claims. One of the incorporators and the principal stockholder was Green Flake, one of two black slaves in the original party of pioneers who arrived in the Salt Lake Valley in July 1847.[104] The company worked the mine until 1880, when it was reorganized as the Old Evergreen Mining and Tunnel Company, which continued to work the property well into the twentieth century.[105] With the early presence and importance of this mine, it was perhaps inevitable that the mountain in which it was worked would be named after it. Indeed, as early as May 1871 it was called Evergreen Mountain, and later that year Evergreen Hill.[106] And it has been known by the name of Evergreen ever since.

Lake Solitude is located at the head of Mill F South Fork, but it is more closely linked with Brighton because of the popular trail around Mount Evergreen from Silver Lake. The first name used for this body of water was Lizard Lake, probably an indication that there were salamanders in it, as there were in many other Wasatch lakes. Salamanders were often called electric dogfish, so their presence is supported by the fact that another prospector called it Electric Lake.[107] In June 1872 a young man by the name of Oscar B. Young, a son of Brigham and Harriet Cook Young, came into Mill F South Fork and filed a claim for the Zoah Tunnel, located only a few hundred feet from the lake. He very modestly called it Oscar Lake.[108] Oscar Young must have been a loquacious fellow and spread the news of his lake's name, because no less than seven other claims referred to Oscar Lake that year and four more in the years following. In 1873 Young was involved in a claim in City Creek Canyon, but after that he left the mining scene. His name went with him and the

lake returned to being nothing more than the little lake in Mill F South
Fork. When recreationists started to frequent Brighton they discovered
this lake cradled among the cliffs. What better name to give it than Lake
Solitude? The name appeared first in 1899 and has been used ever since.[109]

Twin Lakes has been mentioned. There were originally two lakes in
this high basin, but they were combined into the single Twin Lakes Reservoir when the dam was built in 1915.

Mount Millicent, towering over Brighton to the southwest, is an impressive mountain of granite, so it should be no surprise that the early
miners called it Granite Mountain.[110] A writer in 1880, impressed by its
granite bulk situated in the center of the Brighton basin, said it was called
Mount Eyrie, named by a young lady who climbed it and saw an eagle
circling overhead.[111] It has been claimed that its present name was in
honor of Millicent Godbe Brooks, a daughter of William S. Godbe, early
Salt Lake City merchant and miner.[112] He was one of the men advocating
the opening of Utah mines, against the wishes of President Brigham
Young. His ideas led to the Godbeite movement and his expulsion from
the Mormon Church. He went on to a highly successful, if somewhat
rocky, career in mining. He was involved in mines at Alta, in the South
Fork of Big Cottonwood, and in Mill F South Fork. In 1883 he built a
cottage at Brighton's mountain retreat, after which time the Godbe family figured heavily in Brighton society.[113] In his mining interests, Godbe
certainly worked closely with Charles P. Brooks, a civil engineer who did
a great deal of work in mines in the Wasatch, and it is likely Brooks was a
close friend of the Godbe family. He married Millicent Godbe on 28 September 1876. After the Godbe cottage was built, the Charles P. Brooks
family shared a prominence in Brighton society with the Godbes. Millicent Godbe Brooks died on 27 September 1889, leaving her husband and
three daughters. Two years later, on 15 December 1891, Brooks married
Miriam Godbe, Millicent's sister, and the Brooks and Godbe families continued their close association. As the Brooks daughters grew older, their
names frequently appeared in Brighton society columns. It is not difficult
to imagine that they might have named the mountain after their mother,
certainly with the approval of their close-knit family. A reference to Mount
Millicent appeared in 1899, after which time its name was generally accepted.[114] William S. Godbe died on 1 August 1902 at his son's cottage in
Brighton.[115]

Mount Wolverine, although higher than Mount Millicent, is hidden by
the latter's summit and is not generally visible from Brighton. Two different claims, both named Wolverine, were located on the slopes of this

peak. Since they were posted a year apart, they may have referred to the same location, but one was posted in the Big Cottonwood district, while the other was in the Little Cottonwood district.[116] The claims were not heavily worked, for they failed to appear in any later mining news, but it is likely they were the source of the name of this mountain. The Mount Wolverine name appeared as early as 1896, when a Brighton society column reported a large party had climbed the summit.[117]

Mount Tuscarora is adjacent to Mount Wolverine, between it and Catherine Pass. Tuscarora is the name of a tribe of Iroquoian Indians formerly located in North Carolina, now in New York state and Ontario. How it got assigned to a peak in the Wasatch remains a mystery. There were no known mining claims with this name. The first use of the name that has been found was on the USGS Cottonwood Special Map of 1907.

Sunset Peak, southwest of Catherine Pass, and neighboring Pioneer Peak both rise above Lake Catherine so prominently that they must have been named by the time early Brighton vacationers were enjoying the area. Both names appear on the USGS Cottonwood Special Map of 1907, but a 1908 Salt Lake County map called the former Monument Peak. Pioneer Peak was included in a list of peaks and their elevations that was posted in the Brighton post office in 1903.[118]

Lake Phoebe was the first in a group of lakes to be found when one followed the trail from Brighton to Catherine Pass. It was lost in 1915 when a dam combined it and Lake Mary into one body of water. The dam was originally called the Lakes Phoebe–Mary Dam, but the Phoebe name was subsequently dropped and is virtually unknown today. Phoebe, the smaller of the two lakes, was once known as the Fairy Pool. George M. Ottinger, one of the artists who frequented the area, was supposed to have named the lake after his second wife, the former Phoebe Nelsen. The name was in use as early as 1888. Ottinger went on to become superintendent of the city waterworks and fire chief, but in 1890 he retired from those duties to devote full time to his art.[119]

Lake Mary was as popular among recreationists in the nineteenth century as it is today; but in those days before the building of the dam the lake was considerably smaller. Alfred Lambourne wrote that it was first known as Granite Lake, which would be a reasonable name since it was surrounded by granite cliffs and had a granite island in the center. However, in 1872 New York artist Hartwig Borneman was visiting Utah Territory and spent some time at Brighton's hotel with local artist friends. When he first saw the lake "his exclamations of delight are said to have awakened the echoes." He then gave his wife's name to the lake, and it has ever since been known as Lake Mary.[120] While in Utah Borneman painted a portrait of Brigham Young. At least two examples of his work are in the LDS Museum of Art and History in Salt Lake City.

Lake Martha, immediately above Lake Mary, is much different from its neighbor, being surrounded by meadow and forest rather than stone. It has an island, ringed with white boulders and crested with trees, that the artists likened to a crown. The lake was named after Alfred Lambourne's mother at the same time Borneman named Lake Mary.[121]

Lake Catherine is the highest lake in this basin. Like several other lakes in the Wasatch, it has no surface outlet. This feature was used to describe the location of one of the earliest mining claims in the area: at "a small lake with no outlet."[122] It is also the site of some of William S. Brighton's earliest mining claims, one of which was the Catherine Lode, adjacent to the lake and covering its southeastern shore, and the parallel Brighton Lode, covering the center of the lake.[123] Catherine Brighton was said to have considered this lake the most beautiful of all. When Hartwig Borneman was at the hotel and was going to the upper lake to paint she asked him to name the lake after her. Her son, William H. Brighton, went along to carry the artist's equipment and was said to have watched Borneman paint her name on a rock.[124] That may be true, but the Catherine Claim had been recorded two years before Borneman was there, and was later surveyed for patent. The mineral survey may have done as much to establish the lake's name as the artist painting a name on a rock. The Lake Catherine name was not documented until 1890 when it was used in a claim located nearby.[125]

Dog Lake is about two-tenths of a mile southeast of the Lake Mary dam. It is surrounded by meadow and marsh and provides a suitable habitat for salamanders, which are the source of its name. The lake was called Dog Lake as early as 1870.[126]

Lake Annette, a little-known body of water almost too small to be called a lake, doesn't even appear on the USGS map of the area. Lambourne called it a lakelet or a pool. It was named by Henry Culmer after his wife-to-be, Annette Wells, in 1875. Culmer wrote that he and Lambourne once spent a whole day in a vain search for this elusive body of water.[127] It rests on a shelf below and to the north of Lake Mary, not far from the trail between Lake Mary and Twin Lakes. In recent years it has gathered a mass of downed trees and is filling with silt, giving it a rather cluttered and unkempt appearance. Not too many years ago, however, while those trees were still standing, the pool was a beautiful, clean, quiet haven from the hustle and bustle of the outside world, probably looking very much as it did when Culmer gave it its name.

Clayton Peak and Mount Majestic are one and the same, rising to the southeast of Brighton. The official name, Clayton Peak, appears on current USGS maps. It is said to have been named after Prof. J. E. Clayton, a mining engineer of considerable repute who lived in Salt Lake City from 1872 to the mid-1880s and spent considerable time in and about mines in the Wasatch. Before that he was engaged in mining for nearly thirty-six years, first in Alabama and Georgia, then in California, Nevada, and Utah. He was heavily involved in the Emma Mine for its British owners and was called as a witness in the congressional Emma Mine investigation.[128] While he certainly visited Brighton, perhaps frequently, there is no evidence that he ever was a resident of that summer community. A reference to Mount Clayton did appear in 1882, while he was still a Utah resident; however, the Brighton residents of the 1880s and 1890s chose to give it their own name, Mount Majestic. The name was in use as early as 1896, and remains in popular usage today.[129]

Scott Hill is some distance from the Brighton community but was a popular destination of recreationists there. In 1870 a group of men filed a claim on the Scott Lode, named after four of the claimants: Joseph L., John W., Ephraim, and Simmeon Scott. The mine was an immediate success and became a recognized point of reference. As early as the following year the hill on which it was located was known as Scott Hill.[130] The four men of the Scott family disappeared from the scene within a year, but they left their name behind for others to use to this very day.

CHAPTER 14

Little Cottonwood Canyon—Names and Places

LITTLE COTTONWOOD CANYON has about 27.7 square miles of watershed, making it considerably smaller than Big Cottonwood Canyon but still the third largest in size among the major Wasatch canyons draining into Salt Lake County. As explained earlier, the canyon took its name from its stream, a stream that has been known as the Further Cottonwood, South Cottonwood, and Little Cottonwood. The latter name appeared as early as January 1851. Today the all-season Utah Highway 210 extends about eight miles up the canyon to Alta, with another 2.5 miles of dirt road going into Albion Basin, the latter being open only during the summer months. Highway milepost markers are used in this chapter to locate places under discussion. Milepost zero is located at the intersection at the mouth of Big Cottonwood Canyon, where Highway 210 begins. The fork at the mouth of Little Cottonwood Canyon, where Highways 209 and 210 join, is at mile 3.9.

*

Ferguson Canyon. While not a part of Little Cottonwood Canyon, Ferguson Canyon does lie between Big and Little Cottonwood and is included here since it is on the way to the latter canyon. It is a short, steep canyon draining into the valley, its mouth being little more than a quarter mile south of the mouth of Big Cottonwood Canyon. Its name is said to be taken from Isaac Ferguson, an early pioneer who lived and worked in the vicinity of Big Cottonwood Canyon. There is a story that Isaac Ferguson discovered some ore while dragging logs from the canyon, had a sample assayed, and found it to be rich in gold. When he told Brigham Young what he had found, Young told him to keep it a secret and cover the discovery. When Ferguson died, the location of the gold deposits died with him, never to be found since.[1] While popular history feeds upon this kind of story, there is little substance to it. Gold fascinates people and, as we shall see, Ferguson Canyon isn't the only local place where its lure has been felt. In 1912 the Blue Point company was working a group of fourteen claims in Ferguson Canyon and reported its ore was averaging $8.80 a ton in gold.[2] This may have inspired the folktale related above, but it was not enough to keep the company going.

Actually Isaac Ferguson and his family were rather active in the mining scene. Between 1870 and 1872 they filed at least eighteen claims; but most were near the head of Big Cottonwood Canyon, none in what would be known as Ferguson Canyon. After 1872 the Fergusons quit prospecting. In 1876 the Ferguson Canyon name appeared when a petition was presented to the county court (not by Ferguson) for the right to build a road into the canyon and control its timber resources.[3] That same year, Isaac Ferguson moved his family to Whipple's mill in Big Cottonwood and hauled logs out of Whipple Fork. He did the same again the following spring but left in June, leaving nothing behind except his name in the canyon adjacent to Big Cottonwood.

Ferguson Canyon received its first mining claim in 1877, with only two more filed in the nineteenth century. The first decade of the twentieth century saw much more activity, with twenty-one claims filed. As has been noted, a company was working a group of claims in 1912, but the place never became a noted center of mining, or, for that matter, of anything else.

The next drainage south of Ferguson Canyon, a fairly shallow and short one, was known around the turn of the century as Dry Canyon, Wide Hollow, or White Hollow. The latter name probably referred to the granite butte that stands near the head of the hollow. Known today as Hounds Tooth, it was called by the miners Granite Peak, Granite Bluff, Granite Needle, and Granite Knob.

*

Little Willow Canyon. This canyon drains into the valley about midway between Big and Little Cottonwood Canyons, at mile 2.2 on the highway. It has two forks, the north fork extending nearly three and one-half miles to the north side of Twin Peaks, climbing nearly 5,000 feet in that distance. This is but one of many Willow Canyons and Willow Creeks found all over the West, named, of course, for the willow trees growing along their streambeds. There is a Willow Creek draining from the west side of Lone Peak, south of Little Cottonwood Canyon, that provided sustenance for residents in the extreme southern end of the valley, especially the budding community of Draper. It is likely Little Willow gained its qualifier to differentiate between it and the other creek farther south, as well as in recognition of its smaller flow of water. The southern Willow Creek was named as early as 1850, but the Little Willow name did not appear until 1865. At that time John J. Thayn applied to the county court for a grant for the use of the canyon "known as Little Willow Creek Kanyon, situate betwixt the two Cottonwoods." He was given the grant, but he returned six months later to give it back to the court, saying the condition of the canyon would not justify his building a road.[4]

When the mining boom took off, very few prospectors ventured into Little Willow Canyon; there were only six claims posted for that area during the 1870s. In December 1875 one Henry Standish petitioned the county court and received the right to charge toll on a road he made in Little Willow.[5] Several years earlier Standish had held a contract to haul ore from Emmaville, near the mouth of Little Cottonwood Canyon, to Salt Lake City. Now he put his freighting experience to good use, and with the help of seven men and twelve mules he hauled a steam sawmill, boiler and all, several miles up Little Willow Canyon. The difficult access probably discouraged him from operating the mill very long, for three years later the road was in a state of disrepair and the mill had been abandoned. Before Standish vacated the premises, he did a little prospecting and filed one claim in the canyon.[6]

In the early twentieth century Little Willow became a favored route for hikers and climbers to ascend Twin Peaks. As early as 1916 the YMCA was sponsoring climbs. The participants would ride the streetcars to Holladay, then hike to a camp in Little Willow, climb the peaks, and return the same way.[7] A number of years after that the Physical Education Department at the University of Utah conducted similar hikes. Under the direction of Professor J. R. Griffiths, the parties would take trucks to the mouth of Little Willow, where they would begin their climb. Usually two nights were spent in the canyon, with the summit(s) being reached during the intervening day.[8] However, not everyone chose this rather strenuous route. In 1916, shortly after the YMCA party had been on the summit, Dr. Charles G. Plummer and Otto R. Peters, manager of the Brigham Street Pharmacy, climbed the peaks by way of Lake Blanche, an equally strenuous route, which rewarded the good doctor with two broken ribs and other painful injuries when he fell during the descent. In the early spring of 1920 five young men of the Tahoma Club of Salt Lake repeated Plummer and Peters' route and left an excellent description of their hardships and joy in reaching the summit.[9]

Today Little Willow Canyon is known as Deaf Smith Canyon. The source of the latter name is not known, but it appeared sometime between 1952 and 1962. The 1952 edition of the USGS Draper, Utah, 7.5-minute quadrangle shows the canyon as Little Willow, while the 1962 printing has it labeled Deaf Smith Canyon. It is an unusual name, one that is not heard often. For that reason, one cannot overlook the fact that the state of Texas has a Deaf Smith County, adjacent to Amarillo. The county was created in 1876 and named after Erastus (Deaf) Smith, who played an important role in the Texas Revolution, gaining the lasting respect of his fellow Texans. However, Smith was not a native Texan. He was born in Duchess County, New York; but when he was eleven his parents immigrated to Mississippi Territory. In 1821, when he was in his early thirties, he moved to Texas and made it his home. At that time he was totally deaf, hence his nickname.[10] To carry this diversion one step farther, it is necessary to note that in 1847 and 1848 a number of pioneers came into the Salt Lake Valley from Mississippi, where they had relocated from New York. And Smith was a very common name among the pioneers. However, Erastus Smith died in November 1837, long before the Mormon migration, and if there was any connection between him, or his family, and any of the immigrants, it has failed to come to light. Was there another Deaf Smith? And, if so, who was he?

*

Gold City. Curiously, there was little prospecting in Little Willow Canyon before 1893. Then, in less than twenty years, nearly 200 claims were filed in and around the canyon, probably triggered by the hope of finding gold. Two men, Leonard Hilpert and John Ruebhausen, started an intense prospecting effort in the area, both in Little Willow and on the west-facing slopes north and south of it. They filed some thirty-five claims in less than ten years and induced others to follow. One of the others was George F. Dalton, who with several associates filed the Gold Hill and Hidden Treasure Claims in May 1894.[11] They were located north of Little Cottonwood Canyon and south of Little Willow Canyon. When Dalton found surface indications that led him to believe a mineral vein was hidden below, he applied for a patent on the claims. Uncertain which mining district had jurisdiction over the site, he posted another claim for the Gold Hill, the more promising of the two, in the Big Cottonwood Mining District in July of the same year.[12]

Throughout 1894 the two claims were explored, and at least six more were posted in the immediate vicinity. In November 1895 a body of gold-bearing ore was found, pieces of which assayed incredibly high, reported to be as much as $1,800 per ton and averaging $290. Dalton's backers formed a corporation known as the New State Mining and Milling Company. They tried to keep their discovery a secret, but when the corporation was announced and men were put to work on the property, the news was out. It was published on 20 January 1895.[13] Two days later it was reported that as many as 500 mining men had visited the office of Wolstenholme & Morris, coal dealers on First South Street, to see the magnificent specimens of gold quartz from the new discovery.

William J. Wolstenholme and Richard P. Morris, two of Dalton's backers, were secretary-treasurer and president, respectively, of the New State Mining Company. Although it was the middle of January and the mountains were covered with snow, many people went to the slopes north of Little Cottonwood to see the mine, and some went to prospect and post their own claims. One witness said the hillside had the appearance of having been trodden by a flock of sheep, with trails stretching in every direction, radiating from the discovery shaft on the Gold Hill Claim. The Denver and Rio Grande Western Railway company took part in the excitement by announcing a special "Excursion to The Gold Fields in Little Cottonwood, . . . fare but $1 for the round trip."[14]

On 26 January the growing gold fever received another impetus when the Cottonwood Gold Mining and Milling Company was incorporated to work four claims north and east of the New State property. Two men from the New State company also were principals in this new company, one of them being George F. Dalton.[15] Reports poured in of new discoveries, some extending as far north as Ferguson Canyon, near Big Cottonwood. Prospectors brought samples of ore into the city for assay and display. A stage line was started to transport people from the city to the gold fields. And it was reported that another group of men had taken an option on 160 acres of land adjacent to the gold fields and were platting a townsite on it. Both mining companies were building cabins to house their workers, and this, it was said, would form the nucleus of the new town, to be named Gold City. The promoters distributed circulars of the new camp and townsite, reciting the history of Cripple Creek, Colorado, and suggesting that Cripple Creek's history would be repeated at Gold City. The *Salt Lake Tribune,* which had reveled in the excitement thus far, began to express concern. "The contagion proves a little dangerous," it wrote, "for it seems to be overtaking many men who hitherto have lived only on the conservative side of mining ventures."[16]

The Gold City Land and Townsite Company began to do business as soon as the townsite was announced, although it did not file its corporation papers until mid-February. On its first day of business it sold four lots. It also deeded a lot to the Cottonwood Gold Mining Company in consideration of that company building an extensive boardinghouse, 60 by 90 feet in size, there. A grocer and saloonkeeper from the city ordered lumber to build a store and bar in the new town. He opened for business on 3 February even though his shop was not yet finished. Three days later the townsite company announced that twenty-nine lots had been sold and that a lumberyard had been set up by a Sandy firm. Although it was the middle of winter, twelve buildings were being constructed and tents dotted the hillside. The Union Pacific Railroad advertised that the best way to get to the gold fields was to take their train to Sandy, from which place there was only six miles of staging.[17]

During all this time manager Dalton of the New State company continued development of the mine. Surface explorations indicated the vein grew in size as it went deeper, so a tunnel was driven to cut the vein at depth; however, it missed it altogether. A shaft was sunk twenty feet on the vein, and another tunnel was started from that point. As they followed the vein it widened, then split into four separate streaks. They were believed to be feeders to a still larger and more compact body of ore. Then, in mid-March, a day of disappointment came when the miners found the feeders spreading into a conglomerate mass of unprofitable dirt. The bubble had burst. The dreams of $1,800-per-ton ore were turning into the reality of low-grade ore that might, with luck, yield one-tenth that amount. As interest in the gold fields subsided, the Gold City venture collapsed. Within three months suits were being filed against the company for its failure to provide deeds for lots that had been purchased. One purchaser, who had erected a building on his lot, claimed the company could not provide a deed because it had no title, and that the real owners of the property had ordered him to vacate the land.[18]

The New State company continued to work its claims, attempting to recover that which was lost. In mid-summer of 1895 new investors began buying New State stock, including the entire personal holding of George Dalton. The following year Dalton was reported working the Baby Ruth Claim in the south fork of Little Willow. The New State, under new management, worked its claims as late as 1901, but its days of glory were in the past.[19]

Another of the followers of Hilpert and Ruebhausen was Nicholas Schmittroth, who began filing claims on the slopes north of Little Willow in 1893. He was an unlikely man to be prospecting, being a baker by trade. He had emigrated to New York City from his native Germany in 1871, and moved to Salt Lake City in 1890. After his life was over, he was re-

membered for his achievements in founding baking companies, both in
Salt Lake City and elsewhere. His baker's background followed him to the
slopes, for one of his first claims, less than a quarter mile north of Little
Willow Canyon, was called Model Steam Bakery, the name of a bakery
where he worked in the early 1890s.[20] Schmittroth did not take part in
the Gold Hill and Gold City excitement in 1895, perhaps attending to his
bakery interests. But in 1896 he was back, and in the next three years he
posted at least a dozen claims in the area. In 1899 he formed the Jefferson
Gold and Copper Mining Company and deeded his most promising claims
to the new company.[21] Although most of the claims were north of Little
Willow, one, the Jefferson, was adjacent to the Gold Hill claim, and there
the company located its operations. In 1903 it took over the New State
company and became the Consolidated Jefferson Gold and Copper Mining Company.[22] During this period the company built a substantial surface installation, involving a large shaft house, boardinghouse, and a
store called the Jefferson Mercantile Co. But expansion requires capital,
and when the company's stockholders rejected a request to approve a
sizeable loan, Schmittroth resigned and was replaced by one of his former
investors from Omaha.[23] Two years later, in 1908, he formed two more
companies, the Jefferson Extension Mining Company and the Diana Gold
and Copper Exploration Company, to exploit claims to the south of the
Consolidated Jefferson's property. Curiously, one of the assets of the first
company was an option to purchase a quarter section of land, the same
quarter section that had become Gold City some thirteen years earlier.
The two companies were not successful, however, for the charter for the
first was revoked in the spring of 1910, and that for the second in the
spring of 1911, both for nonpayment of license taxes.[24]

The Consolidated Jefferson continued to operate until 1908, when it
was decided to reorganize under the laws of the State of Nevada in order
to make the stock assessable, thereby making it easier to raise money
when the need might arise. It then became the Wasatch-Utah Mining
Company. In the years that followed, the company installed a large mill
to treat the low-grade ore on the site. It continued to operate and be in
the news until 1916, when a new group of people from Chicago took
control, after which time it faded into obscurity.[25]

There was one more act left in this gold-field drama. At the end of
1924 Nicholas Schmittroth appeared again, after more than a dozen years
of absence, to incorporate the Golden Phorphyry Mines Company. Assets
of the company included all of the formerly productive claims between
the two Cottonwoods, including those of the Wasatch-Utah company.

However, most of the news about this company concerned plans for the future and achievements of the past. In the early 1930s the property was being worked by lessees.[26] By that time the national economy was in a serious state of distress, and most operations came to a standstill. The final blow came when the gold field's most ardent supporter, Nicholas Schmittroth, then eighty years of age, died on 29 December 1935.[27]

The gold fields left but little to remember them by. In 1917 the highway between Big and Little Cottonwood Canyons was known as the Gold Road, but that name did not survive.[28] And, at the mouth of Little Willow Canyon (or Deaf Smith Canyon, if you like) there is a residential area known as Golden Hills. Other than that there are only faint memories.

*

The Mouth of Little Cottonwood Canyon. This is the junction of Utah Highways 209 and 210 at the mouth of the canyon—mile 3.9. On the southeast side of the junction is a parking area and a paved nature trail that passes several points of interest. The original canyon road and railroad right-of-way, used today as a popular hiking and biking trail, continues for several miles up canyon from the nature trail and goes as far as the remains of the Columbus Power Plant, described in an earlier chapter. At the upper end of the nature trail is a concrete building housing an abandoned electrical generator and its broken pelton wheel, and in the streambed are shattered remains of a concrete dam. Both are reminders of hydroelectric plants that made use of the power of Little Cottonwood Creek in the past.

In July 1912 the voters in Murray City approved a $60,000 bond issue to erect a municipal hydroelectric power plant near the mouth of Little Cottonwood Canyon.[29] While the plant itself was some distance downstream, its water was taken from the stream where the concrete dam remnants are seen today. The plant operated for nearly two decades before it was relocated and rebuilt, which occurred in 1929 when Salt Lake City was looking for additional sources of culinary water. Having acquired most of the flow from Parleys and Big Cottonwood Canyons in previous years, the city now was looking for water from Little Cottonwood Canyon. However, if water were taken from the stream below the Murray power plant, it could not flow by gravity to the mouth of Big Cottonwood Canyon where it could connect with the Big Cottonwood conduit to be taken to the city. Accordingly, the city negotiated with Murray to move its power plant farther upstream. Salt Lake City paid for the relocation through a June 1931 bond issue that also provided for the Little Cottonwood conduit. The new plant was built at the end of 1931, at which time the old one was abandoned.[30] At that time a new water inlet was constructed about a half mile farther upstream, a new pipeline to the power plant was built, and the old dam was breached. The Murray power plant ran for fifty years before it was again replaced, this time in 1983 by a more modern facility, located on the same site as the old one.[31]

The concrete building housing the abandoned hydroelectric plant belonged to the Whitmore Oxygen Company, today known as Praxair. The company has a long history of electric-power generation in the canyon. It began when Alfred O. Whitmore came to Utah in 1897 to erect a power plant in Provo Canyon to supply power to the Mercur mining camp. With several years experience with the Telluride Power Company in Colorado, he was one of the pioneers in electrical engineering. By 1905 he had moved to Salt Lake City, where he established his own electrical firm; and by 1909 he had expanded his business to include batteries and electric automobiles. To charge the batteries, he installed a small hydroelectric plant in Little Cottonwood Canyon near Coalpit Gulch, a plant that still runs today, albeit with appropriate updating, and is housed in the building along the south side of the highway at mile 5.6. In an attempt to salvage and use the gases given off during the charging process, the Whitmore Oxygen Company was formed.[32] Its first building was built near the stream just below the power plant; but in the 1930s the company moved into larger quarters farther down the canyon. In 1930 Whitmore filed a water appropriation for another power plant to be located above the old Murray dam. When the Murray intake was moved upstream it could not make use of the water that would bypass it for Whitmore's new generator, and legal actions resulted. The matter dragged on for years, the Whitmore company managing to get extensions of time to use its appropriation. Finally, in 1946 a steel pipeline was installed, a fireproof concrete building was constructed, and electrical generators and transformers were on the site. By 1948 the equipment was installed and placed in operation for experimental purposes.[33] It may never have been placed in service before being abandoned, for the company installed a larger, 540-kw hydroelectric generator in its plant located just above the Murray intake. The company later moved into newer quarters in the city, but its two power plants in Little Cottonwood Canyon still operate to provide the company's needs.

*

Granite Quarries. The granite quarries near the mouth of Little Cottonwood Canyon are inescapably linked with the Salt Lake LDS Temple, so much so that the quarries cannot be considered without including that beautiful and impressive structure in downtown Salt Lake City. When the decision was made to build a temple at the general conference of the church on 7 April 1851 it seemed to be generally assumed that the new structure would use sandstone from the Red Butte Canyon quarries east of the city. On that same day it was announced that a railroad had been chartered to run from the Temple Block to the quarries to bring down building materials for a wall around the block and the foundation of the temple.[34] Named The Great Salt Lake City and Mountain Railway, it was not a railroad in the usual sense, as it was intended only to use wooden tracks on which horse-drawn wagons could travel. The route was surveyed and graded and lumber for the tracks and ties was cut at Gardner's mill on the Jordan River and floated down the river to the city. The railroad was never completed, but the roadbed was used by wagons to bring stone into the city. Both the wall around Temple Square and the temple footings were built using this sandstone.[35] The material to be used for the exposed portions of the temple was much debated but was not selected until early in 1855. On 27 April 1855 George A. Smith wrote to Franklin Richards, "It is decided that our Temple will be built of granite, from the quarries between the Cottonwoods."[36] The presence of granite in the Wasatch Mountains had been known for many years. On 14 August 1847 Wilford Woodruff noted in his journal, "Found A mountain of granite good for building runing from 7 to 15 miles from Camp."[37] And after John Brown and his companions climbed Twin Peaks on 20 and 21 August 1847, George A. Smith noted that they had found light-grey granite of all degrees of fineness.[38]

As early as October 1851, at the church's general conference, it was proposed to bring water from Big Cottonwood, Mill Creek, and Big Kanyon [Parleys] Creek to the city for irrigation purposes. A canal route was surveyed and some work was done, but the urgency was not there. Local streams provided enough water for the limited agricultural land in use at the time. By 1855, however, the canal was seen as a means for boating granite from the Cottonwoods to the city. Throughout the next two years considerable effort was expended on the construction of the canal, but when the water was turned into it, it was discovered that the porous soil it ran through at various places allowed the water to escape. For the moment, the canal was a failure.[39] Although it was later repaired and served well as an irrigation canal, it never was used for the transportation of granite. That was done with teams and wagons.

The first granite quarries were at the mouth of Big Cottonwood Canyon. They were not quarries in the usual sense, as rock was not extracted from below the surface. Instead, men used the huge granite boulders that had been dislodged and transported by glacial action in the distant geological past. The quarrymen split and cut them to suit their purposes. On 21 April 1857 foreman B. T. Mitchell with six other men arrived at the head of the Cottonwood canal; the next day they began to build a stone shop. On 27 April, after being joined by ten other men, they began to cut stone.[40] On 23 April young Winslow Farr wrote in his journal, "They are erecting sheds for the stone cutters up to the mouth of Big Cottonwood Kanyon to prepare for the Temple Brothers."[41] William Adams, an immigrant of 1849 who lived in Iron County, was among those who answered a call for stonecutters to come to Salt Lake City and work on the temple. With ten or twelve others, he was sent to Big Cottonwood to cut granite, where he remained until August. In July, when people were heading up Big Cottonwood for the 24 July outing, Wilford Woodruff wrote, "We camped last night in the Big Cottonwood at the place whare the Brethren were Cutting granite for the Temple."[42]

The stonecutters worked through March 1858, but then work on the temple was halted for several years. With the Utah War and the coming of federal troops the temple site was covered with soil to hide all progress. Not until the end of 1861 were the foundation walls uncovered to allow work to resume the following spring.[43] The quarrying of rock, however, was resumed a year earlier, in 1860. At the end of June of that year thirty-five loads of rock were shipped from the Big Cottonwood quarry; however, the following month thirty-one loads were shipped from Little Cottonwood, and there is no evidence that the Big Cottonwood quarries were used after that time. Although quarrying continued throughout the 1860s, it was done at a very slow rate. After the temple foundation walls were uncovered it was found that the foundations had been laid in an unsatisfactory manner. All stones used in the construction were removed and the foundation (footing) was rebuilt. By the time this lengthy process was completed the construction of the new tabernacle had begun. It received most of the attention from workers until 1868, when the bulk of the available working force was sent to work on the construction of the Union Pacific Railroad tracks in Echo and Weber Canyons. It was 1870 before focus again shifted to the temple.

Because of the nature of the quarrying operation the Little Cottonwood quarry site moved periodically. In 1870 the quarry was at the south end of the beaver dam depression, east of the intersection of today's Wasatch Boulevard and Little Cottonwood Road (Utah Highway 209). However, it was soon moved above the fork of the highway in the mouth of the canyon.[44] When the quarry moved, so too did the men and their accommodations. Their small and crude community was known as Granite City. The first Granite City was at the Big Cottonwood quarries, as noted in a time book kept by James Standing in 1857, labeled "Time Book for Stone Cutters at the Head of Cottonwood Canal or Granit City."[45] Granite City then moved south to Little Cottonwood Canyon, where it shifted with the quarry site until its final location at the east end of Glacio Park, near the forks of the highway. The quarries then moved farther up canyon to a site that assumed the name of Wasatch, although many still referred to it as Granite. Local residents continued to use the Granite name to designate the area around the first quarry site at Little Cottonwood, a use that continues to this day.

At that time, as today, there were two roads into the canyon. The first had been used by the Woolley brothers for some fifteen years to haul their lumber to the city. It stayed on the north side of the creek, running through Glacio Park, then turning north to follow a route along today's Danish Town Road and, farther north, Highland Drive. The other road stayed on the south side of the creek and headed west to Sandy, which is the route of today's Little Cottonwood road, Utah 209. The first route was undoubtedly used to haul granite and ore to the city during the 1860s, but in 1870, when the Utah Southern Railroad was operating, it became more expedient to haul to Sandy, where the loads could be transferred to railroad cars.

The similar needs of teamsters hauling ore and mining supplies and those hauling granite caused the area's small communities like Granite City to grow larger and receive considerable prominence and notoriety. By the end of 1871 Granite City had had at least nine houses, because a correspondent reported that two days before Christmas nine houses were blown down.[46] When W. H. Fairfield began running stages from Salt Lake City to Alta, he built a large stable at Granite City, and Messrs. Robinson and Burr opened the Trout House, where a weary traveler could get a square meal.[47] Standish Rood, *Salt Lake Herald* correspondent writing under the pen name of Archibald, said that Granite City "is even harder than the name it bears—such a conglomerate of shanties, old stables, and 'dug-outs' no one outside of this hard place ever saw.... Nature has done an immense amount for this townsite, but its founders spoiled it by building the place."[48]

In the spring of 1873 the Wasatch and Jordan Valley Railroad was completed from Sandy to Granite City, giving that budding village new importance: it no longer was a way-station for the teamsters, it now became a terminus. That didn't necessarily improve it, however; later that summer "Justo" complained that in spite of the place's name, granite had not found its way into the material forming the town: "Board, slab and waney-edged shanties with here and there a half breed dug-out, appear to constitute the available encumbrances on the real estate of the city. It [has] but one slightly curved avenue throughout it, and that like a shepherd dog's tail, very short and stumpy." By this time it was expected that the railroad would be built farther up the canyon to where quarrying operations were then taking place, and there were fears that a rival town would spring up there. But Justo didn't think it would set Granite City back one bit, "as they could couple the town together in an hour and move the thing up in an evening."[49] In March 1874 Archibald noted that "Granite City is very lively, but not quite so gay as a first-class cemetery."[50]

One thing Granite City did produce was a number of pen names. It seemed those who wrote for the newspapers didn't want their real names associated with the city. Archibald and Justo have been mentioned, but one of the most original pen names to appear was "O. Phiddle, D.D."[51]

The granite quarries continued to be worked through the 1880s, by which time most of the stone required for the temple had been cut. The quarries then fell into disuse for a number of years. In 1899 they provided the stone for the foundation walls of the Kearns mansion in Salt Lake City.[52] In 1911, when plans for the state capitol building were being studied, there was some desire to have the structure built with Utah materials. The Little Cottonwood granite was found suitable, although the rock that was chosen was farther up the canyon from the temple quarries. The Utah Consolidated Stone Company, a consortium of four local companies, was formed to handle the quarrying and preparation of the stone.[53] Railroad promoter J. G. Jacobs announced he would build a new railroad line called the Salt Lake and Alta Railroad from Sandy to Wasatch to haul the stone. Quarry operations began in the spring of 1913, and the railroad was completed and ready to haul stone in November.[54] The stone was transported to plants in the city where it was dressed; it then was loaded onto cars that were taken to the capitol building site over the street railways. Stone was also quarried for a building at the University of Utah, and after the Utah State Capitol was completed, granite was provided for the basement of the Idaho State House at Boise.[55] Except for the two major structures in Salt Lake City, all other quarrying was sporadic and for relatively small projects.

*

Wasatch Resort. At mile 4.4 a road turns off to the south to the old Wasatch Resort. When the granite quarry crew moved into the canyon leaving Granite City behind, they established their quarters on the south side of the stream. It was a pleasant spot, nicely shaded by trees, cooled by the stream, and graced by towering cliffs on both sides of the canyon. The workers built a bridge across the stream and spent their leisure time making walkways and gardens, putting up a flagpole and bowery, and building and decorating their own little cottages. As those who were fortunate enough to be invited to visit the spot carried word of it back to the city, Wasatch became a place to visit. By the late 1880s Wasatch had become a resort rather than a domicile for the quarry workers. It also became a settlement of "church" people, with summer residences being constructed for such dignitaries as LDS Church leaders Wilford Woodruff and George Q. Cannon.[56] By 1894 a hotel had been built that later was rated "small but excellent." As its popularity increased, the hotel was enlarged and more cottages were built.[57] The resort lasted until about 1915 or 1916, when it closed its doors. For the next few years it was used as residences for men working for the Little Cottonwood Transportation Company, the railroad from Wasatch to Alta. In 1920 a visitor reported that the old hotel still stood, "but all its former glory of furnishings, crowded lobby and throngs of happy visitors are but a dream." Most of the summer cottages had fallen into ruins.[58] Today there are a number of cottages and homes on both sides of the stream, the large building over the intake for the Murray City power plant, and the fading building that once housed the Whitmore Oxygen Company and still has a hydroelectric generator operating in its lower level. The grand hotel is gone, as are the gardens, fountains, and pathways that made Wasatch Resort so desirable a century ago.

*

Coalpit Gulch. Coalpit Gulch is located on the south side of Little Cottonwood Canyon at mile 5.6. It is a narrow defile whose opening is high above the canyon bottom, offering a beautiful and spectacular waterfall when its stream is running. It takes its name from activities in the main canyon during the early years of the mining boom. When the mines began operating in earnest, they needed coal for their blacksmith shops and steam plants. Coal was not readily available, but charcoal could be used. When smelting plants were erected, they too had the need for coal or charcoal. Late in 1870 the Little Cottonwood Smelting Works was built at the mouth of the canyon by Col. David E. Buel for J. C. Bateman & Co.

and operated first as Buel and Bateman and later as the Flagstaff Smelter. In mid-November Buel announced they would be ready to receive ore and charcoal in two weeks preparatory to commencing extensive operations.[59]

Farther up the canyon, at Tanners Flat, Messrs. Jones and Pardee built a smelter and began operating in July 1871. A *Salt Lake Tribune* reporter stated that the proprietors obtained their coal from a grove of quaking aspen and "spruce pine" a couple of miles above their works. A number of pits were being burned, and no difficulty was anticipated in supplying the two furnaces operating in the canyon with charcoal from this body of timber, at least during the summer season.[60] The men who burned timber to make charcoal were called coal-burners. While their product may have been much desired by those who used it, they often were not held in high esteem by others. As early as May 1871 one Hawkeye, a *Tribune* correspondent, wrote that one species of vandalism practiced in the canyon deserves the most severe and speedy remedy, the destruction of timber of which there is not any surplus. He maintained that for every tree suitable for timbering mines that is put into charcoal another must be brought from some other region to replace it.[61] This was not a complaint unique to Little Cottonwood Canyon, for similar complaints were made in other mining districts throughout the years that followed. Quite aside from the burning of timber that otherwise could be used for mining purposes, this was a time when the U.S. Land Office was becoming active in protecting the timber resources on public lands. In 1872 the United States District Attorney placed a notice in the newspapers:

> Woodchoppers and Coal Dealers Take Notice: All persons cutting timber on United States lands for coal will be prosecuted criminally and the severest penalties enforced. The judicial and military force will be employed to arrest all trespassers on public lands who cut timber for coal purposes. George C. Bates, United States District Attorney, Utah.[62]

In spite of these objections and notices, the coal-burners continued to operate. In 1873 S. J. Sedgwick, in an article in the 19 August *Sale Lake Herald* on his visit to Little Cottonwood Canyon, wrote that the fires and smoke of the burning pits were seen in many places. But the coal-burner's days were coming to an end with the coming of the railroad. When the Wasatch and Jordan Valley Railroad was built to Wasatch at the mouth of the canyon, it began hauling charcoal into the district while it hauled ore out. Soon the smelters were importing charcoal from distant states and the need for the local product diminished. But the pits where the coal was burned remained for many years and gave their name to the gulch hanging high over the canyon.

*

Hogum Fork. On the south side of Little Cottonwood Canyon, at milepost 7, a drainage drops steeply into the main canyon. This is Hogum Fork, whose name dates from the early days of the mining boom in the canyon. At that time, when there were many, literally hundreds, of teams hauling equipment and supplies up and large quantities of ore down the canyon, there was need for places to pause and allow the teams to recruit. One of these was near the confluence of the Hogum Fork stream and Little Cottonwood Creek. In April 1874 a Land Office surveyor who was running a survey line along Little Cottonwood Creek noted that this place was a thriving community. "It contains," he wrote in his notebook, "a store, a harness shop, a blacksmith shop, a boarding house, two saloons and a wash house. Also a large stable. The population of the town is probably 60."[63] He didn't say whether this was a permanent or transient population, but his comments do indicate that it was a town of some importance. The name of this place was Hog-Um. How long this settlement remained is not known, but it has long since vanished, leaving behind only its name to identify the fork whose stream still flows down to the old town site.

Incidentally, the place where the town of Hog-Um was located is relatively level. The surveyor added a comment to his notes saying that the place was known as White's Flat, named after Robert White, the first settler. In June 1874 one J. B. White filed a mining claim for the Maybird Gold, Silver and Lead Mining Company, located one and one-half miles southeast of Whitesville. And in August 1875 Robert White was one of three men filing a claim on the Snowdrop Lode, albeit farther up the canyon.[64]

The area where Hog-Um was located may be reached from a parking area at mile 6.5, where a road goes down toward the canyon bottom. This was an access road to the power plant, used after the present highway was built and the old canyon road was abandoned. When it meets the old canyon road, which parallels the stream, that road can be followed up canyon a short distance to the site of the old Columbus Power Plant. The road disappears under debris pushed down when the present highway was constructed, but it can be picked up again a little farther up the canyon and followed until it joins the present highway at Tanners Flat. The site of the Woolley sawmill is directly across the stream from the access road, where roads and trails used by the loggers can be found.

Hogum Fork is difficult of access; the approach from the canyon bottom is very steep and choked with vegetation. It is much easier to come down the drainage than to go up it, but even that is a challenge. The upper part of the fork contains an enormous boulder field, about a mile by a mile and a half at its maximum extent, surrounded on three sides by vertical granite cliffs. The only reasonable point of entry is a low spot on the ridge between Hogum Fork and Maybird Gulch, just north of the Pfeifferhorn. And this point is a long way from any easily accessible trailhead. For these reasons, Hogum Fork has enjoyed relative isolation, is seldom visited, and is viewed only by those climbing the Pfeifferhorn or hiking the ridge between the Pfeifferhorn and Thunder Mountain.

In 1937, however, this isolated and little-known spot was thrust into the limelight of publicity. The drama began in the very early morning hours of Tuesday, 15 December 1936, when a Western Air Express flight, a Boeing 247 twin-engine transport carrying four passengers and a crew of three from Los Angeles to Salt Lake City, with an intermediate stop in Las Vegas, was droning through the darkness toward its destination. At 3:27 a.m. the pilot had checked in by radio at Milford, reporting "Everything O.K." The plane then continued on its northerly course toward Utah Lake, where the crew would pick up the southern leg of the Salt Lake range station and fly the "beam" to their destination airport.

In 1936 the range station was the most modern of aerial navigation devices; it consisted of two low-frequency radio transmitters whose signals were sent in predetermined directions. One transmitted, in Morse Code, a repeated letter A, dot-dash, the other a repeated letter N, dash-dot. Midway between the directions of the two transmissions, the signals merged to give a steady tone, the "beam," which could be followed directly to the range station. In the case of the Salt Lake City station, located immediately north of the airport, the southern leg, or beam, was oriented slightly east of due south, taking it down the Salt Lake Valley and across Utah Lake. While the range stations were a significant aid to aerial navigation, especially at night and in foul weather, they suffered an equally significant disadvantage, that being their transmission at low frequencies. The Salt Lake station's transmitter operated at 227 kilocycles. Today's AM radio stations operate roughly between 550 and 1600 kilocycles, and listeners know how bad weather can cause a great amount of static on AM radio. The low-frequency range stations suffered the same problems, which became considerably worse at the very time that the signals were most needed—during very bad weather.

During the early morning hours of Tuesday, 15 December, the weather over northern Utah was not good. There was a low overcast preceding a midwinter storm. In the cockpit of the Western Air Express airliner, veteran pilot S. J. Samson and copilot William Bogen listened intently through the static coming out of their headsets for the south leg of the Salt Lake range. When they reached it they would alter their course to the north and follow the beam to the Salt Lake airport, the destination they were destined never to reach.

When nothing was heard from or about the flight by daybreak, the plane was assumed to be down and a search was begun. Two Western Air Express airplanes flew over the route between Salt Lake and Las Vegas, and two army airplanes out of Salt Lake also joined the search, but all were hampered by bad weather and poor visibility. Reports that the airplane had been heard flying over Pleasant Grove and Alpine caused the search to focus on the area around and above Alpine; but, in spite of both air and ground searches, nothing was found. Other reports from locations farther south and west caused the search to be expanded until it covered most of the air route between Salt Lake City and Milford. The ground search was augmented by up to 400 men from several Utah camps of the Civilian Conservation Corps, as well as soldiers from Fort Douglas. On Sunday the airline company offered a $1,000 reward for information leading to the discovery of the missing airplane, causing many volunteers to join the search. Famous aviatrix Amelia Earhart got involved when she flew into Salt Lake City on Monday and on a hunch briefly searched north toward Brigham City. By the following Wednesday, with the airplane missing for over a week, more than one thousand military and CCC men were reported on the search. They were joined by four marine corps airplanes from San Diego and two army bombers from March Field. Then, on Christmas Day, a snowstorm forced most of the efforts to be halted. More storms followed until by the end of the month the search was virtually stalled by deep snows and cold weather. The marine corps airplanes and the army bombers were ordered back to their home bases as soon as weather would permit. By 5 January the air search had dwindled to but one airplane. The reward offer had expired the previous day, removing the incentive for volunteers to brave the winter weather. The search had lost its momentum. For all the effort that had gone into it, no one had a clue as to where the missing airliner might be.[65]

The mystery remained for the rest of the winter and following spring. Then, late in June 1937, Alpine residents Frank Bateman and Fred Healy discovered some weather-stained letters in the snow high in Dry Creek Canyon above Alpine. The letters were identified as having been aboard the missing airliner. The search began anew, this time centered in Dry Creek Canyon, where many people, including the Western Air Express chief pilot, all along had believed the airplane would be found. Again men from CCC camps and soldiers from Fort Douglas were made available for the search, and aircraft combed the area. More letters were found, and the search centered on the mountainsides around Lake Hardy, below and between Lone Peak and Chipman Peak. A permanent base camp, Millset, was established in the upper reaches of Dry Creek Canyon. Then, on Sunday, 6 June, two groups of men inspecting the ridge above the camp simultaneously came upon small bits of wreckage, including a torn mail sack. The events of that early morning in December began to unfold. When the pilots were unable to pick up or recognize the south leg of the Salt Lake range station, they continued flying northeast, over Alpine and toward the high granite ridge that marks the southern limits of the Little Cottonwood Canyon drainage. The airplane plowed into the ridge only a few feet from the top and was catapulted up and over the vertical cliffs on the north side to fall some 800 feet onto the snow-covered slopes of Hogum Fork. In all likelihood, the incident triggered an avalanche that cascaded down to cover the wreckage, making it impossible to spot during the subsequent air searches. In fact, even at this late date, almost six months after the accident, the wreckage was still hidden from sight.[66]

The difficulty of reaching the wrecked airplane and its victims, if and when they would be found, was becoming obvious. The cliffs separating the site from the base camp were a formidable obstacle, yet the alternative, an approach from Little Cottonwood Canyon to the north, seemed equally forbidding. Two men explored the route through White Pine Fork, crossing into Red Pine Fork, then across the hog-back ridge into Maybird, and up that gorge until they could cross into the head of Hogum. These men actually reached the site of the wreckage and found a quantity of mail and a portion of the instrument panel, which they brought back with them. Another party, composed of men from the Provo CCC Camp, succeeded in hiking directly up Hogum Fork and declared their route the shortest and most feasible, but authorities rejected it in favor of taking horses up the White Pine trail. In the end that route was abandoned because horses couldn't be taken even close to the site. Ultimately, access was via the ridge, with ropes and a windlass used to lower and raise searchers and equipment. In the process more bits and pieces were found in recesses in the cliffs, but the main wreckage defied discovery. The search was further hindered and endangered by rain and snow squalls and repeated avalanches and falling rocks that cascaded down upon the slopes below. Several times the search was suspended due to the dangers to the men, but even then a crew remained at the camp and made daily inspections from the ridge.[67]

On Wednesday, 16 June, melting snow exposed a portion of the fuselage. Men with shovels began digging to uncover the airliner's cabin, only to find the sides intact but the bottom missing. No bodies were found. Falling rocks and small avalanches continued to bombard the workers, causing the search to be suspended again. Finally, on Friday, 2 July, the first body was exposed, that of Mrs. John F. Wolfe, a passenger. It was hoisted to the top of the cliffs and taken down to Alpine, as were all bodies recovered later. Four days later, the body of her husband was found. Mrs. Wolfe had been an employee of Western Air Express in Seattle for four years and had left her post on the first of December to marry John Wolfe, a Chicago businessman and aviation enthusiast. The couple was en route to Chicago from their Mexican honeymoon at the time of the accident. Now that a pattern had been established, Hogum Fork slowly gave up the rest of the victims it was hiding beneath its snows. On Thursday, 15 July, the 37-year-old pilot, S. J. Samson, of Los Angeles, was found. He had been the fifth-oldest pilot in Western Air Express service seniority, and he left a wife and seven-year-old daughter. The next day passenger Carl Christopher, 37, a businessman from Dwight, Illinois, was given up. Christopher was the only victim who did not have some connection with aviation. Another body, that of passenger Henry W. Edwards, 31, of Minneapolis, was exposed on Tuesday, 3 August. He was a radio superintendent for Northwest Air Lines. The following day 26-year-old stewardess Gladys Witt's body was returned. Only the 29-year-old copilot, William Bogen, of Los Angeles, remained missing. Hogum Fork held him for another week, finally giving his body up on Wednesday, 11 August.[68]

With all the victims recovered, a formal investigation was made by a board of inquiry headed by Department of Commerce officials. Frank J. Eastman, Western Air Express station manager at Salt Lake City and director of the search effort that had extended over eight months, said that he had permission from the Forest Service to use dynamite to blast the cliffs above Hogum Fork, removing forever any evidence of the plane's wreckage.[69] It is not known whether this was done, but Nature has a way of taking such matters into Her own hands. Each winter the snows press down to rearrange the boulder fields and avalanches cascade down, bringing rocks with them. But while they might cover, they also might uncover. So even today, more than sixty years later, bits and pieces of the Western Air Express wreckage can be found in the upper reaches of Hogum Fork.

*

Maybird Gulch. Maybird Gulch is the next drainage east of Hogum Fork, at about mile 7.4. As is the case with all drainages on the south side of the canyon, this one is high above the canyon bottom and drops steeply at its lower end, near its confluence with Little Cottonwood. In June 1874 the Maybird Gold, Silver and Lead Mining Company claim was filed, located on the south side of Little Cottonwood Canyon, one mile southwest of Tanner's hotel.[70] At that time the drainage was known as Miner's Gulch, and the creek was called Miner's Creek as late as 1882. The Maybird mine was worked, relocated, and used as a reference into the twentieth century. The first use of the name Maybird Gulch was found in the Mineral King claim in 1901.[71]

Pfeifferhorn. The Pfeifferhorn is the peak at the south end of the ridge between Maybird Gulch and Hogum Fork. Its very distinctive horn shape may be seen from many high points throughout the Wasatch, but despite its prominence it does not appear in any historical records. Some early maps show it as the Little Matterhorn, but today it is universally known as the Pfeifferhorn.[72] This came about as the result of its being named by members of the Wasatch Mountain Club as a memorial to one of their own. Charles (Chick) Pfeiffer was a very popular, well-known member and president of the Wasatch Mountain Club. His friends were astounded when, in March 1939, he was discovered dead in his shoe repair shop in Salt Lake City, still wearing his ski clothes after a skiing trip to Brighton. Some time later, when a group of mountain club climbers were ascending the Little Matterhorn, they decided that the peak should be named after their friend. They improvised the name Pfeifferhorn, a name that has survived for sixty years and probably will survive for many more.[73]

*

Tanners Flat. At milepost 8 is the Tanners Flat Campground. This was an important station in the canyon during the mining days. It also seemed to be a transition point between very heavy snows and lighter snows, such that teamsters using sleighs could come down to this point, and teamsters using wagons could come up this far. Or, if the wagons were immobilized by the snow, Tanners Flat was a place where the ore could be stockpiled until wagons could get there. This also was the terminus of the aerial tramway put in by the Continental-Alta company, and later rebuilt by the Michigan-Utah company people. The flat got its name from the Nathan Tanner family. There were at least six Tanners in the area in 1870: Nathan, Sr.; Nathan, Jr.; J. W.; M. H.; Stewart; and Alva. They set up a camp on the flat and posted two claims in the area.[74] By early spring of 1871 buildings were being erected and a telegraph office was installed, giving Tannersville, as it was now known, communication with the outside world. One of the buildings was a hotel or inn where a traveler could get a meal. As early as 26 April Samuel A. Woolley had stopped at Tanners for a meal. J. W. Tanner was running the inn, while Nathan Tanner, Jr., had a contract to haul Emma Mine ore and was putting fifty teams on the job.[75]

In March 1871 Jones and Pardee began construction of their smelter at Tanners Flat. They erected a building forty feet square and two stories high, which was to hold two furnaces. At the end of April they were hauling charcoal imported from California to their furnace, and by mid-June they were in operation, producing bullion from Emma Mine ore.[76] In the middle of July the furnace was running with charcoal produced by burning aspen and pine in Little Cottonwood Canyon. The furnace, incidentally, had a lining of sandstone from Red Butte Canyon. In spite of its apparently successful operation, the furnace was idle in October. At the end of 1871 it was sold to the Wellington Mining and Smelting Company, and it produced a small amount of bullion from that company's ore for the next six months.[77] The furnace doesn't appear to have been used after the middle of 1872.

In September 1872 J. W. Tanner's inn caught fire from an unknown source and was destroyed in the conflagration. Tanner and his family escaped but were left destitute.[78] Apparently they left the area and took the rest of the family with them, for none of the Tanners appear in any canyon news or mining recorder entries after that time. The following year a new hotel, the Mountain House, was erected and kept by Cy Iba, who had been the proprietor of the Grand Hotel at Alta. He ran the hotel for two years. Fritz Rettich ran the Mountain House in 1875; after that time nothing more was heard of the hotel.[79] By that time the tramway had been built past Tannersville. Perhaps the village was being left behind in the march of progress.

*

Red Pine Fork. Red Pine Fork is a drainage on the south side of Little Cottonwood Canyon about a half mile east of Maybird Gulch and directly across from Tanners Flat. In the early days of the mining era it was known as Upper Slide Canyon, referring to the logging slide located there to bring logs down from the heights for the Woolley sawmill a short distance down Little Cottonwood Canyon. The Upper Slide name probably first was used by loggers and picked up by miners, who used it as early as October 1864, and continued to use it as late as July 1870.[80] Samuel A. Woolley, who was running the Little Cottonwood sawmills at this time, began calling the mountain at the slide the Red Pine Mountain. That name also was picked up by miners and was applied to the slide, Red Pine Slide, and to the entire fork.[81] The falls created by the stream as it drops into the canyon were as spectacular then as they are today; they were mentioned in a number of mining claims and were the inspiration for the Niagara Claim, situated "near the falls from which this lode takes its name." The hill on the southeast side of the drainage, Woolley's Red Pine Mountain, was briefly known as Niagara Hill.[82] Woolley called the stream flowing out of the fork Crapo Creek, after John Crapo, one of his loggers.[83] However, no one else seemed to accept that name. Miners called this fork Tanners Gulch as a result of Tannersville and Tanners Flat being located near the confluence of the Little Cottonwood and Red Pine streams, but that name was only briefly used.[84] Today the Tanners Gulch name is applied to a very steep drainage on the north side of Tanners Flat.

*

White Pine Fork. In 1866 Samuel A. Woolley called the hillside next to this fork White Pine Mountain. The name was picked up by miners and applied to the logging slide located there, White Pine Slide. The fork also picked up the White Pine name, but it did not appear until 1891.[85]

*

Snowbird Area. The Snowbird Ski and Summer Resort extends from its entry one at mile 9.9 to entry four at mile 10.7. The resort was based on an accumulation of patented mining claims and took its name from the Snowbird Claim, located at the portal of the Wasatch Drain Tunnel.[86] Although its very name is steeped in history, the resort made use of historical names only during its very first years. Its first ski runs were primarily located in Gad Valley, a broad drainage on the south side of the canyon directly above the lower parking lot. The name, used by Samuel A. Woolley as early as June 1869 when he was looking for a site for a new sawmill, was one of the miners' early contributions, since a gad is a pointed steel bar used to break up or loosen ore.[87] Peruvian Gulch, at mile 10.9, takes its name from the Peruvian Lode, which was located there in June 1870 by James Wall, the original recorder of the Mountain Lakes Mining District, and his associates.[88] His lode received much attention and immediately became a point of reference for other claims. One year later the Peruvian Gulch name was in use. Wall worked the mine sporadically for many years, and he was still at it in 1892. However, the Peruvian was a fairly small producer. In a 1902 estimate it was suggested that the mine had generated only $50,000 in ore during all its years of operation.[89] But its name lives on and is used for a lift at Snowbird and a lodge at Alta. The gulch hosted many other claims, including a group of six at its upper end, under the Bullion Divide, the ridge between Mount Baldy and the American Fork Twin Peaks. They were known as Iron Blosam Numbers 1 through 6, and that name was used for the first lodge built at Snowbird Resort.[90] On the slopes west of the Snowbird tram was the Harpers Ferry Mine and tunnel, a name used for two of the resort's ski runs.[91]

*

Alta Area. Previous chapters have dealt with Alta and its origins, and some names were mentioned there. There are many more, however, most being reflections of names from the past that should be noted, and some have stories that should be told.

Mount Superior, with its great concave face that poured an endless succession of snow avalanches upon the canyon road below, is the symbol of modern-day Alta. Its name came from the Superior Lode, located on the east side of Superior Gulch in the early days of Little Cottonwood mining.[92] The mine gave its name to the gulch and the mountain above it.

Hellgate, the narrow part of the canyon between Snowbird and Alta, with blue and white limestone cliffs towering above, takes its name from Fritz Rettich's early-twentieth-century Hellgate group of claims.[93] The dump that can be seen on the north side of the highway, directly below the cliffs, is from the Frederick Tunnel, originally driven in the early 1870s to work the Frederick and Crown Prince Claims on the mountainside directly above, but over the years extended to great depth, eventually connecting with the Columbus-Rexall Tunnel, which extended to the Cardiff workings under Cardiff Fork. It came under Rettich's control and was used to work his Hellgate claims. He, and later his son, Hugo, worked the property until the latter's death in 1934.[94]

The entire slope on the north side of Alta was known as Emma Hill. Although the Emma Mine covered only a very small part of the hill, its prominence in the district spread in all directions. Portions of the hill had other names as well, reflecting the importance of mines that were in the immediate area. The slopes above the Frederick Tunnel were known as Frederick Hill, or Frederick Mountain. Directly below Cardiff Pass was Toledo Hill, named after the Toledo Mine.[95] Flagstaff Mountain, the high point on the ridge north of Alta, was named after the Flagstaff Mine. If one ascends the peak, it will be seen that it is not a true summit but only the intersection of the divide between the two Cottonwoods and the Reed and Benson Ridge, and that the latter rises somewhat higher than the high point of Flagstaff Mountain. From Alta it appears as the summit of a mountain, however, and so it was named. Emily Hill, named after the Emily Mine, was directly below Flagstaff Mountain.[96] The north slopes above Grizzly Gulch were, and still are, known as Davenport Hill, named after the Davenport Mine.[97]

Flagstaff Hill was the slope surrounding the Flagstaff Mine, somewhat east of Flagstaff Mountain.[98] Over the years several tramways served the Flagstaff mine. On the north slopes, almost due north of the Alta Lodge, lie the remnants of the most recent one, a wire-rope gravity tramway. The area around the lower terminal shows signs of a road that circled an ore bin. The cables, now on the ground, still run up the slope to the mine dump. This tramway was constructed in 1926, when the Flagstaff was part of the Mineral Veins Coalition Mining Company. The cable and various attachments were hauled over from the Pittsburg Mine in American Fork Canyon.[99]

On the south side of the canyon the ridge directly above the Peruvian Lodge was known as Peruvian Ridge, named after the mine that gave Peruvian Gulch its name. The north end of the ridge was known as Lexington Hill, after the Lexington Mine. General P. E. Connor gained control of the mine in 1873, one of his few mining ventures in the Wasatch Mountains. He filed a location for the Lexington Tunnel the following year, but nothing was done until Col. Elijah Sells joined him. Another tunnel claim was filed in 1880, and after that time it became the Sells property, although the tunnel was known as the Lexington Tunnel as late as 1918. In 1921 George H. Watson took over the mine, after which time both the Sells and Lexington identities were lost.[100] Lexington Hill was also known as Alpha Hill, after the Alpha Mine, and Peruvian Hill, after the Peruvian Mine.[101]

Collins Gulch has been mentioned before, it having taken its name from Charles H. Collins, who ran the Collins group of claims in that drainage. At the head of the gulch is a summit known as Mount Baldy. It is a natural name, for the summit has few trees and from below appears totally bald. It was called Bald Mountain as early as 1871.[102] On the east side of Collins Gulch the Rustler Mine was located on the ridge.[103] It left its name behind for the North and West Rustler slopes, and the noted High Rustler ski run. The north end of that ridge, overlooking the old town of Alta, was known as both Iris and Emerald Hill, named after mines, and both names were used into the twentieth century.[104]

Several other claims left their mark on present-day Alta. The Germania Claim was on the east slope of Bald Mountain, extending down to the pass between Baldy and Sugarloaf. This was a fairly recent claim, having been recorded in 1905 by the Alta Germania Mining Company, an organization soon to fall under the control of George H. Watson. The name came to be used by one of Alta's early ski lifts.[105]

The Greeley Claim, filed in 1884, was located high on Emerald Hill. It was one of Alta's better mines during the late-nineteenth-century slump in mining. In November 1893, when there were not more than twenty miners employed in Alta, three of them were working at the Greeley. The mine eventually fell under Watson's control and lost its identity in his conglomerate.[106] Its name, however, survived in Greeley Bowl, a favored powder skiing slope where the mine once was located.

The Neversweat group of claims was located on the east side of the Albion Basin stream. When the first ski lift was built in the basin in the 1960s, it crossed the Neversweat claims and took their name. Since that time, however, the Neversweat lift has been renamed and is now known as the Albion lift.[107]

*

Patsey Marley Hill. This hill is east of Alta and south of Grizzly Gulch. The name came from the Patsey Marley Claim, filed in May 1870. Had this claim been like so many others in the area, it would have been soon forgotten; but it did yield some good ore, enough to build a reputation and implant its name firmly in the minds of contemporary miners. And so the hill on which the claim was located became known as Patsey Marley Hill. It probably would not surprise readers to learn that one of the ten locators of the claim was a man by the name of Patsey Marley.

Patrick Marley was born in Ireland in 1840 or 1841. When he left his family home he went to London for a period of time, then emigrated to the United States. It is not known what drew him to Utah, but he arrived in the territory early in 1870 and soon gained a considerable reputation as both a pugilist and a miner. The first recorded evidence found of Marley's presence in Utah was the recording of the Patsey Marley claim on 23 May 1870, it having been "dated at Curtis & Spaffords cabin."[108] Levi Curtis and W. W. Spafford, the owners of the cabin, were two of the locators of the Grizzly Claim, filed just six weeks earlier. At this time Marley was about thirty years old; he must have had a considerable knowledge of prospecting and mining matters, for six weeks later the *Salt Lake Herald* mentioned that "Patsey Marley brought into town ninety six pounds of bullion, the result of three quarters of an hour run of his newly built smelting works."[109] This is a bit of a surprise, as mining histories mention nothing of this smelter, but the mining recorder's books do, for the following year there was an entry referring to "a furnace known as the Marley Furnace."[110] And, if Marley really brought bullion rather than ore into town, he must have used some sort of furnace, however crude it may have been. The mine was becoming known. A letter from Salt Lake City was published in the *Missouri Democrat* claiming Marley had a number of men at work for him and that the claim was estimated to be worth $100,000.[111] The following February it was reported that the mine was bonded by western parties for thirty days; that is to say, the parties had an option to buy. As often happened when an option was taken on a mine, the discovery of a rich deposit in the mine was reported. Then the mine's production of news worth printing ended, but its brief days of glory assured that it would be remembered by the name it gave to the hill where it was located.

Patsey Marley remained in the area, although his name appeared on only two other claims in the district, neither one amounting to anything. But he made more news with his other profession, that of a pugilist. Some sparring matches were arranged, "mills" they were called, and others were spontaneous. The latter more often than not ended in police court or jail. One such event, on election day in August 1871 at Central City, started as a friendly match but quickly escalated to involve rocks, then pistols. It ended with one man being shot in the stomach and dying early the following morning. In 1873 Marley and an associate, Matt Brennan, opened a saloon at 7 Commercial Street in downtown Salt Lake City, where they taught boxing, foiling, and some gymnastics.[112] The saloon was host to incidents that involved arrests and appearances in court. In October 1873 Patsey was back at Alta where he engaged in a series of matches at Nick Dramer's Hall in the Grand Hotel. He was willing to take on anyone willing to challenge him. He gave a good account for himself in a series of matches, but he admitted he was getting too fat to fight and only put the gloves on for the exercise it gave him.[113]

In spite of his skirmishes with the law, Marley was a very popular fellow. The press loved him for the stories he gave them, such as someone posting his name on the Central House bulletin board as candidate for the presidency of the United States, with other prominent citizens as candidates for various other offices.[114] And when a desperado by the name of George Curran assumed an alias of Patsey Marley, the newsmen happily published Patsey's denial that he was not the same man. "Patsey," they wrote, "has attended a few Irish weddings but has not and never expects to disturb other people's rights or property."[115] They then started calling Curran "Patsey Marley No. 2" and for the next two years followed and reported his misdeeds ranging from Salt Lake City west into Nevada.

Patsey Marley himself drifted west and for some time engaged in mining in the Tooele and Stockton areas, then moved north into Idaho. Little was heard from him in local circles and it seemed he had drifted into oblivion. But then he returned to Utah, took up some claims in Farmington Canyon, and worked them in spite of his advancing years. In July 1905 he was brought to Holy Cross Hospital in Salt Lake City in a most enfeebled condition, disabled by inflammatory rheumatism. He was destitute, but he still had some friends. D. P. Felt, editor of the Davis County *Argus,* undertook to contact Marley's mining friends to help him. Marley was back in the hospital at least two more times in the years that followed, but he was not easily put down. In 1912 he was still working his claims near Farmington. He struggled along until mid-1916, when his health became so seriously impaired that he had to move to Salt Lake City. After a period of time in the county infirmary he was moved to the county hospital, where he died on 12 December. Someone who knew not only him but also his dire financial situation stepped forward and had a brief article published in the *Salt Lake Tribune.* It said that unless his friends subscribed to a fund to pay for a funeral, he would be buried in a pauper's grave. Within a week's time funeral arrangements were made, and on Friday, 22 December 1916, funeral services were held, attended by a few old friends. One man, dressed in tattered clothing, tiptoed into the chapel and stood in a far corner. Several times he was seen to wipe his eyes with the sleeve of his tattered coat. It was later learned that he was an old friend of Marley, one who was in similar financial circumstances and who wished to remain inconspicuous while he bid his friend farewell.[116] Patsey Marley now rests in an unmarked grave in Mount Calvary Cemetery.

*

Albion Basin. This is the basin to the south of Alta, known in the early days as the South Fork of Little Cottonwood Canyon. The name comes from the Albion Mining Company, but it has a long history behind it. In August 1870 four men recorded the Wellington Lode, located a little more than a quarter mile northwest of Cecret Lake. The Wellington Mining and Smelting Company was incorporated in San Francisco to run the mine.[117] The company bought the Jones and Pardee smelter at Tanners Flat and used it for a short time to reduce its ores. It drove several tunnels into the lode and built a number of buildings to support its operations. The mine produced rich ore for a number of years, but then the quality degraded and operations were discontinued. The mine was turned over to lessees, who ran it until the late 1870s. During that time the Wellington came into the news as the result of two tragedies. The first was in April 1872 when a heavy snow avalanche came down from the summit of Wellington Hill, the mountain above the mine. It missed the mine buildings but crossed the trail to the mine, where the day shift of miners, eight in number, were on their way to their daily toil. All were carried some five hundred feet down the hill. The mine workers were not buried deeply and were able to extricate themselves, but superintendent H. H. Murray died of suffocation before he could be found.[118]

In the afternoon of 28 December 1876 another snowslide came down the mountain, this time striking the boardinghouse, crushing the building and burying the occupants: four men, one woman, and one child. On New Year's Day a miner came over the pass from American Fork Canyon and went down to Alta. On his return in the afternoon he noticed that the dwelling had been covered by a snowslide. He hastened back to Alta to give the alarm. A large group of volunteers went to the scene and worked late into the night, but when they found neither victims nor survivors they gave up and returned to their homes for the night. Unknown to them, two men were still alive under the surface. They could hear the would-be rescuers and even recognized the voices of their friends. They shouted but were not heard. When the search party left, they gave up all hope, and one of them became delirious. The next morning the search party returned, and, when probing with a long iron rod, struck the roof of the house. They were surprised to hear a knocking in return. Quickly sinking a shaft some twenty feet through the snow, they found the two survivors. J. W. Brown was delirious and died several days later, but John Varcoe was in good spirits and declared that moment to be the happiest in his life. Three bodies were recovered that same day, but that of the final victim was not found for nearly seven months, on 23 July 1877.[119]

In 1880 a group of claims was filed for the region around the abandoned Wellington Tunnel. One of them was the Albion, which was patented and whose name was then applied to the old Wellington Mine.[120] Little was done with it, however, until 1886, when it began to be worked by lessees. In 1898 the Albion Mining Company was incorporated to take over the property. It was worked with considerable vigor and became one of the regular producers in the Alta mining camp. By 1915 George H. Watson was on the Alta scene and became manager of the mine. The Albion was soon rumored to be involved in a big merger of Alta properties. It took several years, but in 1919 the Albion was consolidated with several other properties to form the Albion Consolidated Mining Company, with George H. Watson as president.[121] Two years later Watson organized the South Hecla Mines Company, which absorbed the Albion Consolidated company.[122] From that point the mine lost its identity and was carried along in Watson's corporate shell game. But its name did survive in the many Albion patents that remain on the mineral survey maps, and it became the source of the name of the Albion Basin.

*

Cecret (Secret) Lake. This lake with its peculiar name is located high in the Albion Basin, between and below Devils Castle and Sugarloaf Peak. When prospectors started posting claims in this vicinity, they called this small body of water Mountain Lake. In fact, one of the first claims in its vicinity was the Mountain Lake Lode, posted in August 1866. In 1871 it was also called Little Cottonwood Lake when the Flora and Flora No. 2 mining claims were recorded.[123] Those two claims crossed at the Flora shaft, high on the slope toward Sugarloaf Peak, but the first claim extended toward the south shore of the lake. The Flora claims were among the very earliest to be surveyed for patent, and U. S. Mineral Monument No. 5 was located at the shore of the lake in the process. This linked the Flora claims, the Mineral Monument, and the lake in such a way that it was perhaps inevitable that the lake would eventually take the Flora name. It first appeared in 1885 and continued to be used well into the twentieth century. Salt Lake County used the name on its county maps as late as 1937.[124] Before the Flora name came into use, Alfred Lambourne and Henry Culmer were tramping the Wasatch and naming various lakes. Lambourne called this little body of water Lake Minnie after Minnie Williamson and Minnie Shanks, a mother and daughter.[125] The mother was Wilhelmina Marie Williamson, whom Lambourne married on 13 September 1877. They had a daughter, Minnie, who later married Joseph Shanks, well after the time Lambourne bestowed the name upon the lake. The Lake Minnie name was used in the same 1885 claim that used the Flora Lake name, but it was not as well accepted and went out of use by the early twentieth century. Lambourne, incidentally, had a special attachment to this lake. He made at least four paintings of it and used it as the base of action in two of his books.[126]

Another name that was used briefly around 1890 was Cases Lake, named after George W. Case, an old-time miner who had been in the area since 1870. In 1878 the North Pole Claim was filed, the mine being located south of the lake near the cliffs of Devils Castle.[127] A cabin was built near the shore of the lake to house those working the claim. Several years later, in 1885, Case used the then-abandoned North Pole cabin while working in the vicinity, and the structure became known as Case's cabin. It was natural that some miners then called the body of water next to it Cases Lake.[128] But this name was not generally accepted either and was soon forgotten.

The Cecret name has its origin in four claims, Cecret Numbers 1 through 4, filed in 1905. Cecret No. 2 encompassed all of Flora Lake. There is no doubt that these claims were supposed to be named Secret, for the following year the Secret Mining and Milling Company was incorporated to work them.[129] And in 1906 and 1907 the company filed at least ten additional claims, all of them using the Secret name. When claims were recorded, the mining recorder always copied the claim notice exactly as it was given to him. This caused some curious and amusing entries, for many of the miners and prospectors were marginally literate and spelled words the way they sounded to them; hence, Cecret for Secret. However, the four Cecret claims went beyond the usual misspellings; they were atrociously written. For instance, the claim description for "Cecret in Little Cotonwod," was written as: "Runing 3 hendered feet in a suothwestrley Drection to 1 stak N1 from ths Discovry thens 15 hundred to stak N2 thens 600 hendred feet to stake N3 thens 15 Hudred feet to Stak N4 thens 600 hundred feet to Plase of Beging."[130] This must have been an embarrassment to the company after it was formed, for in 1907 the company's treasurer, who incidentally was one of the original claimants, filed three amended notices whose descriptions were much more legibly written.[131] When claims were surveyed for patent, the mineral surveyor always used the name of the claim as it appeared in the mining recorder's books; thus the Cecret name was preserved for posterity.[132]

The Secret Mining and Milling Company operated for about ten years before suffering hard times. A lawsuit filed in 1916 alleged that the company was insolvent and unable to pay its debts due to dissension and lack of harmony among its directors. In 1919 the company announced it had resumed development work, but within a month it modified its capitalization, increasing its number of shares from 600,000 to one million but decreasing the par value from one dollar to ten cents a share. With that move the company entered its final plunge into oblivion. In 1923 its corporate charter was revoked because it was no longer paying its license tax.[133] The company, however, left its name behind, for Flora Lake on its property became Secret Lake. That name was used in the 1939 edition of the USGS Cottonwood, Utah, map, and continued to be used in the 1955 edition of the Brighton, Utah, map, as well as the 1980 edition of the Salt Lake City, Utah–Wyoming map. However, the Cecret Lake name has been used by the Forest Service on its trail signs and maps.

*

There are many other names whose origins defy discovery. The crags overlooking Cecret Lake, for instance, are an excellent example. Known today as Devils Castle, the rugged cliffs are so imposing that the miners must have had a name for them. Yet, none is mentioned in a claim notice. The trail that zig-zags up the slope between Devils Castle and Sugarloaf, portions of which can still be seen, was mentioned often; it went to the Pittsburg Mine on the other side. But for Devils Castle, never a word. There was a Devils Slide claim, but it was in one of the steep gullies on the north slopes of Bald Mountain, one of today's Baldy Chutes.[134] Alfred Lambourne, in his book *Jo, A Christmas Tale of the Wasatch,* called it the Tower of Babel, but that name appears in neither mining records nor newspaper files. Perhaps he was using his poetic license. Yet, twenty years later he wrote, "The writer was little more than a boy when he first looked upon the high mountain lake of the Wasatch, and that great mountain at its back, by some called the Tower of Babel, and by others the Devil's Peak."[135] The names probably came from recreationists like Lambourne, many of whom summered at Brighton. The USGS used the Devils Castle name on its Cottonwood Special Map of 1907 and on every map it has issued since that time.

Notes

Note to Preface

[1]The Hartzog inscription is on the ridge of the Red Butte, above the mouth of Red Butte Canyon. In spite of my interest in this artifact and the man who carved it, J. B. Hartzog has eluded all my attempts at discovery. Since other inscriptions from the same period were made by soldiers from Camp Douglas, Hartzog may have been one of their number.

Notes to Introduction

[1]This map has been published by the Utah State Historical Society. It may also be found in David E. Miller, *Utah History Atlas*.

[2]Richard D. Poll et al., eds., *Utah's History*, 87; John W. Van Cott, *Utah Place Names*, 24.

[3]Van Cott, *Utah Place Names*, 392.

[4]Poll, *Utah's History*, 57–58; Miller, *Utah History Atlas*, map 14.

[5]Poll, *Utah's History*, 66.

[6]Miller, *Utah History Atlas*, map 15.

[7]Poll, *Utah's History*, 631.

[8]Ibid., 66.

[9]Jedediah Smith, *The Southwest Expedition of Jedediah S. Smith*, 40.

[10]Ibid., 194–95.

[11]Miller, *Utah History Atlas*, map 18.

[12]Charles Kelly, *Salt Desert Trails*, 8.

[13]Osborne Russell, *Journal of a Trapper*, 113–22.

[14]Ibid., 123.

[15]Poll, *Utah's History*, 73–76; Miller, *Utah History Atlas*, map 23; John Charles Frémont, *Memoirs of My Life*, 228.

[16]J. Roderick Korns, *West From Fort Bridger*, 10; Kit Carson, *Kit Carson's Own Story of his Life*, 60–65; Frémont, *Memoirs*, 431.

[17]Korns, *West From Fort Bridger*, 24.

[18]Ibid., 36.

[19]Ibid., 37–38.

[20]Ibid., 39.

[21]Ibid., 191; James Clyman, *James Clyman, Frontiersman*, 229.

[22]Korns, *West From Fort Bridger*, 72.

[23]Jacob Wright Harlan, *California '46 to '86*, 43.

[24]Korns, *West From Fort Bridger*, 198–99, note 16.

[25]Ibid., 203.

[26]Ibid., 201, note 20.

[27]William Clayton, *William Clayton's Journal*, 307; Korns, *West From Fort Bridger*, 204, note 28.

[28]Clayton, *Journal*, 319.

[29]Journal History of the Church of Jesus Christ of Latter-day Saints, hereafter noted as Journal History, entry for 9 August 1847; Wilford Woodruff, *Wilford W. Woodruff Journal, 1833–1898*, entry for 14 August 1847.

[30]Journal History, 22 August 1847.

[31]Clayton, *Journal*, 307.

[32]Salt Lake County Court, Minute Book A, 31 July 1852; *Deseret News Weekly*, 2:103 (6 November 1852).

[33]Lorenzo Brown, "Diaries, 1853–1859," entry for 1 April 1853.

[34]Lorenzo Dow Young, "Diary of Lorenzo Dow Young," 165.

[35]Journal History, 28 October 1849, 10 February 1848; "Historical Department Journals, 1844–1890," entry for 20 May 1850; Sir Richard Burton, *The City of the Saints*, 346.

[36]Salt Lake Stake High Council Minutes, entry for 22 April 1848; Journal History, 7 January 1851.

[37]"History of the Church," entry for 3 August 1847.

[38]John Brown, *Autobiography of Pioneer John Brown, 1820–1896*, 82–83.

[39]For instance, see Alexis Kelner, *Skiing in Utah*, and Clyde B. Hardy, "The Historical Development of Wasatch Trails."

[40]Journal History, 20 August 1847.

[41]Ibid., 22 August 1847.

[42]Ibid.

NOTES TO CHAPTER 1

[1]Levi Jackman, "Letter from Salt Lake City, 1850."

[2]John R. Young, "Reminiscences of John R. Young."

[3]William Clayton, *William Clayton's Journal*, 309.

[4]Journal History, 23 July 1847.

[5]Clayton, *Journal*, 318.

[6]Ibid., 324.

[7]Journal History, 28 July 1847.

[8]Clayton, *Journal*, 309.

[9]Ibid., 337.

[10]C. H. Wendel, *The Circular Sawmill*, 5.

[11]Robert S. Bliss, "The Robert S. Bliss Journal," 392, 394; Lewis Robinson to D. H. Wells, 1 March 1848, D. H. Wells Collection, bx. 1, fd. 4.

[12]Louis C. Hunter, *A History of Industrial Power in the United States, 1870–1930, Volume 1: Waterpower*, 16–17.

[13]Salt Lake Stake High Council Minutes, 1847–48, 13 October 1847.

[14]Ibid., 4 October 1847.

[15]Delila Gardner Hughes, *Life of Archibald Gardner*, 42. Mrs. Hughes was a granddaughter of Archibald Gardner.

[16]Robert Gardner, *Robert Gardner, Jr. 1819–1906, Utah Pioneer 1847*, 19–20.

[17]Salt Lake Stake High Council Minutes, 1847–48, 5 February 1848.

[18]Hughes, *Life of Archibald Gardner*, 43.

[19]"Epistle of the High Council to Brigham Young," Journal History, 6 March 1848.

[20]Salt Lake Stake High Council Minutes, 1847–48, 1 July 1848.

[21]*Acts, Resolutions and Memorials*, approved 4 February 1852.

[22]Journal History, 2 June 1849.

[23]Hughes, *Life of Archibald Gardner*, 43; Journal History, 13 August 1851, 15 August 1851.

[24]Salt Lake County Court, Minute Book A, 7 June 1852.

[25]Hughes, *Life of Archibald Gardner*, 65.

[26]This was John Neff, who had built a grist mill on Mill Creek stream near today's 2700 East street.

[27]Orrin Porter Rockwell was a son-in-law of John Neff, so had good reason to be at the Neff home.

[28]Gardner, *Robert Gardner, Jr.*, 24–25.

[29]For instance, see Clyde Brian Hardy, "The Historical Development of Wasatch Trails in Salt Lake County"; Craig Jay Brown, "The Allocation of Timber Resources Within the Wasatch Mountains of Salt Lake County—1847–70"; and Asa R. Bowthorpe, "History of Pioneer Sawmills and Local Canyons of the Salt Lake Valley."

[30]Selina O. Stillman, "Mill Creek Canyon," in Daughters of Utah Pioneers, *Heart Throbs of the West*, vol. 3, 5–6, and *Treasures of Pioneer History*, vol. 6, 400–42. Selina Osguthorpe Stillman was a daughter of John Osguthorpe, who came to Utah and settled at the mouth of Mill Creek Canyon in 1853. For the rest of his life he was active in the canyon and had firsthand knowledge of the mills and people involved, and probably relayed this information to his daughter. However, she was only fourteen years old when he died in April 1884. In December 1897 she married Samuel S. Stillman, a member of another pioneer Mill Creek family. He died in 1933. Her recollection of the Mill Creek mills was prepared many years later and published by the Daughters of Utah Pioneers in 1941, when she was seventy years old. See obituaries for John Osguthorpe,

Deseret News Weekly, 33:201 (16 April 1884) and 33:208; Samuel S. Stillman, *Deseret News,* 23 May 1933; and Selina O. Stillman, *Deseret News,* 4 October 1953 and *Salt Lake Tribune,* 4 October 1953.

[31]Salt Lake Stake High Council Minutes, 1847–1897, 21 October 1848.

[32]Journal History, 10 March 1849.

[33]*Deseret News,* 4 May 1892.

[34]Irene Black Wrigley, "Life of Samuel Thompson."

[35]Mary Jane Mount Tanner, *A Fragment: The Autobiography of Mary Jane Mount Tanner,* 63–66. Mary Jane Tanner was the daughter of Joseph and Elizabeth Mount. Also see entry for Joseph Mount in Susan Lindsay Ward Black, *Membership of the Church of Jesus Christ of Latter-day Saints, 1830–1848.*

[36]*Deseret News Weekly,* 1:31 (6 July 1850).

[37]Salt Lake County Court, Minute Book B, 7 June 1859.

[38]"Nancy Areta Porter Mattice," in Daughters of Utah Pioneers, *Our Pioneer Heritage,* vol. 13, 463–67.

[39]Journal History, 22 December 1849.

[40]Wilford Woodruff, *Wilford Woodruff's Journal, 1833–1898,* 27 January 1851.

[41]Salt Lake County Court, Minute Book A, 12 April 1852.

[42]Ibid., 13 and 14 June 1853.

[43]Ibid., 21 September 1853.

[44]See Henry Eyring Bowman, "Autobiography of Henry Eyring Bowman."

[45]Salt Lake County Court, Minute Book B, 13 May 1867. For more about Porter and his families see Joseph Grant Stevenson, *Porter Family History.*

[46]Salt Lake County Court, Minute Book A, 8 December 1854.

[47]Ibid., 2 October 1855.

[48]Salt Lake County Court, Minute Book B, 7 June 1859.

[49]Ibid., 21 June 1860.

[50]*Deseret News,* 17 January 1894.

[51]See Edmund Ellsworth, "Account Books, 1850–1877."

[52]See Edmund Ellsworth, "Autobiography."

[53]See Selina O. Stillman, "Mill Creek Canyon."

[54]Salt Lake County Court, Minute Book B, 13 May 1867.

[55]*Deseret News Weekly,* 16:320 (13 November 1867).

[56]*Deseret News,* 5 February 1869. The term South Mill Creek was often used to differentiate it from North Mill Creek Canyon in Davis County.

[57]See *Gazetteer of Utah and Salt Lake City Directory, 1874.*

[58]Salt Lake County Court, Minute Book B, 4 December 1865.

[59]Ibid., 4 June 1866.

[60]"John Thayn," in *Our Pioneer Heritage,* vol. 5, 197–99.

[61]Lloyd Leon Peterson, "The Pioneer Ancestors of Lloyd L. Peterson."

[62]Salt Lake County Court, Minute Book C, 4 May 1875.

[63]*Salt Lake Herald,* 6 July 1877.

[64]Salt Lake County Court, Minute Book A, 8 December 1854.

[65]Lorenzo Brown, "Diaries, 1853–1859," 30 September 1853.

[66]Salt Lake County Court, Minute Book A, 26 March 1855.

[67]*Deseret News Weekly,* 6:88 (21 May 1856).

[68]See Stillman, "Mill Creek Canyon."

[69]Salt Lake County Court, Minute Book B, 15 October 1873.

[70]*Salt Lake Tribune,* 29 November 1899

[71]*Utah Territory Census Schedules 1880,* Schedule 3: Products of Industry.

[72]"Peter Ranck," in *Our Pioneer Heritage,* vol. 5, 45.

[73]*Salt Lake Tribune,* 12 November 1879 (italics added).

[74]*Journal of Discourses,* vol. 4, 303–304, 6 April 1857.

[75]Journal History, 30 May 1852.

[76]Ibid., 7 April 1857; Brigham Young to George Taylor et al., 30 May 1857, Brigham Young Letterbook 3:578.

[77]Samuel Amos Woolley, "Diaries 1846–1899," 6 June 1857.

[78]Leonard Arrington, *From Quaker to Latter-day Saint,* 418–19.

[79]Salt Lake County Court, Minute Book A, 7 June 1852.

[80]Ibid., 31 July 1852.

[81]Ibid., 13 June 1853.

[82]Ibid., 15 November 1853.

[83]Ibid., 2 October 1855.

[84]Ibid., 1 September 1856.

[85]Ibid., 27 March 1860.

[86]Ibid., 16 October 1860.

[87]*Deseret News Weekly,* 10:260 (17 October 1860).

[88]Salt Lake County Court, Minute Book B, 24 January 1862.

[89]Ibid., 2 September 1861.

[90]Ibid., 19 June 1863, 7 March 1864.

[91]Ibid., 6 August 1867.

[92]Ibid., 15 October 1873.

[93]Salt Lake County Court, Minute Book C, 13 October 1877, 9 June 1879.

[94]Journal History, 22 October 1848.

[95]An Ordinance, for the purpose of controlling the wood and timber in the first Kanyon south of Mill Creek, 4 December 1850, in Dale L. Morgan, *The State of Deseret,* 195–96.

[96]See Joseph L. Rawlins, *The Unfavored Few: The Autobiography of Joseph L. Rawlins.*

[97]Journal History, 20 November 1851.

[98]Salt Lake County Court, Minute Book A, 17 March 1852.

[99]See, for instance, ibid., 30 January 1854, 7 September 1855.

[100]Ibid., 21 September 1853, 15 November 1853.

[101]Ibid., 30 January 1854, 7 September 1855.

[102]Ibid., 3 March 1856, 4 March 1856.

[103]Ibid., 4 June 1860, 5 June 1860.

[104]See Helen Greenland, *Williams, Pugh and North Family Biographies.*

[105]Edmund Ellsworth, "Account Book, 1850–1877."

[106]Water claims by A. V. Taylor, Joseph E. Taylor, and W. B. Stout, 5 August 1898, Salt Lake County Recorder, Water Claims Book B, 107–8.

[107]Salt Lake County Court, Minute Book B, 6 June 1864.

[108]*Deseret News Weekly,* 10:336 (19 December 1860). See also *Utah Territory Census Schedules, 1880.*

Notes to Chapter 2

[1]Salt Lake High Council Minutes, 1847–49, 22 April 1848.

[2]Ibid., 13 May 1848.

[3]Journal History, 17 October 1848; John Brown, *Autobiography of Pioneer John Brown, 1820–1896,* 102.

[4]Journal History, 24 February 1849.

[5]Salt Lake County Court, Minute Book A, 31 July 1852, 12 August 1852.

[6]Ibid., 21 September 1852.

[7]Of all Wasatch Mountain canyons draining into the Salt Lake Valley, Big Cottonwood has the second largest watershed, at 48.47 square miles. City Creek has 19.15 square miles, Parleys 50.14 square miles, Mill Creek 21.3 square miles, and Little Cottonwood 27.72 square miles. See Table VIII in C. T. Carnahan and C. T. Wright, *Report Covering Public Health Aspects of Salt Lake City's Water Supply, with Reference to Watershed Developments.*

[8]Salt Lake County Court, Minute Book A, 20 September 1853.

[9]*Deseret News Weekly,* 2:104 (6 November 1852).

[10]Journal History, 6 February 1853.

[11]Ibid., 28 May 1853. It may be nothing more than coincidence, but in 1853 Brigham Young undertook to have his private residences, the Lion House and the Beehive House, constructed next to his newly built church office—another demand on the meager supply of lumber.

[12]*Deseret News,* 29 May 1897; H. L. A. Culmer, "Mountain Scenery of Utah," *The Contributor* 13, n. 4 (February 1892): 175.

[13]Big Cottonwood Lumber Company Ledgers, Balance Sheet, March 1859.

[14]"Big Cottonwood Lumber Company Ledger A, November 1854–March 1863," 205, 303, 311.

[15]Frederick Kesler may well have been the most influential millwright in Utah Territory. For more about this man and his work see the well-researched histories by Allen D. Roberts, "Pioneer Mills and Milling," and "The Chase Mill," in Daughters of Utah Pioneers, *An Enduring Legacy,* vol. 7, 85–192. Also see Kimberly Day-Holm, "The Importance of Frederick Kesler to the Early Economic History of Utah."

[16]Brigham Young to John M. Bernhisel (Washington, D. C.), 17 July 1856, Brigham Young Letterbook (hereafter designated BYLB) 2:867–77; Brigham

Young to John Taylor (New York), 28 July 1856, BYLB 2:890–96; Brigham Young to George A. Smith, 30 July 1856, BYLB 2:882–86.

[17]*Deseret News Weekly,* 6:165 (30 July 1856).

[18]Henry L. A. Culmer, "Mountain Scenery of Utah," 175.

[19]*Deseret News,* 29 May 1897, 13.

[20]Big Cottonwood Lumber Company Ledger A, November 1854–March 1863, 841–43; Big Cottonwood Lumber Company Balance Sheet, March 1859.

[21]*Daily Union Vedette,* 1 August 1867.

[22]Wasatch Placer, 16 April 1874, Big Cottonwood Mining District Book B, 679; Brown & Sanford Ditch & Water Company, in Water Claims Book A, 110, Salt Lake County Recorder's Office.

[23]Celest Claim, 13 January 1876, Big Cottonwood Mining District Book C, 219.

[24]Lorenzo Brown, "Diaries, 1853–1859," entry for 25 January 1855.

[25]Journal History, 28 May 1853.

[26]*Deseret News Weekly,* 6:161–62 (30 July 1856).

[27]Big Cottonwood Lumber Company Ledger A, November 1854–March 1863, 17–19.

[28]Brigham Young to J. W. Cooley, 23 April 1855, BYLB, 2:107.

[29]Brigham Young to Feramorz Little, 23 April 1855, BYLB, 2:108–9.

[30]Balance Sheet, 25 July 1855, in Big Cottonwood Lumber Company Miscellaneous Papers, 1855, 1862.

[31]John William Cooley, in Black, *Membership of the Church of Jesus Christ of Latter-day Saints;* 1860 census listing, in J. R. Kearl, et al., *Index to the 1850, 1860 & 1870 Census of Utah—Heads of Household.*

[32]George Laub, "Reminiscences and Journals, January 1845–April 1857."

[33]Balance Sheet, 18 February 1862, Big Cottonwood Lumber Company, Miscellaneous Papers, 1855, 1862.

[34]This was at a time when a man might earn two dollars per day.

[35]About 660 feet.

[36]Brown, "Diaries, 1853–1859," 26 January 1855.

[37]"The Utah War, Journal of Albert Tracy 1858–1860," *Utah Historical Quarterly* 13, (1945): 89.

[38]Lone Tree Claim, 7 May 1888, Hot Springs Mining District Book C, 144; also 25 May 1909, Book H, 596.

[39]Great Ochre Spring Claim, 24 September 1898, Big Cottonwood Mining District Book I, 19.

[40]Brown, "Diaries, 1853–1859," 26 January 1855.

[41]Ibid., 28 January 1855.

[42]Ibid., 25 February 1855.

[43]Lorenzo Brown, *Journal of Lorenzo Brown, 1823–1900,* entry for 19 April 1855.

[44]Ibid., 18 June 1855.

45Ibid., 6 September 1855, 27 September 1855.

46Ibid., 19, 20 October 1855.

47Ibid., 7 December 1855.

48Brigham Young, Letter 29 September 1855, LDS Church History Archives.

49Heber C. Kimball to Wm. H. Kimball, 29 February 1856, in "History of the Church."

50Journal History, 17 April 1856.

51Ibid., 21, 22 May 1856; Winslow Farr, "Diary May 1856–September 1899," 22 May 1856.

52Journal History, 8 July 1856.

53"History of the Church," 18 July 1857.

54Ibid. 28 June 1856.

55Ibid., 12 July 1856.

56*Deseret News Weekly*, 6:161–62 (30 July 1856).

57Ibid., 6:165 (30 July 1856).

58Ibid., 6:333 (24 December 1856).

59Peter Sinclair, "Diary 1856 January–1862 March," 15 and 16 December 1856.

60*Deseret News Weekly*, 5:192 (22 August 1855).

61"Maxfield History"

62Articles of Agreement, 13 October 1855, BYLB, 2:427–28.

63Frederick Kesler, Diary, 17 April 1857, in "Frederick Kesler Papers."

64Phineas Wolcott Cook, "Reminiscences and Journal, 1844–1886."

65D. H. Wells to Hon. George Peacock, 25 March 1859, BYLB, 5:91–92.

66Kesler addenda to letter, Brigham Young to James H. Hart, 30 November 1855, BYLB, 2:485.

67Kesler, Diary, 13 July 1857.

68Journal History, 22 July 1857.

69Ibid., 19 July 1857.

70*Deseret News Weekly*, 7:165 (29 July 1857).

71Hosea Stout, *On the Mormon Frontier, The Diary of Hosea Stout*, 22 and 23 July 1857.

72Brown, "Diaries, 1853–1859," 23 July 1857.

73Stout, *On the Mormon Frontier*, 23 July 1857.

74"Utah Stake Minutes, 1857–1858," 26 July 1857, in Edyth J. Romney, "Typescript Collection"; Wilford Woodruff, *Wilford Woodruff's Journal, 24 July 1857*.

75Kesler, Diary, 4 August 1857.

76Ibid., 5 August 1857.

77Winslow Farr, "Diary, May 1856–Sept. 1899," 24 July 1857.

78Kesler, Diary, 6 August 1857.

79Map: Township No. 2 South, Range No. 3 East, Salt Lake Meridian, 22 December 1877, No. 443, Salt Lake City BLM office.

80Daniel H. Wells to Family, 2 May 1858, D. H. Wells Collection, Bx. 2, Fd. 5.

[81]Journal History, 26–29 July 1858; Woodruff, *Journal,* 26–29 July 1858.

[82]See Elizabeth Cumming, "The Governor's Lady, A Letter From Camp Scott, 1857," *Utah Historical Quarterly* 22 (1955): 163–73; *Deseret News Weekly,* 8:118 (8 September 1858); Stout, *On the Mormon Frontier,* 26–28 August 1858.

[83]Big Cottonwood Lumber Company Ledger A, 451.

[84]Horace Greeley, *An Overland Journey From New York to San Francisco in the Summer of 1859,* 208–9.

[85]*Deseret News Weekly,* 10:173 (1 August 1860).

[86]Sir Richard Burton, *The City of the Saints,* 348.

[87]Journal History, 12 August 1860.

[88]*New York Herald,* 5 October 1860; Journal History, 5 October 1860.

[89]Woodruff, *Journal,* 23 July 1860.

[90]Brigham Young to Orson Pratt, 27 December 1860, BYLB, 5:657–59.

[91]Brigham Young to Bishop Peter Maughan, 12 July 1861, BYLB, 5:829; Elias Smith, "Elias Smith; Journal of a Pioneer Editor, March 6, 1859–September 23, 1863," 10 July 1861.

[92]Brigham Young to James Street, 27 August 1861, BYLB, 5:872; also see Zebulon W. Jacobs, "Reminiscences and Diaries, 1861–1914." Jacobs was one of the teamsters hauling poles for the western line.

[93]*Deseret News Weekly,* 11:189 (23 October 1861); Smith, "Elias Smith; Journal of a Pioneer Editor," 18 October, 24 October 1861.

[94]Kesler, Diary, 5 March 1861.

[95]Balance Sheet, 18 February 1862, Big Cottonwood Lumber Company, "Miscellaneous Papers, 1855, 1862," item 2.

[96]Letters, James Jack to D. H. Wells, 9 May 1864, 23 June 1864, D. H. Wells Letterbook, 1855–1868, in D. H. Wells, "Collection."

[97]Journal History, 8 October 1861.

[98]Kesler, Diary, 5 November 1862; Warren Foote, "Autobiography and Journals 1837–1903," 174.

[99]Kesler, Dairy, 5 November 1867, 12 November 1867. *Salt Lake Tribune,* 13 June 1899.

NOTES TO CHAPTER 3

[1]Salt Lake County Court, Minute Book B, 1 January 1870, 14 May 1872.

[2]Deeds Book F, 32–33, 29 June 1872, Salt Lake County Recorder.

[3]Liens Book A, 430–33, 28 February 1874, Salt Lake County Recorder.

[4]Frank Esshom, *Pioneers and Prominent Men of Utah,* 786.

[5]J. R. Kearl, C. L. Pope, and L. T. Wimmer, *Index to the 1850, 1860 and 1870 Census of Utah: Heads of Household.*

[6]Salt Lake County Court, Minute Book B, 5 June 1871, 4 June 1872.

[7]*Salt Lake Tribune,* 12 June 1872; *Salt Lake Herald,* 24 November 1872.

[8]*Salt Lake Herald,* 10 July 1874, 16 March 1875.

[9]Judith B. Butters, *History of the Butlerville Area,* 7.

[10]*Salt Lake Tribune,* 19 March 1875, 20 March 1875, 25 March 1875.

[11]Nelson Wheeler Whipple, "Autobiography and Journal, 1819–1885," 377; *Salt Lake Tribune,* 8 June 1876, 9 June 1876.

[12]Memorandum for Cottonwood Cañon, 17 February 1863, Brigham Young Collection, Box 106, Fd. 18.

[13]Salt Lake County Court, Minute Book B, 6 June 1871.

[14]See, for instance, Pheebe Lode, 22 September 1870, Big Cottonwood Mining District Book A, 99; Mill B Lode, 31 October 1870, Big Cottonwood Mining District Book B, 37; Pheeby Lode Relocated, 21 October 1870, Book B, 37; Mansure Lode, 6 December 1870, Book B, 49; Marietta Lode, 6 December 1870, Book B, 49; Andrew Jackson Lode, 22 September 1871, Book B, 342; Mountain Springs Lode, 26 March 1872, Book B, 353.

[15]Brigham Young to Joseph A. and John W. Young, 28 March 1864, Brigham Young Letterbook (BYLB) 6:850–52.

[16]See *Salt Lake City Directory and Business Guide for 1869.*

[17]*Salt Lake Tribune,* 23 August 1874.

[18]*Salt Lake Tribune,* 29 September 1875.

[19]Deeds Book L, 372, 6 March 1877, and 435–36, 13 March 1877, Salt Lake County Recorder.

[20]Deeds Book N, 805–6, 6 January 1879, Salt Lake County Recorder.

[21]Phone Claim, 23 January 1904, Big Cottonwood Mining District Book I, 410.

[22]Whipple, "Autobiography and Journal," 181.

[23]George H. Taylor, "Autobiography of George H. Taylor," 40; *Salt Lake City Directory, 1867.*

[24]*Salt Lake Daily Reporter,* 23 June 1868. The Eighth Ward included the area between Third and Fifth South Streets, and between East Temple (State Street) and Third East.

[25]*Salt Lake City Directory and Business Guide for 1869;* Salt Lake Herald, 5 September 1873, 14 April 1880.

[26]Daughters of Utah Pioneers, *Tales of a Triumphant People, A History of Salt Lake County, Utah, 1847–1900,* 107.

[27]*Deseret News,* 15 June 1899.

[28]*Salt Lake Herald,* 28 June 1871, 3 April 1875.

[29]James Jack to D. H Wells, 23 June 1864, Letter Book, 1865–1868, D. H. Wells Collection.

[30]Whipple, "Autobiography and Journal," 359–60; *Salt Lake Herald,* 25 September 1873.

[31]James Jack to D. H. Wells, 13 October 1864, Letter Book, 1865–1868, D. H. Wells Collection.

[32]Ibid.; Articles of Agreement, 14 January 1865, Letter Book, 1865–1868, D. H. Wells Collection. C. B. Hawley mentioned the "large saw that was injured by the falling of your mill," in B. Snow to D. H. Wells, 25 November 1866, Incoming Letters 1866, D. H. Wells Collection; and N. W. Whipple mentioned

the roof "had fallen in by the heft of the emence [*sic*] amount of snow." See Whipple, "Autobiography and Journal," 163–65.

[33]Whipple, "Autobiography and Journal," 163–65.

[34]Ibid., 377.

[35]Ibid., 199.

[36]James Jack to D. H Wells, 13 October 1864, Letter Book, 1865–1868, D. H. Wells Collection.

[37]Salt Lake County Court, Minute Book B, 5 June 1866.

[38]BYLB 10:16, 5 March 1867.

[39]Salt Lake County Court, Minute Book B, 5 March 1867.

[40]Memorandum for Cottonwood Cañon, 17 February 1863, Brigham Young Collection, Box 106, Fd. 18.

[41]See *Salt Lake City Directory, 1867*.

[42]*Salt Lake Telegraph & Telegram,* 7 December 1867

[43]J. H. Beadle, *The Undeveloped West; or Five Years in the Territories,* 338–39; *Salt Lake Herald,* 28 September 1871, 5 October 1871, 25 September 1873.

[44]An Act to establish the Office of Surveyor-General of Utah . . . , 21 February 1855, U.S. Statutes at Large 10 (1855): 611.

[45]An Act to create the Office of Surveyor-General in the Territory of Utah, and establish a Land Office . . . , 16 July 1868, U.S. Statutes at Large 15 (1868): 91.

[46]*Salt Lake Tribune,* 2 October 1874.

[47]*Salt Lake Tribune,* 8 November 1874.

[48]*Salt Lake Tribune,* 10 November 1874.

[49]*Salt Lake Herald,* 29 September 1876.

[50]*Salt Lake Tribune,* 20 June 1875. United States Marshal George Maxwell has been described as "a fire-eating anti-Mormon." See Robert Joseph Dwyer, *The Gentile Comes to Utah,* 90.

[51]See *Salt Lake Herald,* 3 July 1875 and *Salt Lake Tribune,* 25 June 1875.

[52]*Salt Lake Tribune,* 29 June 1875. The ad was published daily until 23 July.

[53]*Salt Lake Tribune,* 13 July 1876.

[54]*Salt Lake Tribune,* 9 July 1875. This article was headlined "'Yours In Krist'—or—The One-eyed Pirate of the Wasatch."

[55]*Salt Lake Herald,* 28 August 1875, 28 September 1876.

[56]*Salt Lake Tribune,* 10 September 1876.

[57]*Deseret News,* 27 September 1876.

[58]*Salt Lake Herald,* 28 September 1876.

[59]*Salt Lake Tribune,* 10 September 1876, 27 September 1876, 28 September 1876, 29 September 1876, 30 September 1876; *Salt Lake Herald,* 28 September 1876, 29 September 1876, 30 September 1876; *Deseret News,* 27 September 1876, 28 September 1876, 29 September 1876.

[60]*Salt Lake Herald,* 7 October 1876, 8 October 1876; *Deseret News,* 9 October 1876.

[61]*Salt Lake Herald*, 11 October 1876; *Deseret News*, 9 October 1876, 10 October 1876.

[62]*Salt Lake Herald*, 14 October 1876, 17 October 1876; *Salt Lake Tribune*, 19 October 1876. The *Salt Lake Tribune* article gives the bond value as $18,000, but that must have been an error. When the case was resolved in court the award was $1,800. See *Salt Lake Tribune*, 3 June 1877.

[63]*Salt Lake Herald*, 10 October 1876.

[64]*Salt Lake Herald*, 10 October 1876, 11 October 1876, 12 October 1876; *Salt Lake Tribune*, 11 October 1876.

[65]*Salt Lake Tribune*, 27 October 1876.

[66]*Salt Lake Tribune*, 31 October 1876.

[67]*Salt Lake Tribune*, 1 November 1876.

[68]*Salt Lake Tribune*, 3 June 1877.

[69]*Salt Lake Tribune*, 22 June 1877.

[70]See *Utah Directory and Gazetteer for 1879–80.*

[71]L. L. Despain to D. H. Wells, 7 August 1882, Incoming letters, 1877–79, Daniel H. Wells Collection.

[72]Silver Cloud Claim, 31 July 1889, Big Cottonwood Mining District Book E, 130.

[73]Copper King Claim, 3 January 1898, Big Cottonwood Mining District Book I, 2.

[74]*Salt Lake Herald*, 6 June 1872; *Salt Lake Tribune*, 5 June 1872, 6 June 1872.

[75]*Salt Lake Herald*, 14 June 1873; *Salt Lake Tribune*, 14 June 1873.

[76]*Salt Lake Herald*, 3 September 1874.

[77]*Salt Lake Tribune*, 3 September 1875.

[78]Timber receipt, in F. W. Armstrong Collection, Box 2, Armstrong & Bagley Lumber Company.

[79]*Salt Lake Tribune*, 3 September 1875, 7 September 1875, 9 September 1875, 22 September 1875.

[80]Albert F. Potter, "Diary of Albert F. Potter," entry for 28 July 1902. Also see Charles S. Peterson, "Albert F. Potter's Wasatch Survey, 1902."

[81]*Deseret News*, 15 June 1899; *Salt Lake Tribune*, 16 June 1899.

[82]*Salt Lake Tribune*, 22 September 1875.

[83]Michigan No. 1 Claim, 28 May 1901, Big Cottonwood Mining District Book I, 139.

[84]Whipple, "Autobiography and Journal," 200.

[85]Ibid., 202–4, 352.

[86]Surveyor's notes for T2S R2E, Book A239, p. 234, 3–7 June 1902, BLM office, Salt Lake City.

[87]Mortgages Book E, 287, 5 March 1874, Salt Lake County Recorder.

[88]Nelson Wheeler Whipple, "Journal, 1877 Jun.–1878 August," 24 June 1877.

[89]Ibid., 26 December 1877, 28 December 1877, 1 January 1878, 6 January 1878.

[90]Most of the material on Whipple comes from Nelson Wheeler Whipple, "Autobiography and Journal of Nelson Wheeler Whipple, 1819–1885." Obituary information is from the *Deseret News,* 6 July 1887, 20 July 1887.

[91]Mary L. Lode Addenda, Big Cottonwood Mining District Book F, 23 May 1895, 229.

[92]Whipple, "Journal, 1877 Jun.–1878 August," 27 June 1877.

[93]Charlotte E. Gilchrist, "Pen Picture from Silver Lake Glen . . . ," 23 August 1882.

[94]*Deseret News,* 5 July 1883; *Salt Lake Herald,* 6 July 1883, 7 July 1883; *Salt Lake Tribune,* 6 July 1883, 7 July 1883.

[95]*Deseret News,* 3 October 1884; *Salt Lake Herald,* 3 October 1884; *Salt Lake Tribune,* 4 October 1884.

[96]*Salt Lake Herald,* 7 October 1884.

[97]Asa R. Bowthorpe, "History of Pioneer Sawmills and Local Canyons of Salt Lake Valley." In 1870 Bowthorpe's father, William, then only nineteen years old, was one of Daniel H. Wells's lumber haulers in the canyon. Asa was born on 7 April 1885. He spent much of his life in and around the canyons of the Wasatch and gathered considerable knowledge of events there. However, his history, compiled late in his life, about 1960, suffers many of the problems and pitfalls of oral histories. It makes interesting reading but as a primary source should be used with care. Bowthorpe died on 28 March 1973.

[98]Butler Mill Site, 18 April 1872, Big Cottonwood Mining District Book B, 354.

[99]Golden Curry Mine Claim, 10 May 1880, Big Cottonwood Mining District Book D, 31.

[100]*Salt Lake Tribune,* 13 November 1888.

[101]Bear Trap Claim, 13 September 1900, Big Cottonwood Mining District Book I, 118.

[102]Potter, "Diary," entry for 26 July 1902.

[103]Bob Lode, 15 November 1870, Big Cottonwood Mining District Book B, 41; Gold Belt Lode, 1 April 1895, Big Cottonwood Mining District Book F, 215.

[104]Rocky Mountain Lode, 15 March 1871, Big Cottonwood Mining District Book B, 92.

[105]Willow Patch Lode, 17 July 1895, Big Cottonwood Mining District Book F, 266; Willow Lake Lode, 13 August 1895, Book F, 307; Willow Creek Lode, 18 August 1895, Book F, 309. Kuck's name, pronounced Cook, is variously spelled Cook, Kucke, or Kuk. The spelling used in the homestead patent is used here.

[106]See, for instance, Salt Lake County Recorder Deeds Book 5F, 528, 4 November 1909.

[107]Homestead Patent 9598, issued to Julius A. Kuck, approved 4 April 1904, BLM office, Salt Lake City.

[108]Mortgages Book 5F, p.528 and Book 6D, p.348–9; Deeds Book 11C, 355, Salt Lake County Recorder.

[109] *Salt Lake Tribune,* 10 April 1937.

[110] *Deseret News,* 29 September 1938; *Salt Lake Tribune,* 29 September 1938, 1 October 1938, 5 October 1938.

NOTES TO CHAPTER 4

[1] Journal History, 15 January 1850.

[2] Ibid., 7 January 1851; Dale L. Morgan, "The State of Deseret—1847–1849," 199–200.

[3] Warren Foote, "Autobiography and Journals 1837–1903," 29 November 1851.

[4] *Deseret News Weekly,* 3:28 (19 February 1853).

[5] Journal History, 7 April 1852.

[6] Ibid., 3 October 1852.

[7] Ibid., 7 April 1859; Elisha Drown Clapp, "Notebook, 1880–1883."

[8] Susan Lindsay Ward Black, *Membership of the Church of Jesus Christ of Latter-day Saints, 1830–1848.*

[9] Salt Lake County Court, Minute Book A, 26 March 1853.

[10] Ibid., 13 June 1853, 14 June 1853.

[11] Ibid., 16 November 1853.

[12] Journal History, 28 June 1854; John M. Woolley, "Diaries, 1854–1859," 10 July 1854.

[13] Journal History, 29 February 1856.

[14] Surveyor's notes for T3S R2E, Book R137, April 1874, BLM office, Salt Lake City.

[15] *Deseret News Weekly,* 5:205 (5 September 1855), 5:228 (26 September 1855).

[16] *Deseret News Weekly,* 6:376 (28 January 1857), 7:152 (15 July 1857).

[17] *The Mountaineer,* 2 June 1860; 9 June 1860; *Deseret News Weekly,* 10:112 (6 June 1860).

[18] Samuel A. Woolley, "Diaries 1846–1899," 16 June 1864.

[19] Ibid., 18 June 1864.

[20] Preston W. Parkinson, *The Utah Woolley Family,* 286.

[21] *Salt Lake Daily Telegraph,* 18 August 1864, 20 August 1864; *Deseret News Weekly,* 13:377 (24 August 1864); Samuel A. Woolley, "Diaries," 16–19 August 1864.

[22] Samuel A. Woolley, "Diaries," 20 August 1864.

[23] See Isaac Groo, "Autobiography."

[24] Land Office Map, T3S, R2E, SLM, 5 August 1871, No. 472, BLM office, Salt Lake City.

[25] Samuel A. Woolley, "Diaries," 30 March 1868, 31 March 1868.

[26] *Deseret News,* 28 March 1912, 1 April 1912.

[27] *Deseret News,* 19 October 1915.

[28] Robert Earl Despain, *Robert Henry Despain Family, 1875–1967,* 28–37; Solomon Joseph Despain, "Autobiography."

[29]Samuel A. Woolley, "Diaries," 18 July 1865, 22 May 1866, 8 May 1867, 8 July 1867.

[30]Salt Lake County Court, Minute Book B, 10 April 1866.

[31]Samuel A. Woolley, "Diaries," 1 September 1866.

[32]*Deseret News Weekly,* 16:191 (12 June 1867). The notice ran in every issue until 28 August 1867.

[33]Samuel A. Woolley, "Diaries," 18 May 1869.

[34]Ibid., 1–3 August 1869, 20 August 1869, 27 August 1869; *Deseret News Weekly,* 18:317 (11 August 1869).

[35]Samuel A. Woolley, "Diaries," 16 February 1866, 18 March 1866.

[36]Ibid., 9 April 1868, 21 April 1868, 28 April 1868.

[37]Ibid., 2 June 1869.

[38]Black Bessie Claim, 26 June 1871, Little Cottonwood Mining District Book A, 315; Samuel A. Woolley, "Diaries," 25 June 1872.

[39]There were many claims referring to Hawley's mill, the first being the Barrington Claim, 26 June 1872, Little Cottonwood Mining District Book B, 243. The latest reference called it "the Turbine mill, the same being known as the Hawley mill," this being the Blossbery Claim, 26 July 1875, Book C, 238. See Hawley Flat Nos. 1 through 4 claims, 7 March–3 July 1907, Little Cottonwood Mining District Book G, 344, 345, 354, 355. Also see English Bank No. 5 Claim, 2 April 1908, Book M, 79.

[40]Samuel A. Woolley, "Diaries," 27 January 1871.

[41]Ibid., 18–21 April 1871.

[42]*Salt Lake Herald,* 15 September 1872, 20 December 1872.

[43]*Salt Lake Herald,* 21 August 1872. Also see 3 September 1872 and *Salt Lake Tribune,* 28 December 1872.

[44]*Salt Lake Herald,* 7 August 1873.

[45]*Salt Lake Herald,* 29 October 1872, 31 January 1873.

[46]Samuel A. Woolley, "Diaries," 30 September 1873.

[47]Ibid., 30 May 1872.

[48]Ibid., 30 May 1875.

[49]Parkinson, *Utah Woolley Family,* 283–84.

[50]Harkers Canyon was named after Joseph Harker of Taylorsville, the bishop of the West Jordan Ward. In 1858, when Harker was released as bishop, Archibald Gardner was chosen to take his place. See John W. Van Cott, *Utah Place Names,* 177; and Clarence Gardner, "Autobiography and Journals, 1895–1957."

[51]Samuel A. Woolley, "Diaries," 5 August 1872, 3 December 1872.

[52]Salt Lake County Court, Minute Book C, 24 March 1875.

[53]Samuel A. Woolley, "Diaries," 26 April 1875. E. T. City was at the south end of the Great Salt Lake in Tooele County; in 1923 its name was changed to Lake Point. See Van Cott, *Utah Place Names,* 121.

[54]*Salt Lake Herald,* 20 July 1875.

[55]Samuel A. Woolley, "Diaries," 25 June 1875, 6 and 7 July 1875, 10 July

1875; Leonard J. Arrington, *From Quaker to Latter-day Saint: Bishop Edwin D. Woolley,* 474–75.

[56]Samuel A. Woolley, "Diaries," 27 May 1873.

[57]Ibid., 8 June 1876.

[58]Delila Gardner Hughes, *Life of Archibald Gardner,* 104.

[59]Ibid., 105.

[60]*Salt Lake Herald,* 5 December 1876, 13 December 1876; *Salt Lake Tribune,* 5 December 1876, 4 February 1877, 16 February 1877.

[61]*Salt Lake Tribune,* 27 February 1877; Hughes, *Life of Archibald Gardner,* 105.

[62]*Salt Lake Tribune,* 5 August 1877.

[63]Chief Joseph Claim, Tessie Claim, 17 August 1877, Little Cottonwood Mining District Book C, 425, 426.

[64]*Salt Lake Herald,* 30 April 1875.

[65]Salt Lake County Court, Minute Book C, 13 July 1878.

[66]Ibid., 3 June 1879.

[67]Hughes, *Life of Archibald Gardner,* 110.

[68]Potter, "Diary," 27 July 1902. On this same day Potter visited Alta and wrote, "Here is an old mining camp that has been worked until both the ore and the timber were pretty well exhausted. The stumps show it to have been well forested originally but every tree (and seedling) has been cut. It certainly is a picture of forest destruction and I do not wonder that the town was once destroyed by a snowslide coming down the denuded mountain side. As I remarked . . . , it would be hard to find a seedling big enough to make a club to kill a snake."

NOTES TO CHAPTER 5

[1]For instance, see Leonard Arrington, "Abundance from the Earth; The Beginnings of Commercial Mining in Utah"; Robert G. Raymer, "Early Mining in Utah"; B. S. Butler, G. F. Loughlin, V. C. Heikes et al., *The Ore Deposits of Utah;* F. C. Calkins and B. S. Butler, *Geology and Ore Deposits of the Cottonwood– American Fork Area, Utah;* and Laurence P. James, *Geology, Ore Deposits and History of the Big Cottonwood Mining District.* Laurence P. James, *Silver Mountain,* a history of mining in the Wasatch, is being completed.

[2]*Union Vedette,* 27 November 1863.

[3]Connor had been promoted to brigadier general on 29 March 1863 for "heroic conduct and brilliant victory on Bear River." *Deseret News Weekly,* 12:313 (1 April 1863).

[4]T. B. H. Stenhouse, *Rocky Mountain Saints,* 713.

[5]Edward W. Tullidge, *History of Salt Lake City and its Founders,* 697–98. It is interesting that Tullidge presents the Ogilvie story, since the rest of his discussion of mining was taken nearly verbatim from Stenhouse.

[6]Fred B. Rogers, *Soldiers of the Overland,* 15, 111.

[7]Connor to R. C. Drum, Asst. Adjutant General, U. S. Army, 21 July 1864, cited in *Tullidge's Quarterly Magazine* 1:185; it is also found in Tullidge, *History of Salt Lake City*, 328–30. Connor's efforts were described in a letter written by George A. Smith, First Councilor to President Brigham Young: "A daily paper is being published in the Camp called the 'Vedette,' an abusive, filthy sheet, which is doing its best to break up the 'Mormons' and humbug mankind generally, by making them believe that the country is full of gold mines." Journal History, 11 March 1864.

[8]Rogers, *Soldiers of the Overland*, 112.

[9]*Union Vedette*, 4 December 1863.

[10]*Union Vedette*, 22 July 1864.

[11]Mountain Lakes Mining District Book B, "Mining Laws of the Mountain Lakes Mining District as amended at the miner's meeting held in Little Cottonwood Canon August 30, 1869 . . ." pasted inside back cover of book.

[12]*Salt Lake Herald*, 22 February 1874.

[13]*Salt Lake Tribune*, 1 January 1892. James Wall enlisted in the California Volunteers at San Francisco on 26 September 1861. He was mustered out as a sergeant at Camp Douglas on 4 October 1864. See Richard H. Orton, *Records of California Men in the War of the Rebellion 1861–1867*.

[14]Big Cottonwood Mining District Book M, 26; notices of miners' meetings, 17 March 1870, 10 May 1870.

[15]*Salt Lake Tribune*, 24 August 1872; *Salt Lake Herald*, 25 August 1872.

[16]Big Cottonwood Mining District Book M, 54, 10 July 1876.

[17]*Daily Union Vedette*, 30 August 1866.

[18]Hot Springs Mining District Book A, 9 December 1870.

[19]*Salt Lake Herald*, 31 January 1871.

[20]*Salt Lake Tribune*, 1 January 1880. No recorder's books have been found for the Mill Park District.

[21]*Salt Lake Herald*, 6 July 1873.

[22]Stenhouse, *Rocky Mountain Saints*, 716; John R. Murphy, *Mineral Resources of the Territory of Utah*, 4.

[23]*Croffett's Western World*, cited in *Salt Lake Tribune*, 16 November 1871.

[24]*Salt Lake Tribune*, 2 April 1876.

[25]U.S. Statutes at Large 17 (1872): 91–96.

[26]See *Salt Lake Telegram*, 24 January 1929. A *New York Sun* article, quoted in *Salt Lake Tribune*, 19 March 1876, said Silas Brain was "an old soldier." No other reference has come to light to support this. Brain was not one of Connor's California Volunteers.

[27]*Daily Union Vedette*, 27 August 1866. For use of the M.D. title see *Daily Union Vedette*, 10 October 1866.

[28]Mountain Lake Mining District Book A, 120, 12 May 1866.

[29]Samuel A. Woolley, "Diaries," 1 September 1866.

[30]*Daily Union Vedette*, 25 August 1866.

³¹Douglas, General Smith, Volcano, Gen. Scott, and General Washington Lodes, 28 August 1866, Mountain Lake Mining District Book A, 153–55.

³²*Daily Union Vedette,* 10 October 1866.

³³*Salt Lake Tribune,* 2 April 1876.

³⁴*Salt Lake Tribune,* 17 November 1871.

³⁵See *Salt Lake City Directory, 1867.*

³⁶Water Power Relocation, 25 June 1867, Mountain Lake Mining District Book A, 190.

³⁷*Mining & Scientific Press,* 9 October 1869.

³⁸*Emma Mine Investigation,* James E. Lyon testimony, 400; *Engineering & Mining Journal,* 12 February 1876; *Salt Lake Herald,* 10 October 1879, 20 August 1880.

³⁹Relocation notice, 7 April 1870, Little Cottonwood Mining District Book A, 36; Pandora Claim, 11 June 1882, Little Cottonwood Mining District Book D, 156.

⁴⁰*Salt Lake Tribune,* 8 August 1874.

⁴¹General Grant Lode, 6 September 1865, Mountain Lake Mining District Book A, 71; Great Eastern Lode, 10 November 1865, Book A, 93; and Great Western Lode, 10 November 1865, Book A, 94.

⁴²Emma Claim, 7 April 1870, Little Cottonwood Mining District Book A, 35.

⁴³*Salt Lake Tribune,* 3 August 1872, 19 March 1876.

⁴⁴*Engineering & Mining Journal,* 13 May 1876.

⁴⁵For instance, see W. Turrentine Jackson's two articles: "The Infamous Emma Mine: A British Interest in the Little Cottonwood District, Utah Territory," and "British Impact on the Utah Mining Industry." See also chapter 8, "The Emma Silver Mining Company, Limited: A Case Study," in Clark C. Spence, *British Investment and the American Mining Frontier, 1860–1901;* and *Emma Mine Investigation.*

⁴⁶Warren Hussey testimony in *Emma Mine Investigation,* 587. Correspondent John Morgan mentioned this sale in his lengthy, anti-mining letter to the editor, *Deseret News,* 18 July 1870.

⁴⁷*Gazetteer of Utah and Salt Lake City Directory, 1874,* 302. Hussey's banking activities are described in C. James Wall, "Gold Dust and Greenbacks"; and Leonard J. Arrington, "Banking Enterprises in Utah, 1847–1880."

⁴⁸*New York Times,* 18 February 1884, 5.

⁴⁹Warren Hussey testimony, in *Emma Mine Investigation,* 588.

⁵⁰James M. Day testimony, ibid., 444.

⁵¹Spence, *British Investment and the American Mining Frontier,* 153–54.

⁵²Warren Hussey testimony, in *Emma Mine Investigation,* 592.

⁵³*Salt Lake Herald,* 25 May 1872.

⁵⁴*Salt Lake Herald,* 5 June 1872, 16 June 1872; *Salt Lake Tribune,* 6 June 1872, 8 June 1872, 14 June 1872.

[55] *Salt Lake Herald,* 21 June 1872, 13 July 1872; *Salt Lake Tribune,* 21 June 1872, 24 June 1872, 4 July 1872, 12 July 1872, 18 July 1872.

[56] Spence, *British Investment and the American Mining Frontier,* 165–66.

[57] Trenor W. Park testimony, in *Emma Mine Investigation,* 460.

[58] *Salt Lake Tribune,* 8 August 1871.

[59] Flagstaff Claim, Little Cottonwood Mining District Book A, 43.

[60] *Salt Lake Herald,* 10 June 1870.

[61] *Salt Lake Herald,* 2 September 1871, 27 February 1872; *Salt Lake Tribune,* 30 November 1871.

[62] In his paper, "British Impact on the Utah Mining Industry," W. Turrentine Jackson always put Buel's title in quotation marks.

[63] *Salt Lake Herald,* 19 April 1871.

[64] Nicholas Groesbeck Morgan, *Our Groesbeck Ancestors in America,* 20–21.

[65] See Jackson, "British Impact on the Utah Mining Industry."

[66] *Deseret News,* 31 March 1900.

[67] *Salt Lake Tribune,* 17 July 1875, 4 January 1876, 8 December 1876.

[68] *Salt Lake Herald,* 10 October 1879.

[69] Arrington, "Banking Enterprises in Utah," 321–23; *Spokane Falls Illustrated,* 14–15, 46, 49.

[70] Daniel Sylvester Tuttle, *Missionary to the Mountain West,* 57, 250, 364, 392, 395.

[71] Davenport Claim, 19 July 1870, Little Cottonwood Mining District Book A, 88; also recorded 29 June 1870, Big Cottonwood Mining District Book A, 18.

[72] *Salt Lake Herald,* 27 October 1871, 9 January 1872.

[73] Great Eastern Lode, Great Western Lode, 10 November 1865, Mountain Lake Mining District Book A, 93, 94.

[74] *Salt Lake Herald,* 20 December 1872.

[75] *Salt Lake Tribune,* 3 February 1873, 21 April 1873; *Salt Lake Herald,* 11 March 1873; *Salt Lake Mining Review,* 15 June 1929, 11–12.

[76] *Salt Lake Herald,* 9 May 1873; *Salt Lake Tribune,* 1 January 1880; *Mining and Scientific Press* 46 (19 May 1883): 345.

[77] *Salt Lake Herald,* 27 June 1873; *Salt Lake Tribune,* 16 June 1873.

[78] *Salt Lake Tribune,* 17 December 1873.

NOTES TO CHAPTER 6

[1] John Morgan correspondence in *Deseret News,* 22 June 1870, 7 July 1870. At this time Morgan was running the Morgan Commercial College, housed in a Main Street building in Salt Lake City owned by Morgan's father-in-law, Nicholas Groesbeck. While Morgan was strongly anti-mining in his writings, he had kind words to say for Groesbeck's Flagstaff Mine, perhaps because Groesbeck had recently given the Morgans a block of ground on the north side of First South between West Temple and First West Streets, where a building

was constructed sufficiently large to accommodate the growing college as well as Morgan's family. See Arthur Richardson, *The Life and Ministry of John Morgan.*

[2]*Salt Lake Herald,* 18 March 1871. The Deseret Telegraph Company was formed to provide telegraph service between territorial communities. It was organized on 21 March 1867, although most of the line was in operation before that time (*Deseret News,* 3 July 1867). The line from Salt Lake City to the Little Cottonwood mines was started in mid-February 1871 and reached the mouth of the canyon before the end of that month. Heavy snows delayed construction up the canyon.

[3]*Salt Lake Tribune,* 27 April 1871, 30 April 1871.

[4]*Salt Lake Herald,* 17 June 1871.

[5]*Salt Lake Tribune,* 14 November 1871.

[6]*Salt Lake Tribune,* 19 June 1871, 30 June 1871.

[7]*Salt Lake Herald,* 8 November 1871.

[8]*Salt Lake Herald,* 27 November 1872.

[9]*Salt Lake Herald,* 16 October 1872.

[10]Albert F. Potter, "Diary of Albert F. Potter," entry for 27 July 1902.

[11]*Salt Lake Herald,* 15 October 1872.

[12]*Salt Lake Tribune,* 26 September 1872.

[13]*Salt Lake Herald,* 23 October 1872.

[14]*Salt Lake Herald,* 10 June 1871, 21 June 1871, 28 June 1871; *Salt Lake Tribune,* 29 December 1871.

[15]*Salt Lake Herald,* 2 September 1871, 27 February 1872.

[16]*Salt Lake Herald,* 8 October 1872; *Salt Lake Tribune,* 26 September 1872, 12 November 1872, 31 January 1873.

[17]The Vallejo tramway carried forty-seven small buckets, each able to carry 100 pounds of ore, and two large ones. Its completed cost was $14,000, well above its projected cost of $10,000. See *Salt Lake Herald,* 12 July 1872, 21 August 1872, 18 October 1872, 23 October 1873; *Salt Lake Tribune,* 19 August 1872, 26 September 1872, 16 January 1873.

[18]*Salt Lake Herald,* 18 October 1872, 19 October 1872, 23 October 1872. It might be noted that the young lady's male companion chose not to accompany her on her ride up the tramway.

[19]*Salt Lake Herald,* 9 January 1873.

[20]*Salt Lake Herald,* 9 May 1873.

[21]*Salt Lake Herald,* 30 June 1874, 1 July 1874, 11 July 1874.

[22]*Salt Lake Herald,* 20 April 1873; *Gazetteer of Utah and Salt Lake City Directory, 1874,* 78–80.

[23]*Salt Lake Herald,* 3 June 1874; *Salt Lake Tribune,* 24 April 1874, 20 May 1874, 2 May 1875. The bath was still operating three years later, then under the management of another barber, Felix Reinhold; see *Salt Lake Tribune,* 3 January 1877.

²⁴*Salt Lake Tribune*, 23 September 1873.

²⁵*Salt Lake Herald*, 17 July 1873.

²⁶*Salt Lake Tribune*, 19 March 1912. There have been suggestions that Chinese workers were used to build the Wasatch and Jordan Valley tramway to Alta in the early 1870s, and one the tramway's stone walls that still stands along today's highway is sometimes called the "China Wall." However, no firm evidence has been found by the author to substantiate these beliefs.

²⁷*Salt Lake Tribune*, 6 May 1873, 28 May 1873.

²⁸*Salt Lake Tribune*, 26 June 1873, 27 June 1873, 14 July 1873; *Salt Lake Herald*, 13 July 1873.

²⁹*Salt Lake Herald*, 12 October 1873; *Salt Lake Tribune*, 19 October 1873.

³⁰The first issue of the *Cottonwood Observer*, published on 14 July 1873, has not survived, but the article was mentioned in the *Salt Lake Tribune*, 16 July 1873. The *Observer*'s second issue, dated 19 July 1873, has survived and contains a lengthy article on the townsite subject.

³¹*Salt Lake Herald*, 23 May 1871.

³²*Salt Lake Tribune*, 5 October 1871; *Salt Lake Herald*, 8 October 1871.

³³Nagler filed at least fourteen claims, the first of which was the Independence Claim, 25 July 1869, Little Cottonwood Mining District Book A, 27.

³⁴*Salt Lake Tribune*, 21 July 1873.

³⁵See *U.S. Statutes at Large* 11, 292.

³⁶Salt Lake County Recorder, Deeds Book H, 287, 26 July 1873.

³⁷Salt Lake County Recorder, Deeds Book F, 592, 1 February 1873.

³⁸See B.A.M. Froiseth, "Froiseth's Map of Little Cottonwood Mining District and Vicinity."

³⁹Subdivision Plats: Alta City, A:47, 23 July 1873. Salt Lake County Recorder's Office.

⁴⁰Deed for lot in Block 27, Alta City, 25 March 1871, Salt Lake County Recorder Deeds Book E, 55–56.

⁴¹*Salt Lake Herald*, 16 July 1873; *Salt Lake Tribune*, 16 July 1873.

⁴²*Salt Lake Herald*, 1 August 1873.

⁴³*Salt Lake Herald*, 6 August 1873.

⁴⁴Deeds: John W. Heines and Cecelia A. Heines (formerly Amanda Brown) to Samuel J. Brown, 4 October 1873, Salt Lake County Recorder Deed Book H, 708–9; Samuel J. Brown to Alexander F. Bell, 4 October 1873, Salt Lake County Recorder Deed Book H, 707–8; Samuel J. Brown to Samuel G. Anderson, 17 December 1873, Salt Lake County Recorder Deed Book H, 893–94.

⁴⁵Deed: Alexander F. and Elizabeth A. Bell to Samuel J. Lees, 31 December 1873, Salt Lake County Recorder Deed Book H, 921–22.

⁴⁶Deed: Samuel J. Brown and Samuel G. Anderson to Samuel J. Lees, 6 January 1874, Salt Lake County Recorder Deed Book H, 912–13.

⁴⁷Deed: Samuel J. Lees to Benjamin G. Raybould, trustee, 28 November 1884,

Salt Lake County Recorder Deed Book Y, 871–73. For more on the Walker brothers and their many business activities see Jonathan Bliss, *Merchants and Miners in Utah: The Walker Brothers and Their Bank.*

[48] *Salt Lake Tribune,* 4 January 1874.

[49] *Salt Lake Herald,* 23 June 1874.

[50] *Salt Lake Tribune,* 16 November 1873, 7 January 1874.

[51] *Salt Lake Tribune,* 26 March 1874.

[52] *Salt Lake Tribune,* 25 June 1874.

[53] *Salt Lake Tribune,* 8 September 1874.

[54] *Salt Lake Tribune,* 25 February 1875, 9 March 1875, 25 March 1875, 22 April 1875.

[55] *Salt Lake Herald,* 8 July 1874, 11 July 1874; *Salt Lake Tribune,* 30 July 1874, 7 August 1874, 15 December 1874.

[56] Deed: S. J. Lees to James P. Schell, Lots 10–12, Block 30, 13 September 1873, Salt Lake County Recorder Deeds Book I, 682.

[57] *Salt Lake Herald,* 16 September 1873, 8 October 1873, 28 October 1873; *Salt Lake Tribune,* 7 October 1873.

[58] *Salt Lake Herald,* 28 October 1873, 19 December 1873.

[59] *Salt Lake Tribune,* 17 January 1874, 23 January 1874.

[60] *Salt Lake Tribune,* 20 May 1874, 2 July 1874.

[61] *Salt Lake Herald,* 14 March 1875; *Salt Lake Tribune,* 3 March 1876, 13 March 1877; Salt Lake County Court Minutes, Book C, 5 September 1876, 19 September 1876.

[62] *Salt Lake Tribune,* 27 January 1900, 28 January 1900.

[63] *Salt Lake Herald,* 20 July 1875, 17 May 1876, 17 August 1876; *Salt Lake Tribune,* 8 September 1874, 22 July 1875, 8 September 1877, 27 July 1878. For Mrs. Davis's school see *Salt Lake Herald,* 30 April 1875, 30 May 1875, 10 May 1876, 17 August 1876, and *Salt Lake Tribune,* 13 May 1875. Her husband was of the Alta assaying firm of Bennett & Davis.

[64] *Salt Lake Herald,* 26 August 1873, 14 October 1873; *Salt Lake Tribune,* 7 August 1874.

[65] *Salt Lake Tribune,* 29 January 1874, 11 February 1874. The shovel races were described in the *Alta Independent,* quoted in the *Salt Lake Tribune,* 14 May 1873.

[66] *Salt Lake Tribune,* 9 July 1874.

[67] *Salt Lake Herald,* 7 July 1875, 11 July 1875; *Salt Lake Tribune,* 17 June 1875, 4 July 1875, 7 July 1875.

[68] *Deseret News,* 6 April 1870.

[69] *Salt Lake Herald,* 11 June 1870, 17 June 1870, 29 June 1870; *Deseret News,* 1 August 1870.

[70] *Salt Lake Herald,* 14 August 1870, 16 August 1870, 17 August 1870, 18 August 1870; *Deseret News,* 15 August 1870, 17 August 1870, 18 August 1870.

[71] *Salt Lake Tribune,* 9 August 1871, 10 August 1871, 12 August 1871.

[72] *Salt Lake Tribune,* 5 November 1872; *Salt Lake Herald,* 6 November 1872.

[73] *Salt Lake Herald,* 9 February 1873.

[74] *Salt Lake Herald,* 1 April 1873.

[75] *Salt Lake Herald,* 2 December 1873.

[76] *Salt Lake Herald,* 11 February 1874, 12 February 1874, 6 September 1879; *Salt Lake Tribune,* 11 February 1874, 13 February 1874, 18 February 1874, 5 September 1879.

[77] *Salt Lake Tribune,* 2 June 1873.

[78] *Salt Lake Herald,* 14 September 1875.

[79] Brigham Young to Prest. H. S. Eldredge, 16 February 1871, Brigham Young Letterbook (hereafter designated BYLB) 12:529–34.

[80] Brigham Young to Gen. Thomas L. Kane, 16 April 1971, BYLB 12:641–45.

[81] *Salt Lake Tribune,* 9 May 1871.

[82] *Salt Lake Herald,* 5 September 1871.

[83] Brigham Young to John W. Young, 14 August 1872, BYLB 13:180–81; Brigham Young to Wm. C. Staines, 24 October 1872, BYLB 13:239–41.

[84] Brigham Young to John W. Young, 8 May 1871, BYLB 12:670–73; *Salt Lake Herald,* 3 August 1871, 4 August 1871.

[85] Brigham Young to D. M. Stuart, 4 August 1871, BYLB 12:782–84.

[86] See Roxie N. Rich, *History and People of Early Sandy,* 63–64; D. McKenzie to Chas. F. Raymond, 9 March 1872, BYLB 12:995.

[87] *Salt Lake Herald,* 28 September 1871; Letter, Brigham Young to Willard Young, 17 October 1871, BYLB 12:878–79; Journal History, 5 November 1872 (*Deseret News* Weekly, 21:618)

[88] *Salt Lake Tribune,* 31 May 1872.

[89] *Salt Lake Herald,* 5 November 1872; Journal History, 5 November 1872 (*Deseret News Weekly* 21:618).

[90] *Salt Lake Herald,* 1 March 1873, 3 April 1873.

[91] Journal History, 5 April 1873 (*Deseret News Weekly* 22:166); Brigham Young to Willard Young, 14 April 1873, BYLB 13:332–35; Brigham Young to A. Carrington, 19 April 1873, BYLB 13:340–43.

[92] *Salt Lake Herald,* 17 April 1873; *Salt Lake Tribune,* 11 April 1873, 21 April 1873.

[93] *Salt Lake Herald,* 27 April 1873; *Salt Lake Tribune,* 29 April 1873.

[94] *Salt Lake Herald,* 21 September 1873; *Utah Mining Gazette,* 27 September 1873.

[95] *Salt Lake Herald,* 24 June 1875; *Salt Lake Tribune,* 25 June 1875, 26 June 1875.

[96] *Salt Lake Tribune,* 26 October 1875.

[97] *Salt Lake Tribune,* 1 January 1878; *Salt Lake Herald,* 14 September 1875. Portions of the stone wall still can be seen at some places along the Little Cottonwood highway.

[98] *Engineering and Mining Journal* 24 (8 September 1877): 186–87.

[99]*Salt Lake Herald*, 18 September 1879.

[100]Charlotte E. Gilchrist, "Pen Picture from Silver Lake Glen, Big Cotton-wood Canyon, Utah," 12 August 1880 entry.

[101]*Salt Lake Tribune*, 26 September 1878.

NOTES TO CHAPTER 7

[1]Journal History, 24 January 1850, 12 April 1850.

[2]*Salt Lake Herald*, 27–30 December 1872, 14 June 1873; *Salt Lake Tribune*, 27, 30, 31 December 1872, 16 June 1873.

[3]*Salt Lake Herald*, 13, 14, 17 January 1875; *Salt Lake Tribune*, 13, 14, 17, 19 January 1875.

[4]*Salt Lake Herald*, 14 January 1875.

[5]*Salt Lake Herald*, 13 January 1875; *Salt Lake Tribune*, 13 January 1875, 19 January 1875.

[6]*Salt Lake Herald*, 21 January 1875; *Salt Lake Tribune*, 21 January 1875, 22 January 1875.

[7]Alfred Lambourne, *Jo, A Christmas Tale of the Wasatch*. Lambourne later re-wrote the book but told the story in verse in *Plet, A Christmas Tale of the Wasatch*.

[8]*Salt Lake Tribune*, 22 January 1875, *Salt Lake Herald*, 21 January 1875.

[9]*Salt Lake Tribune*, 25 February 1875.

[10]*Salt Lake Herald*, 18 March 1875; *Salt Lake Tribune*, 18 March 1875, 24 March 1875.

[11]*Cottonwood Observer*, 26 July 1873; *Salt Lake Herald*, 27 July 1873; *Salt Lake Tribune*, 27 July 1873, 29 July 1873.

[12]*Salt Lake Herald*, 2 December 1873.

[13]Archibald overestimated the weight of 430 gallons of water; it is more like 1.75 tons. *Salt Lake Herald*, 30 June 1874; *Salt Lake Tribune*, 10 June 1874.

[14]*Salt Lake Tribune*, 2 July 1874.

[15]*Salt Lake Tribune*, 20 November 1874.

[16]*Salt Lake Tribune*, 9 June 1875, 3 July 1875.

[17]*Salt Lake Herald*, 2 August 1878, 3 August 1878; *Salt Lake Tribune*, 2 August 1878, 3 August 1878, 4 August 1878, 6 August 1878, 8 August 1878, 15 August 1878.

[18]*Salt Lake Tribune*, 8 September 1878.

[19]*Salt Lake Tribune*, 6 August 1878, 15 October 1878.

[20]*Salt Lake Tribune*, 17 September 1878, 26 September 1878; *Salt Lake Herald*, 21 September 1879.

[21]*Salt Lake Tribune*, 15 October 1878.

[22]*Salt Lake Herald*, 26 October 1878; *Salt Lake Tribune*, 5 August 1875, 3 May 1876.

[23]*Salt Lake Herald*, 25 September 1879; *Salt Lake Tribune*, 1 January 1880; *Utah Directory and Gazetteer, 1879–80*, 183–85.

[24]*Salt Lake Tribune*, 15 October 1878, 23 September 1879.

[25] *Deseret News*, 14 January 1881, 17 January 1881, 20–21 January 1881; *Salt Lake Herald*, 15–16 January 1881, 18–20 January 1881, 22 January 1881; *Salt Lake Tribune*, 15–16 January 1881, 18–20 January 1881, 22 January 1881.

[26] *Salt Lake Herald*, 27 May 1881, 19 June 1881.

[27] *Salt Lake Tribune*, 18 May 1881. The sloping roof is mentioned in the *Salt Lake Tribune*, 15 February 1885.

[28] *Salt Lake Tribune*, 17 June 1882; O. J. Hollister letter in *Salt Lake Tribune*, 13 August 1882.

[29] *Salt Lake Tribune*, 17 August 1882.

[30] *Salt Lake Tribune*, 1 January 1880.

[31] *Salt Lake Tribune*, 22 February 1884.

[32] The March 1884 avalanche was reported in the *Deseret News*, 12 March 1884; *Salt Lake Herald*, 11 March 1884, 13–15 March 1884; *Salt Lake Tribune*, 11–13 March 1884, 15–16 March 1884, 20 March 1884; *Park Record*, 15 March 1884.

[33] *Salt Lake Tribune*, 19 August 1884, 27 August 1884, 12 September 1884, 7 November 1884.

[34] McDaniels was presumed dead according to the *Salt Lake Tribune*, 18 February 1885. No other mention of him was found. Initial reports had two Chinese men missing; see *Salt Lake Herald* and *Salt Lake Tribune*, 15 February 1885. Later reports said one was still not found and Charlie Foh was dead, *Salt Lake Herald* and *Salt Lake Tribune*, 18 February 1885. In the *Salt Lake Tribune*, 19 February 1885, one Chinese man was still reported missing. In the last report of the incident found, *Salt Lake Herald*, 21 February 1885, he still had not been found.

[35] The 1885 Alta disaster was reported in the *Salt Lake Herald*, 15 February 1885, 17–18 February 1885; *Salt Lake Tribune*, 15 February 1885, 17–21 February 1885, 27 February 1885. For the life of John Ford and more about Jeremy Reagan see Charles L. Keller, "Tales of Four Alta Miners."

[36] *Salt Lake Tribune*, 6 June 1885.

[37] *Salt Lake Tribune*, 1 August 1885.

[38] *Salt Lake Herald*, 18 February 1885.

[39] *Salt Lake Tribune*, 6 June 1885.

[40] *Salt Lake Tribune*, 29 September 1885, 1 January 1886.

[41] *Salt Lake Tribune*, 13 July 1886, 21 July 1886, 29 August 1886.

[42] *Salt Lake Herald*, 12 May 1891; *Salt Lake Tribune*, 30 August 1889, 8 May 1891, 17 May 1891, 30 May 1891, 2 June 1891, 10 June 1891, 25 June 1891, 26 June 1891.

[43] *Salt Lake Tribune*, 5 May 1899.

[44] *Salt Lake Tribune*, 1 January 1896.

[45] Albert F. Potter, "Diary of Albert F. Potter," 27 July 1902.

[46] *Deseret News*, 6 November 1903. While it is not known when the tramway tracks were removed, a mineral survey map made in July 1904 shows the

tramway still running through the area of the claims. Plat of Hellgate and Hellgate Nos. 2–4 Lodes, Mineral Survey 5282, BLM office, Salt Lake City.

[47] *Salt Lake Tribune,* 26 August 1888; *Park Record,* 11 August 1888, 18 August 1888, 25 August 1888.

[48] *Salt Lake Tribune,* 30 December 1900, 12 August 1903, 20 August 1903; *Salt Lake Mining Review,* 30 August 1903.

[49] C.W. Lockerbie, "Sidelight Stories of Romantic Alta."

[50] *Salt Lake Tribune,* 4 March 1899, 7 March 1899.

[51] Veitch and Jones Claim, Little Cottonwood Mining District Book G, 283.

NOTES TO CHAPTER 8

[1] *Engineering and Mining Journal,* 16 May 1903, 6 June 1903; Columbus Consolidated Mining Co., 9 April 1902, Salt Lake County Corporate File 3498, in Utah State Archives, series 3888.

[2] *Engineering and Mining Journal,* 13 and 20 June 1903; 4 and 25 July 1903; 26 September 1903; 3 October 1903; *Salt Lake Tribune,* 13 September 1903.

[3] *Engineering and Mining Journal,* 4 July 1903, 26 May 1904.

[4] *Salt Lake Tribune,* 6 July 1904; *Engineering and Mining Journal,* 14 July 1904; *Salt Lake Mining Review,* 15 July 1904.

[5] *Engineering and Mining Journal,* 10 November 1904, 24 November 1904.

[6] *Salt Lake Tribune,* 8 September 1905, 23 November 1905, 31 December 1905.

[7] *Salt Lake Tribune,* 12 November 1905; *Engineering and Mining Journal,* 14 October 1905, 23 December 1905, 14 April 1906.

[8] *Salt Lake Mining Review,* 15 June 1911; *Engineering and Mining Journal,* 1 July 1911.

[9] *Engineering and Mining Journal,* 30 November 1912.

[10] *Salt Lake Mining Review,* 30 November 1912, 30 January 1913, 28 February 1913; *Engineering and Mining Journal,* 12 October 1912, 23 November 1912, 30 November 1912, 8 February 1913.

[11] *Salt Lake Tribune,* 7 October 1919, 4 August 1920.

[12] *Salt Lake Tribune,* 4 November 1915, 12 December 1915; *Salt Lake Mining Review,* 15 November 1915, 30 December 1915. There have been suggestions that this pipeline brought water from the Hogum Fork stream. Indeed, a section of pipe leading over the ridge on the south side of the stream above the power plant still exists today, but its route or destination is not obvious. See Robert Marvin to L. James, 27 February 1964, 23 July 1964, in Laurence P. James, "Little Cottonwood Canyon Collection."

[13] *Salt Lake Tribune,* 31 October 1917; *Engineering and Mining Journal,* 17 November 1917.

[14] *Salt Lake Tribune,* 24 July 1923; *Salt Lake Mining Review,* 30 July 1923, 15 September 1925.

[15] *Salt Lake Mining Review,* 15 October 1943, 15 April 1944.

[16]Laurence P. James, *Geology, Ore Deposits and History of the Big Cotton-wood Mining District*, 66.

[17]*Engineering and Mining Journal*, 1 February 1913; *Salt Lake Tribune*, 8 September 1914; *Salt Lake Mining Review*, 15 September 1914. The power plant property was sold by Utah Power and Light Company via quitclaim deed on 20 October 1943, according to information received from C. Burton of Utah Power on 23 December 1998.

[18]*Salt Lake Mining Review*, 15 July 1903, 15 August 1903, 30 October 1903.

[19]*Salt Lake Tribune*, 12 September 1903, 13 September 1903; *Salt Lake Mining Review*, 30 September 1903.

[20]*Salt Lake Tribune*, 17 September 1904, 20 September 1904, 18 October 1904, 30 October 1904, 29 November 1904, 15 December 1904, 7 April 1905; *Salt Lake Mining Review*, 30 September 1904, 30 November 1904, 30 April 1905.

[21]*Salt Lake Tribune*, 26 October 1904, 30 October 1904, 23 November 1904, 29 November 1904, 15 December 1904.

[22]*Salt Lake Tribune*, 5 February 1905, 9 February 1905, 11 March 1905, 22 March 1905, 24 March 1905, 2 April 1905; *Salt Lake Mining Review*, 30 April 1905.

[23]*Salt Lake Herald*, 22 January 1906; *Salt Lake Tribune*, 22 January 1906,; *Salt Lake Mining Review*, 30 April 1906.

[24]For the Unity Mines Corporation see *Salt Lake Mining Review*, 30 August 1907. Jacobson's announcement was in the *Salt Lake Tribune*, 4 July 1908.

[25]For the sale of the City Rocks property and formation of the new company see *Salt Lake Tribune*, 28 May 1905, 31 August 1905, 2 August 1906, 11 August 1906, 30 December 1906, 4 August 1907; City Rocks Mining Co., 31 July 1906, Salt Lake County Corporate File 4404, Utah State Archives. The City Rocks tramway is reported in the *Salt Lake Tribune*, 11 August 1906, and *Engineering and Mining Journal*, 18 August 1906. For the Utah Mines Coalition suit see *Salt Lake Tribune*, 16 May 1910 and *Salt Lake Mining Review*, 15 November 1911. The Michigan-Utah formation and merger is in *Engineering and Mining Journal*, 22 October 1910, 11 March 1911, 9 September 1911, 16 September 1911, 27 January 1912, 10 February 1912, 9 March 1912, 15 June 1912; *Salt Lake Mining Review*, 30 December 1911; *Salt Lake Tribune*, 31 December 1911.

[26]*Salt Lake Mining Review*, 15 May 1912, 15 June 1912; *Salt Lake Tribune*, 9 June 1912, 23 June 1912, 2 August 1912, 1 September 1912, 8 September 1912, 21 September 1912, 27 September 1912.

[27]*Engineering and Mining Journal*, 11 November 1905, 7 July 1906, 6 October 1906; *Salt Lake Tribune*, 22 June 1906.

[28]*Engineering and Mining Journal*, 3 June 1911; *Salt Lake Tribune*, 10 June 1913, 13 June 1913. The railroad from Sandy to Wasatch was originally built as

a 36-inch narrow gauge road, but in 1891 the Denver & Rio Grande Western Railroad converted it to standard gauge. See *Salt Lake Tribune*, 8 May 1891, 17 May 1891, 10 June 1891.

[29]*Salt Lake and Alta Railroad Co.*, 25 August 1913, Salt Lake County Corporate File 7138, Utah State Archives; *Salt Lake Tribune*, 26 August 1913.

[30]*Salt Lake Tribune*, 16 November 1913; *Engineering and Mining Journal*, 29 November 1913; *Salt Lake Mining Review*, 30 November 1913.

[31]*Salt Lake Tribune*, 6 June 1914, 26 June 1915; *Salt Lake Mining Review*, 15 September 1915, 30 September 1915, 30 November 1915; *Engineering and Mining Journal*, 13 November 1915.

[32]*Salt Lake Mining Review*, 29 February 1916, 30 May 1916. The Pearson company is also mentioned in the *Salt Lake Tribune*, 11 October 1916, 21 December 1916, 4 June 1917, 14 June 1917; and in the *Salt Lake Mining Review*, 30 December 1916, 15 March 1917, 30 June 1917. The Little Cottonwood Transportation Company name first appeared in the *Salt Lake Tribune*, 13 September 1916.

[33]*Salt Lake Tribune*, 28 October 1916, 5 November 1916.

[34]It is not clear why the first locomotive was not placed into service. The *Salt Lake Tribune*, 21 December 1916, said the Shay was being equipped with a new and distinctive type of track brake designed by Yorston.

[35]Harry Hartwell of New York, vice-president of the company, told the Salt Lake County Commissioners that Yorston had been dismissed; see Salt Lake County Commission Minute Book T, 6 July 1917. The *Salt Lake Tribune*, 27 July 1917 and 8 August 1918 provided information on Dunn. Yorston went on to take a contract to build a 3.5-mile electric railroad at Bingham; see *Salt Lake Mining Review*, 15 October 1917. Shand Smith's appointment was reported in the *Salt Lake Tribune*, 8 August 1918. The *Salt Lake Tribune*, 16 October 1917, reported the track within one-half mile of the Columbus Rexall tunnel, about a mile and a half from Alta. The slow progress of the railroad up the canyon was probably due to the rate at which the Emigration Canyon Railroad was dismantled. That project had been delayed by numerous protests of Emigration Canyon residents who didn't want to lose the rail service.

[36]*Salt Lake Tribune*, 30 November 1917, 4 May 1918, 20 June 1918; *Engineering and Mining Journal*, 13 April 1918, 25 May 1918.

[37]*Salt Lake Tribune*, 31 October 1917, 30 December 1917; *Salt Lake Mining Review*, 15 November 1917.

[38]*Salt Lake Tribune*, 22 April 1917, 14 November 1918; *Engineering and Mining Journal*, 13 April 1918.

[39]Elbert Despain interview, 28 August 1964, in James, "Little Cottonwood Canyon Collection."

[40]*Deseret News*, 6 May 1920; *Salt Lake Herald*, 6 May 1920, 7 May 1920; *Salt Lake Tribune*, 6 May 1920, 8 May 1920.

⁴¹The road work marked one of the first uses of TNT in the mining district, furnished by the government to the state road commission. "Mining operators returning from Alta say that the action of this explosive is marvelous. When it is desired to remove a boulder, a cup-shaped area is made on the surface of the rock with mud. A quantity of TNT, which resembles in appearance brown sugar, is placed in this cavity and exploded. The action of the explosive is so terrific that the largest granite boulders are shattered to a powder," reported the *Salt Lake Tribune*, 5 June 1921. The PUC application is reported in the *Salt Lake Tribune*, 19 August 1921, and *Engineering and Mining Journal*, 3 September 1921.

⁴²George H. Watson & Co., "Market Letter, The Story of Alta."

⁴³*Salt Lake Tribune*, 3 June 1922, 27 August 1922.

⁴⁴Exactly how Watson acquired the railroad is not clear. Years later Robert Marvin suggested that Minor C. Keith of the United Fruit Company gave it to Watson to get it off his, Keith's, hands. Since neither Keith nor the United Fruit Company was mentioned in any reports on the railroad, that suggestion may seem farfetched, but there is a coincidental event. When Shand Smith retired from the railroad in September 1922 he published a statement saying that any business with the company had to be carried through the headquarters at 17 Battery Place in New York City, which was the address of the United Fruit Company and Minor C. Keith's office. It is not known when Keith gained control of the Little Cottonwood Transportation Company, but a good guess would be either when Yorston was relieved in 1917 or when Shand Smith took over in 1918. Keith had many years of experience in building and operating narrow gauge railroads in Central America and in his later years, some said in his dotage, he invested in a number of small railroads in the United States. Presumably the LCTC was one of them. See Robert Marvin to L. P. James, 19 January 1967, in James, "Little Cottonwood Canyon Collection"; *Salt Lake Tribune*, 27 August 1922; New York City Directory, 1922; Watt Stewart, *Keith and Costa Rica, A Biographical Study of Minor Cooper Keith*.

⁴⁵Elbert G. Despain interview, 18 August, 1964, in James, "Little Cottonwood Canyon Collection."

⁴⁶*Deseret News*, 4 June 1928; *Salt Lake Tribune*, 4 June 1928, 5 June 1928, 13 June 1928; Elbert G. Despain interviews, 17 August 1964, 28 August 1964, in James, "Little Cottonwood Canyon Collection." The Clays family had a long and tragic association with the Peruvian mines at Alta. J. P. Clays's father, Valentine F. Clays, became a director and superintendent of the Peruvian Mining Company before the turn of the century. In 1900 he and his son J. P. became president and vice-president, respectively. On 31 January 1911, after a decade of active participation in the company's operation, V. F. Clays was killed in an avalanche while working at the mine. His sons, J. P. and H. W., continued to run the company and work the mine until 20 March 1920, when H. W. and a

co-worker, John Howry, were killed in an avalanche at the mine, reminiscent of the event some nine years earlier. Their bodies were not recovered until 6 June. J. P. Clays carried on, running the company until the accident on the tramway in June 1928 took his life. See *Salt Lake Tribune,* 14 June 1898, 7 July 1900, 1 February 1911, 21 March 1920, 7 June 1920.

[47]Robert Marvin to L. James, 30 January 1965, 19 January 1967, in James, "Little Cottonwood Canyon Collection." Marvin claimed that money was not available to remove the rails using cutting torches, but they discovered that if they nicked the ball of the rail with a torch, allowed it to cool, then struck the rail with a double jack it would snap off at the nick. In that way they cut sixteen miles of rail into thirty-inch pieces and sent them to the American Mine & Smelter plant to be used as flux.

[48]The small locomotive, Number 3, was Lima c/n 1672, built 5 April 1906, 24-inch gauge, 8x8 cylinders, 26½ inch drivers. It and Engine Number 2 were built for the Silver City, Pinos Altos & Mogollon Railroad in New Mexico. Both were converted to 36-inch gauge before they arrived at Wasatch. The larger locomotive, Number 1, was Lima c/n 2194, built for the Santa Barbara Tie & Pole Co. It was completed 14 July 1909, 36-inch gauge, 8x8 cylinders, 26½ inch drivers. At 48,000 pounds it was much heavier than the other two. It went on to be Pioche Pacific Transportation Co. No.3, and Bristol Silver Mines No.3, both in Pioche, Nevada. It ended its career on display at the Last Chance Museum, Las Vegas. The wrecked locomotive, Number 2, was Lima c/n 1673, completed 5 April 1904, 24-inch gauge, 8x8 cylinders, 26½ inch drivers. All three locomotives were bought through W. A. Zelnicker Supply Co., a dealer in St. Louis, Missouri. See Michael Koch, *The Shay Locomotive, Titan of the Timber.*

[49]*Salt Lake Tribune,* 19 December 1936, 7 December 1937.

[50]Robert Marvin to L. James, 30 January 1965, in James, "Little Cottonwood Canyon Collection."

[51]Denver & Rio Grande Western Railroad, "Condensed Profile of the D. & R. G. W. R. R. System," 1 January 1968.

[52]*Salt Lake City Directory,* 1905, 1906; Alta and Hecla Mining and Milling Company, 21 May 1906, Salt Lake County Corporate File 4336, in Utah State Archives, series 3888.

[53]*Salt Lake Mining Review,* 15 March 1908.

[54]*Salt Lake Mining Review,* 15 February 1910, 30 November 1910, 30 December 1916; *Salt Lake Tribune,* 18 August 1910; South Hecla Mining Company, 13 September 1910, Salt Lake County Corporate File 5817, in Utah State Archives, series 3888.

[55]*Salt Lake Mining Review,* 15 April 1923, 15 February 1927, 15 May 1930; Alta Merger Mines Company, 6 April 1923, Salt Lake County Corporate File 10227; Alta Michigan Mines Company, 13 June 1916, Salt Lake County Corporate File 8086, Utah State Archives.

⁵⁶*Salt Lake Mining Review,* 27 September 1932; Emma Silver Mines Company, 3 July 1919, name changed to Alta United Mines Company by stockholder resolution on 26 October 1931, Salt Lake County Corporate File 9266, Utah State Archives.

⁵⁷*Salt Lake Mining Review,* 16 May 1933; Alta Michigan Mines Company, 13 June 1916, name changed to Alta Champion Mining Company by stockholder resolution on 3 May 1933, Salt Lake County Corporate File 8086, Utah State Archives.

⁵⁸Robert Marvin to L. James, 26 April 1964, in James, "Little Cottonwood Canyon Collection." After leaving Watson, Marvin worked on the Columbia Basin project and finally retired to Tolovana Park, Oregon, where he died on 3 August 1969; see *Deseret News,* 4 August 1969.

⁵⁹*Salt Lake Mining Review* 11 May 1937, *Salt Lake Tribune,* 8 October 1938; *Deseret News,* 11 May 1937. Watson's inventions were described in the *Salt Lake Tribune,* 6 November 1949. For Romantic Alta see R. Marvin to L. James, 27 February 1964, in James, "Little Cottonwood Canyon Collection." Also see Alexis Kelner, *Skiing in Utah,* Chapter 4.

⁶⁰*Deseret News,* 1 April 1952; *Salt Lake Tribune,* 1 April 1952, 3 April 1952.

NOTES TO CHAPTER 9

¹Minutes of meeting at Silver Fork, 17 August 1870, Big Cottonwood Mining District Book M, 28.

²First use of the name Rice's Camp was in the Lilly of the West Claim, 30 July 1874, Big Cottonwood Mining District Book C, 21. The name Rice's Flat was used as late as 1 January 1891, Ave Atque Vale Lode, Book E, 321.

³Salt Lake County Court, Minute Book B, 6 June 1871; *Salt Lake Herald,* 24 June 1871.

⁴*Salt Lake Tribune,* 18 August 1871, 15 September 1871, 17 November 1871; *Salt Lake Herald,* 5 October 1871.

⁵*Salt Lake Herald,* 28 September 1871, 5 October 1871

⁶*Salt Lake Tribune,* 18 August 1871, 21 August 1871, 25 June 1872, 29 July 1872.

⁷*Salt Lake Tribune,* 30 July 1872.

⁸CE Patent 730, USA to William Howard, 1 May 1872; CE Patent 731, USA to Stephen G. Sewell, 1 May 1872, BLM office, Salt Lake City; also Salt Lake County Recorder, Deeds Book J, 713, USA to William Howard, 1 May 1872, and Book E, 968, USA to Stephen G. Sewell, 1 May 1872.

⁹Wellington Claim, 6 July 1870, Big Cottonwood Mining District Book A, 21, Congress Claim, 8 July 1870, Book A, 27; Antelope Lode, 18 June 1870, Book A, 12; Highland Chief Claim, 19 January 1871, Book B, 73; Prince of Wales Lode, 4 August 1870, Book A, 47; Wandering Boy Claim, 6 August 1870, Book A, 53.

[10]*Salt Lake Herald,* 12 March 1874, 25 June 1874; *Salt Lake Tribune,* 1 January 1875.

[11]*Salt Lake Tribune,* 13 September 1873, 7 August 1875.

[12]*Salt Lake Tribune,* 3 October 1874, 9 March 1875, 5 August 1875, 28 August 1875, 15 September 1875; Flat Spring Claims, 16 August 1875, Little Cottonwood Mining District Book C, 246.

[13]*Salt Lake Tribune,* 28 August 1875, 1 May 1877; Rosewarne Lode, 22 October 1877, Big Cottonwood Mining District Book C, 389.

[14]Annie Tunnel Claim, 23 April 1874, Big Cottonwood Mining District Book B, 683.

[15]*Salt Lake Tribune,* 19 November 1876, 15 February 1877, 3 June 1877, 29 November 1877; *Salt Lake Herald,* 20 October 1877, 24 September 1879.

[16]*Salt Lake Mining Review,* 18 June 1935.

[17]*Salt Lake Tribune,* 10 September 1913, 29 July 1914, 18 April 1915.

[18]*Salt Lake Tribune,* 7 September 1915, 26 September 1915; *Salt Lake Mining Review,* 15 October 1915, 15 December 1915.

[19]*Salt Lake Tribune,* 6 August 1915; *Salt Lake Mining Review,* 15 August 1915.

[20]Lessees' operations are reported in the *Salt Lake Mining Review,* 12 September 1933, 18 December 1934, 18 June 1935, 16 July 1935, 3 September 1935, 9 June 1936. Bodfish's attempts at reviving the company are reported in the *Salt Lake Mining Review,* 25 May 1937, 31 May 1938, 15 June 1940. Altamina was reported in the *Salt Lake Mining Review,* 31 January 1946. Bodfish's death was reported in the *Salt Lake Mining Review,* 30 September 1946. Also see Laurence P. James and Henry O. Whiteside, "Promoting the Alta Tunnel: The Rise and Fall of F. V. Bodfish."

[21]*Salt Lake Tribune,* 18 August 1920.

[22]Eclipse Mine Claim, 16 September 1877, Big Cottonwood Mining District Book C, 366.

[23]Dugway Lode, 9 August 1892, Big Cottonwood Mining District Book E, 490.

[24]*Salt Lake Tribune,* 1 January 1881, 18 January 1881, 1 January 1882, 4 September 1886, 15 May 1887.

[25]*Salt Lake Tribune,* 3 May 1888, 4 May 1888; *Engineering and Mining Journal,* 12 May 1888.

[26]This latter trail is shown in *Hiking the Wasatch, The Official Wasatch Mountain Club Trail Map for the Tri-Canyon Area.*

[27]Curiously, the Ophir, Reed and Benson, and Excelsior Lodes were recorded twice, first on 6 September 1870 and again a week later on 13 September 1870, but the later records appear first in the Big Cottonwood Mining District Book A on pages 83, 84, and 85, respectively, while the earlier records are on pages 109, 110, and 111. The Brilliant Lode was recorded on 14 September 1870 in Book A, 89.

[28] *Salt Lake Tribune,* 25 April 1871; A. L. Bancroft & Co. map, ca. 1871; Wagon Road & Trail Notice, 30 June 1871, Big Cottonwood Mining District Book B, 174.

[29] *Salt Lake Tribune,* 22 September 1871, 23 September 1871.

[30] Later reports made the claim that initial ore shipments from the Reed and Benson Mine went to Swansea, Wales; see *Engineering and Mining Journal,* 30 August 1902, and *Salt Lake Tribune,* 13 May 1915. However, the quoted quantity, 4,000 tons, and value, one million dollars, of the shipments place these reports in the category of mining legend rather than fact. They likely drew on the much publicized fact that the early Emma Mine shipments went to Swansea, albeit well before the Reed and Benson was producing ore.

[31] Payne & Goodspeed is mentioned in the *Salt Lake Herald,* 13 October 1870 and 16 November 1870. Payne's sale of his part of the mines is recorded in an indenture dated 23 December 1870, recorded on 14 August 1872, Big Cottonwood Mining District Book B, 537. The firm of (Alexander) Majors, (Charles E.) Chapman & Goodspeed is mentioned in the *Salt Lake Herald,* 30 June 1871.

[32] *Salt Lake Herald,* 11 July 1873, 17 July 1873; *Salt Lake Tribune,* 8 October 1873, 6 November 1873, 9 January 1874, 1 January 1875; *Utah Mining Gazette,* 27 September 1873, 13 December 1873.

[33] *Salt Lake Herald,* 11 November 1873, 19 November 1873; *Salt Lake Tribune,* 11 November 1873; *Utah Mining Gazette,* 15 November 1873.

[34] *Salt Lake Tribune,* 23 January 1875.

[35] The avalanche was reported in the *Salt Lake Herald,* 15 February 1878; *Salt Lake Tribune,* 3 February 1878, 16 February 1878, 17 February 1878, 3 March 1878. The tramway was reported operating in the *Salt Lake Tribune,* 1 January 1882, 25 July 1882.

[36] Reed and Goodspeed Tunnel, 13 October 1875, Big Cottonwood Mining District Book C, 188.

[37] *Salt Lake Tribune,* 28 February 1892, 2 March 1892.

[38] *Salt Lake Tribune* 5 August 1894; *Park Record,* 3 November 1894; *Engineering and Mining Journal,* 10 November 1894. Information on the date of Reed's death supplied by the Maine Historical Society Library via Laurence P. James.

[39] *Deseret News,* 21 January 1895; *Salt Lake Herald,* 21 January 1895; *Salt Lake Tribune,* 21 January 1895.

[40] *Salt Lake Tribune,* 15 August 1902; *Engineering and Mining Journal,* 30 August 1902; *Salt Lake Mining Review,* 30 October 1902.

[41] This old road is shown as a trail in *Hiking the Wasatch, The Official Wasatch Mountain Club Trail Map for the Tri-Canyon Area.*

[42] *Salt Lake Herald,* 14 November 1871, 16 November 1871.

[43] *Salt Lake Tribune,* 3 October, 1877, 13 October 1877, 1 January 1878.

[44] *Salt Lake Tribune,* 20 January 1878.

⁴⁵*Salt Lake Tribune,* 1 January 1878.

⁴⁶*Salt Lake Tribune,* 20 January 1878.

⁴⁷*Salt Lake Tribune,* 10 July 1878.

⁴⁸*Salt Lake Herald,* 16 November 1878, 20 November 1878, 24 November 1878; *Salt Lake Tribune,* 21 November 1878, 17 December 1878, 1 January 1879; *Engineering and Mining Journal,* 30 November 1878, 11 January 1879, 8 February 1879. The company's name, Kessler Mining Company, appears in an advertisement in *Engineering and Mining Journal,* 1 March 1879.

⁴⁹*Salt Lake Tribune,* 27 August 1908, 13 September 1908. The first reference calls the mine the East Carbonate, but from the description it is obviously the Carbonate. The East Carbonate Mine was on the east side of the South Fork, at the base of the Reed and Benson Cliffs.

⁵⁰*Salt Lake Tribune,* 9 June 1910, 14 June 1910, 30 July 1910, 7 August 1910, 19 August 1910; *Salt Lake Mining Review,* 30 April 1910, 15 June 1910; *Engineering and Mining Journal,* 25 June 1910, 1 October 1910.

⁵¹*Salt Lake Tribune,* 31 October 1913, 19 June 1914, 16 July 1915, 11 December 1915; *Salt Lake Mining Review,* 15 September 1913, 15 November 1913, 30 April 1916, 30 June 1916; *Engineering and Mining Journal,* 16 May 1914, 1 January 1916.

⁵²James W. Wade, interview, 16 February 1963, in Laurence P. James, "Little Cottonwood Canyon Collection."

⁵³Lloyd Hoskins to L. P. James, 20 January 1963, in James, "Little Cottonwood Canyon Collection."

⁵⁴The Kessler Peak's north route is described in Veranth, *Hiking the Wasatch,* 113.

⁵⁵Price's claims were General Lafayette, 4 December 1901, Big Cottonwood Mining District Book I, 208; Mountain Yueen, 3 September 1902, Book I, 261; Mountain Chief Fraction, 4 November 1904, Book I, 458; and Mountain Chief Extension No.2, 20 December 1904, Little Cottonwood Mining District Book M, 4. Reamer's claim was the Mountain Cheaf, 3 September 1902, Big Cottonwood Mining District Book I, 261. Price and Reamer together filed the Mountain Chief Extension No. 1, 22 August 1903, Big Cottonwood Mining District Book I, 366. Since both claims filed on 3 September 1902 were misspelled, the error was probably the recorder's rather than the prospectors'. Since the name originally filed was used when the Mineral Survey was made, the mining maps show the two misspelled names: Mountain Yueen and Mountain Cheaf. See Laurence P. James, *Geology, Ore Deposits and History of the Big Cottonwood Mining District,* Plate 3.

⁵⁶Little Dora Lode, 5 October 1891, Big Cottonwood Mining District Book E, 417; *Salt Lake Tribune,* 27 July 1892; *Salt Lake Mining Review,* 15 August 1899, 30 August 1899. Thomas Miller ran a brass foundry at 134 South 600 West. Jones was Reamer's father-in-law.

⁵⁷*Engineering and Mining Journal,* 8 May 1920.

[58] *Salt Lake Tribune,* 4 November 1905; *Salt Lake Mining Review,* 30 December 1906; Cardiff Mining and Milling Company, 6 December 1906, Salt Lake County Corporate file 4491, in Utah State Archives, series 3888. In 1903 Price held a job as deputy commissioner in the state Bureau of Statistics. In 1906 he became commissioner, while McMillin, former clerk at the Rocky Mountain Telephone Company, became deputy commissioner; see Salt Lake City directories for 1903–1906.

[59] *Salt Lake Tribune,* 6 June 1908; *Salt Lake Mining Review,* 15 June 1908; *Park Record,* 3 August 1912. Ezra Thompson had been in the freighting business, first in Salt Lake City, then at Park City, while James D. Murdoch had been a machinist, then master mechanic and chief engineer at the Ontario and Daly companies in Park City. In 1900 the two men joined forces to form the Thompson & Murdoch Investment Company in Salt Lake City, and together they later entered numerous other business ventures.

[60] *Salt Lake Mining Review,* 15 March 1910; *Salt Lake Tribune,* 30 April 1910, 12 June 1910, 28 June 1910; *Engineering and Mining Journal,* 16 July 1910.

[61] *Salt Lake Tribune,* 17 June 1911, 12 August 1911, 1 September 1911; *Engineering and Mining Journal,* 26 August 1911.

[62] *Salt Lake Tribune,* 16 October 1914, 18 October 1914, 23 October 1914; *Salt Lake Mining Review,* 30 November 1914.

[63] *Salt Lake Tribune,* 27 October 1914, 30 October 1914, 12 June 1915, 30 June 1915, 4 July 1915, 3 August 1915; *Engineering and Mining Journal,* 23 October 1915.

[64] *Salt Lake Tribune,* 4 April 1915.

[65] *Engineering and Mining Journal,* 26 May 1917.

[66] *Salt Lake Mining Review,* 29 February 1916; Salt Lake County Commission Minutes, Book T, 22 March 1916, 24 March 1916, 25 March 1916, 28 March 1916, 7 April 1916; *Engineering and Mining Journal,* 8 April 1916; *Salt Lake Tribune,* 2 April 1916. Salt Lake County contributed $12,500 toward the expense of the improvement, Salt Lake City $7,500, and the two mining companies $5,000 each.

[67] *Salt Lake Tribune,* 2 April 1916, 11 July 1916, 26 July 1916; *Salt Lake Mining Review,* 15 May 1916, 30 July 1916. Although the reports referred to "caterpillar tractors," it was later clarified that the name caterpillar was a trade name of the Hold Manufacturing Company, which did not provide the tractors used in this experiment; *Salt Lake Tribune,* 2 September 1916.

[68] *Salt Lake Tribune,* 16 August 1916, 3 September 1916, 1 October 1916, 12 October 1916, 3 December 1916; *Salt Lake Mining Review,* 30 October 1916; Salt Lake County Commission, Minute Book T, 8 December 1916.

[69] *Salt Lake Tribune,* 18 July 1917; *Salt Lake Mining Review,* 30 March 1917, 15 January 1918, 30 May 1918; *Engineering and Mining Journal,* 30 July 1921.

[70] *Engineering and Mining Journal,* 4 August 1917; *Salt Lake Mining Review,* 15 November 1917.

[71] *Salt Lake Mining Review,* 15 November 1924, 30 April 1923.

[72] James, *Geology, Ore Deposits and History of Big Cottonwood District,* 65–67.

[73] The ore bin boiler is mentioned in a letter from Lloyd Hoskins to L. P. James, 30 August 1963. The use of dynamite in ore bins is mentioned in a letter from Robert Marvin to L. P. James, 13 November 1967, both in James, "Little Cottonwood Canyon Collection."

[74] *Salt Lake Tribune,* 16 August 1987.

[75] *Salt Lake Mining Review,* 30 April 1915, 30 August 1918, 15 April 1931; *Salt Lake Tribune,* 17 April 1915, 2 May 1915, 8 May 1915, 30 June 1915, 3 April 1931. Also see James, *Geology, Ore Deposits and History of Big Cottonwood District,* 62.

[76] Maxfield Lode, 15 September 1870, Big Cottonwood Mining District Book A, 142.

[77] *Salt Lake Tribune,* 9 May 1873.

[78] Gen. Thomas Lode, 4 May 1871, Big Cottonwood Mining District Book B, 108; Mill A Tunnel Claim, 19 June 1872, Book B, 387; Maxfield Tunnel Claim, 19 June 1872, Book B, 388.

[79] Little Tunnel Claim, 7 April 1877, Big Cottonwood Mining District Book C, 298; Maxfield Mining Company, 15 March 1879, Salt Lake County Corporate File 164, in Utah State Archives, series 3888.

[80] *Salt Lake Herald,* 9 June 1882; *Salt Lake Tribune,* 1 January 1885, 1 January 1886; *Salt Lake Mining Review,* 31 December 1905.

[81] Maxfield Mining Company, 15 March 1879, Salt Lake County Corporate File 164.

[82] Boston Development Company, 8 May 1914, Salt Lake County Corporate File 7331.

[83] *Salt Lake Mining Review,* 15 June 1916, 15 June 1917; *Engineering and Mining Journal,* 17 June 1916. Litigation is reported in the *Salt Lake Mining Review,* 30 October 1916. Also see *Utah Reporter,* 10 September 1917.

[84] Maxfield Mining Company, 2 November 1951, Salt Lake County Corporate File 16040; also see James, *Geology, Ore Deposits and History of Big Cottonwood District,* 51–52.

NOTES TO CHAPTER 10

[1] For instance, see William Clayton, *William Clayton's Journal,* entry for 16 July 1847.

[2] Clayton, *Journal,* 26 July 1847.

[3] Hosea Stout, *On The Mormon Frontier, The Diary of Hosea Stout,* entry for 8 December 1848.

[4] Lorenzo Johnson and John Bagley, both employees of the Big Cottonwood Lumber Company at the time.

[5] Lorenzo Brown, "Diaries 1853–1859," entry for 4 February 1855.

[6] Peter Sinclair, "Diary 1856, January–1862 March," entry for 17 May 1857.

[7]Nelson Wheeler Whipple, "Autobiography and Journal of Nelson Wheeler Whipple," 184–86.

[8]*Utah Gazetteer and Directory of Logan, Provo, Ogden & SLC*, 1884; Salt Lake County Recorder, Deeds Book I, 2, D. H. Wells, SLC Mayor, to W. S. Brighton, 5 April 1873.

[9]See "One of a Family, Two of a Nation: Three Pioneer Families . . . ;" Mary Brighton Timmons, "Address"; Stella Brighton Nielsen, "William Stuart Brighton." Both Timmons and Nielsen were granddaughters of William S. Brighton. Timmons was a daughter of Thomas Bow Brighton, while Nielsen was a daughter of Daniel H. Brighton.

[10]Neils, Titus Extension, and Johnson Claims, 17 October 1868, all recorded in Little Cottonwood Mining District Book A, 23.

[11]Nielsen, "William Stuart Brighton." Mrs. Nielson claimed this information came from her uncle William H. Brighton.

[12]Great Western Lode and Fairy of the Lake Lode, 21 June 1870, Big Cottonwood Mining District Book A, 14–15; Setting Sam Claim, 2 July 1870, Little Cottonwood Mining District Book A, 82; Mountain Lake Lode, 13 July 1870, Big Cottonwood Mining District Book A, 33.

[13]Star of the West Claim, 15 August 1870, Big Cottonwood Mining District Book A, 60; Brighton Lode, 19 August 1870, Book A, 63; Catherine Lode and Day Lode, 6 September 1870, Book A, 78, 79. His cabin was cited in the description of the Pine Tree Lode, 19 July 1872, Book B, 503.

[14]CE Patent 1441, USA to William S. Brighton, 1 November 1875, BLM office, Salt Lake City; see also Salt Lake County Recorder, Deeds Book K, 143–44, USA to Wm. S. Brighton, 1 November 1875.

[15]Salt Lake County Recorder, Mortgage Book D, 748, W. S. Brighton to Elijah Whitaker, 23 January 1873.

[16]*Salt Lake Tribune*, 9 July 1874.

[17]*Salt Lake Herald*, 28 August 1877.

[18]See Nielsen, "William Stuart Brighton."

[19]*Salt Lake Herald*, 28 August 1877.

[20]*Salt Lake Tribune*, 14 August 1878; Salt Lake County Recorder, Deeds Book O, 661–63, W. S. Brighton to Walker Bros., 10 November 1879. The Walkers paid Brighton $250 for the ten acres of land.

[21]Charlotte E. Gilchrist, "Pen Picture from Silver Lake Glen, Big Cottonwood Canyon, Utah." This journal covers the years 1878 through 1883. The following winter, the cabin, Silver Lake Villa, was crushed by heavy snows. Although the cabin was rebuilt, that event marked the end of Gilchrist's journal.

[22]Salt Lake County Recorder, Deeds Book Q, 423–25, Wm. S. Brighton to Dr. W. F. Anderson, 6 October 1880.

[23]CE Patent 2485, USA to Robert A. Brighton, 10 January 1885, BLM office, Salt Lake City; also Salt Lake County Recorder, Deeds Book 2E, p.423–24, USA to Robert A. Brighton, 10 January 1885

²⁴Salt Lake County Recorder, Silver Lake Summer Resort, Subdivision Plat C-57, 2 October 1890.

²⁵*Salt Lake Tribune,* 8 August 1886; Nielsen, "William Stuart Brighton." Mrs. Nielsen claimed her father, Daniel H. Brighton, built this road.

²⁶*Salt Lake Tribune,* 26 July 1890, 28 July 1890; *Deseret News,* 28 July 1890.

²⁷*Salt Lake Tribune,* 17 June 1892.

²⁸Salt Lake County Recorder, Mortgage Book 3N, 443, W. S. Brighton and Catherine Brighton to Valerie E. Stoddard, 1 April 1893.

²⁹Salt Lake County Recorder, Mortgage Book 3O, 337, W. S. and Catherine Brighton to L. H. Sanford, 15 July 1893; Mortgage Book 3O, 338, R. A. and Ellen Brighton to L. H. Sanford, 15 July 1893.

³⁰*Deseret News,* 16 May 1893; *Park Record,* 15 July 1893; *Salt Lake Tribune,* 14 July 1894.

³¹Nielsen, "William Stuart Brighton."

³²*Salt Lake Herald,* 20 July 1894, 21 July 1894; *Salt Lake Tribune,* 21 July 1894, 23 July 1894; *Park Record,* 21 July 1894; *Deseret News,* 23 July 1894; Gilchrist, "Pen Picture from Silver Lake," entry for 31 August 1878.

³³*Salt Lake Tribune,* 26 July 1894, 6 August 1894.

³⁴*Salt Lake Tribune,* 28 April 1895, 29 April 1895; *Salt Lake Herald,* 29 April 1895.

³⁵*Park Record,* 13 August 1887.

³⁶*Salt Lake Tribune,* 26 August 1890.

³⁷Salt Lake County Recorder, Deeds Book 5B, 18–20, Estate of W. S. Brighton to Taylor, Romney and Armstrong Co., Administrator's Deed, 28 March 1896; Deeds Book 4Z, 135–36, Estate of W. S. Brighton to Andrew Howat, Administrator's Deed, 27 March 1896.

³⁸Salt Lake County Recorder, Deeds Book 2P, 537–38, W. S. Brighton to Thomas B. Brighton, 17 March 1886.

³⁹*Park Record,* 9 July 1898; *Salt Lake Herald,* 20 August 1899. Lett was a bookkeeper and clerk, while Lambert, after leaving the hotel business, was in mining. Indeed, between 1899 and 1903 he recorded at least twenty-three claims, all of them in the vicinity of Silver Lake.

⁴⁰*Salt Lake Tribune,* 7 August 1898, 12 August 1898, 30 July 1899.

⁴¹*Salt Lake Tribune,* 26 July 1894, 6 August 1894, 27 July 1896; *Salt Lake Herald,* 9 August 1896, 12 September 1897.

⁴²*Lake at the Head of Cottonwood Canyon—Wasatch Mtn. 1871,* number 15 of some 600 works in Alfred Lambourne, "Catalogue Raisonne."

⁴³H. L. A. Culmer, "Mountain Scenery of Utah," part II, 201.

⁴⁴*Salt Lake Herald,* 6 December 1882, 15 June 1883; *Storm Clearing at Sunset, Silver Lake,* number 376 in "Catalogue Raisonne"; *Salt Lake Tribune,* 24 November 1886.

⁴⁵SeeKennett Alley Culmer, "Some Memories of the Life of H. L. A. Culmer—Artist."

[46]Salt Lake County Recorder, Deeds Book 3Q, 318, R. A. Brighton to James H. Moyle, 23 August 1890.

[47]*Salt Lake Tribune,* 20 July 1896, 27 July 1896.

[48]*Salt Lake Tribune,* 10 August 1900.

[49]*Deseret News,* 16 July 1901; *Salt Lake Herald,* 17 July 1901. A deed was not recorded for the property for another five years. See Salt Lake County Recorder, Deeds Book 8R, 158, Taylor Armstrong Lumber Co. to Alice E. Moyle, 20 November 1906; *Salt Lake Tribune,* 28 July 1901, 3 August 1901.

[50]*Salt Lake Tribune,* 10 August 1902, 24 August 1902, 3 July 1903, 17 July 1904.

[51]*Salt Lake Tribune,* 26 April 1904, 9 June 1912, 1 July 1916.

[52]*Salt Lake Tribune,* 9 June 1912, 30 June 1912.

[53]*Salt Lake Tribune,* 10 September 1916. An advertisement for the King 8 claimed 52 minutes for the Brighton trip; a news article on the same subject quoted 55 minutes. The Hal Twelve was reported in *Salt Lake Tribune,* 1 July 1917.

[54]Salt Lake County Commission Minute Book P, 14 February 1910, records the issuance of Davis's business license. The hotel is referred to in the *Salt Lake Tribune,* 21 July 1912, 2 August 1914 and 20 June 1915.

[55]*Salt Lake Tribune,* 3 March 1929; *Wasatch Rambler* 6, n. 3 (1 September 1926): 16, in Wasatch Mountain Club, "Papers."

[56]*Deseret News,* 22 March 1937; *Salt Lake Tribune,* 22 March 1937.

[57]*Deseret News,* 18 July 1941; *Salt Lake Tribune,* 19 July 1941.

[58]*Deseret News,* 8 October 1959; *Salt Lake Tribune,* 8 October 1959.

[59]*Salt Lake Tribune,* 29 June 1913, 10 August 1919.

[60]*Deseret News,* 20 July 1921, 26 August 1922, 16 July 1923, 21 January 1963; *Salt Lake Tribune,* 20 January 1963; "Brighton LDS Camp History."

[61]Alexis Kelner, *Skiing in Utah,* 156 n.; see also Salt Lake City directories for 1937, 1938, and 1940; *Deseret News,* 1 November 1941.

[62]*Salt Lake Tribune,* 23 August 1945, 24 August 1945.

[63]*Salt Lake Tribune,* 4 September 1955, 11 May 1963, 12 May 1963, 30 May 1963, 5 July 1963, 5 June 1965, 6 June 1965.

NOTES TO CHAPTER 11

[1]Obed C. Haycock, "Electric Power Comes to Utah," 175–77.

[2]*Park Record,* 23 March 1889.

[3]*Salt Lake Tribune,* 1 January 1892, 18 May 1892, 28 May 1892, 2 June 1892, 14 July 1893; *Park Record,* 14 May 1892, 4 June 1892, 17 September 1892.

[4]*Salt Lake Tribune,* 21 January 1895; *Salt Lake Herald,* 21 January 1895; *Deseret News,* 21 January 1895; *Salt Lake Mining Review,* 31 July 1899.

[5]Stairs Water Right, 26 October 1891, Water Claims Book A, 26, Salt Lake County Recorder's Office.

[6]Home Sweet Home and Defiance Westerly Water & Power Claim, 31 March 1892, Big Cottonwood Mining District Book E, 471. For more about

the activities of this controversial mining district recorder see Charles L. Keller, "James T. Monk: The Snow King of the Wasatch."

[7]See Wilford Woodruff, *Wilford Woodruff's Journal*, entries for summer of 1888.

[8]Big Cottonwood Power Company, 29 November 1893, Salt Lake County Corporate File 1079, Utah State Archives, series 3888.

[9]Salt Lake County Court, Minute Book G, 23 and 30 September 1895, 9 and 28 October 1895, 18 November 1895, 2, 14, 16, and 23 December 1895; *Salt Lake Tribune:* 8, 10, and 17 December 1895.

[10]*Park Record,* 20 May 1893; in Big Cottonwood Mining District Book E: Big Cottonwood Creek Water Appropriation, 3 March 1893, 574; Big Cottonwood Creek Water Right, 3 March 1893, 575; Big Cottonwood Creek Water Appropriation, 3 March 1893, 576; and Utah Water & Power Claim, 8 May 1893, 582.

[11]Big Cottonwood Water Power Right, 21 February 1893, 181–82, and Utah Power & Water Claim, 8 May 1893, 187, Salt Lake County Recorder, Water Claims Book A.

[12]Water Right, 20 January 1894, Big Cottonwood Mining District Book E, 646. In July 1893 Gillespie claimed waters flowing in Mill Creek Canyon and in December 1893 claimed the waters of Broads Fork and Mineral Fork. See Water Claims Book A, 196, 216.

[13]*Salt Lake Tribune,* 14 July 1894.

[14]Ibid.

[15]*Salt Lake Tribune,* 26 and 27 April 1895, 1 May 1895.

[16]*Salt Lake Tribune,* 19 December 1895.

[17]*Salt Lake Tribune,* 5 February 1896.

[18]*Salt Lake Herald,* 4 February 1896; *Salt Lake Tribune,* 4 February 1896.

[19]*Salt Lake Herald,* 5 February 1896; *Salt Lake Tribune,* 5 February 1896.

[20]*Salt Lake Tribune,* 25 April 1896.

[21]Salt Lake County Commission Minute Book G, 2, 11 & 18 May 1896; *Salt Lake Tribune,* 28 July 1896

[22]*Salt Lake Tribune,* 19 May 1896; *Deseret News,* 15 June 1896. For detailed descriptions of this power plant see W. P. Hardesty, "The Water Power and Electric Transmission Plant of the Big Cottonwood Power Co.," and "Electrifying Zion: The Big Cottonwood Power Transmission, Salt Lake City, Utah."

[23]*Salt Lake Tribune,* 25 April 1896.

[24]*Salt Lake Tribune,* 7 October 1896, 1 January 1898; *Park Record,* 25 July 1896.

[25]*Deseret News,* 3 March 1916; *Salt Lake Tribune,* 3 March 1916; *Salt Lake Telegram,* 3 March 1916.

[26]Much background on these companies and their mergers and consolidations may be found in "Utah Light and Traction Company: History of Origin and Development," "Utah Power & Light Company: History of Origin and

Development." Interesting instructions on the operation of the Granite generating plant may be found in "Utah Power & Light Co., Instruction to Employees, Power Department."

[27]Salt Lake County Recorder, Deeds Book 4N, 19 January 1894, 396–98; 3 February 1894, 398; 3 February 1894, 398–99.

[28]Mill B Electric Mill Site, water right appropriation and claim, 14 June 1894, Salt Lake County Recorder, Water Claims Book A, 231 (two entries).

[29]Brown and Sanford Ditch and Water Co., 26 April 1884, Salt Lake County Recorder, Water Claims Book A, 110.

[30]Water Right File 57–6, Mill B South Fork, Utah State Division of Water Rights.

[31]The date probably was 1937. In September of that year the state engineer, in a memo reporting a visit to the site, referred to the middle lake as Lake Minnie. See T. H. Humphries memo, 18 September 1937, in Division of Water Resources file 57–41, Mill B South Fork. Lake Lillian was called Lake Minnie in Donald C. McKay, "Ascent of Twin Peaks," *Salt Lake Tribune,* 13 June 1920.

[32]Dam Information Listing, Dams 10200–10202, Utah Division of Water Rights.

[33]See "Charles W. Hardy Collection."

[34]*Salt Lake Tribune,* 27 October 1898, 8 September 1899, 29 November 1899, 13 December 1899.

[35]*Salt Lake Tribune,* 28 November 1900, 7 August 1901.

[36]*Deseret News,* 31 December 1904. In 1906 the Big Cottonwood conduit was connected directly to the Parleys conduit, allowing Big Cottonwood water to flow directly to the city; see *Salt Lake Tribune,* 11 September 1906.

[37]*Salt Lake Tribune,* 12 August 1906, 26 September 1906.

[38]*Salt Lake Tribune,* 16 September 1906, 3 January 1909; Salt Lake County Recorder's Office, Quit Claim Deed 232899, 29 January 1908, Conveyance 232901, 5 February 1908.

[39]*Salt Lake Tribune,* 5 August 1913, 24 September 1915, 3 October 1915, 10 December 1915, 2 January 1916, 19 July 1916; J. T. Hoyt, "Construction of the Lake Mary–Phoebe Dam," *Salt Lake Mining Review,* 15 April 1916.

[40]*Salt Lake Tribune,* 2 October 1914, 22 April 1915, 24 September 1915, 2 January 1916.

[41]*Salt Lake Tribune,* 25 May 1917, 15 July 1919; George Mining Claim, 5 April 1902, Big Cottonwood Mining District Book I, 231; George, U.S. Mineral Survey 5945A, BLM office, Salt Lake City.

[42]*Salt Lake Tribune,* 26 April 1920, 28 April 1920; *Deseret News,* 28 April 1920.

[43]Annual Report of the City Engineer for the year 1926, 47.

[44]*Salt Lake Tribune,* 14 November 1926.

[45]*Salt Lake Tribune,* 8 March 1920.

[46]*Salt Lake Tribune,* 16 December 1926.

[47]*Salt Lake Tribune,* 26 May 1928, 30 May 1928; *Deseret News,* 30 May 1928.

[48]*Deseret News,* 30 March 1929; *Salt Lake Tribune,* 30 March 1929, 31 March 1929, 6 April 1929, 20 April 1929.

[49]*Salt Lake Tribune,* 25 February 1930, 11 March 1930; *Deseret News* editorials, 27 February 1930, 11 March 1930.

[50]*Deseret News,* 1 April 1930, *Salt Lake Tribune,* 2 April 1930

[51]*Salt Lake Tribune,* 13 March 1930, 10 April 1930, 16 April 1930, 17 April 1930, 19 April 1930.

[52]*Salt Lake Tribune,* 17 April 1930, 18 April 1930.

[53]*Salt Lake Tribune,* 1 May 1930.

[54]*Salt Lake Tribune,* 7 May 1930. A total of 9,724 votes had been cast, with only 1,754 being in favor of the issue. This is in contrast with the bond election of 1920 when the issue was soundly defeated but fewer than 3,000 votes were cast; see *Deseret News,* 28 April 1920; *Salt Lake Tribune,* 28 April 1920.

[55]*Salt Lake Tribune,* 12 February 1933, 18 April 1933; An Act For the protection of the municipal water supply of the city of Salt Lake City, State of Utah, 26 May 1934, U. S. Statutes at Large 48 (1938): 808–9.

[56]R. E. Marsell, "Water Supply," in *Geology of Salt Lake County,* 145.

NOTES TO CHAPTER 12

[1]Salt Lake County Commission, Minute Book O, 18 December 1905.

[2]Ibid., 12 June 1906.

[3]*Deseret News,* 4 May 1970. Much of the information on the Mill Creek power plants came from "Utah Power & Light Company: History of Origin and Development," and "Early History of Utah Power and Light Co."

[4]Case files: Mill Creek Power Company, Water Rights 57–15 and 57–16, Utah State Division of Water Resources.

[5]In the spring of 1999 the Forest Service did an extensive grading of the pipeline trail, "improving" it to better accommodate the heavy bicycle traffic it receives, but in the process destroying much of our historical heritage. In a few places the exposed hoops were torn out and discarded on the slopes below, but in most cases they were buried. Only a very few are visible anymore. The steel elbows at the several corners were severed from their deadman restraints and pushed to the trailside, still visible but no longer in their original locations.

[6]C18 517.13, Salt Lake County Engineering Dept., map of streams, pipelines, canals, etc., 17 November 1900, on an aperture card found in Series 5947, Salt Lake County Surveyor Road Location Maps, Utah State Archives.

[7]Journal History, 3 August 1847.

[8]"Epistle to the Saints in Great Salt Lake Valley," ibid., 9 September 1847.

[9]Ibid., 30 June 1848.

[10]Mill Creek Lode, 29 June 1894, Big Cottonwood Mining District Book E, 713. In July 1913 newspaper reports indicated this limekiln was owned by James Langston of the Langston Lime and Cement Comopany. On July 20,

Mr. Langston was killed when his automobile ran off a narrow bridge and overturned into Mill Creek stream after a visit to his limekiln in the canyon. There was no indication that the kiln was active at that time. Langston's company had long operated kilns located north of the city, at 2nd West and 14th North streets. See *Deseret News*, 21 July 1913; *Salt Lake Tribune*, 21 July 1913.

[11]Union Lime and Stone Company, 1 June 1901, Salt Lake County Corporate File 3251, in Utah State Archives, series 3888.

[12]Salt Lake County Commission, Minute Book N, 1 February 1904; *Salt Lake Tribune*, 12 April 1904.

[13]Rattle Snake Lode, 30 September 1891, Big Cottonwood Mining District Book E, 411.

[14]Mill Creek Lode and Rattle Snake Mill Site, 29 June 1894 and 5 November 1894, respectively, Big Cottonwood Mining District Book E, 713, 767.

[15]Salt Lake County Recorder, Deeds Book 6F, 128–30, USA to Wm. H. Dodge and H. K. North, 22 April 1896.

[16]Aquarius Nos. 1–3 Placers, 27 May 1902, Big Cottonwood Mining District Book I, 495–96.

[17]Western Development and Construction Co., 15 October 1903, Salt Lake County Corporate File 3817, in Utah State Archives, series 3888. Quitclaim deeds transferring the placer claims to the company dated from March 1903 but were not recorded until 16 October, the day after the corporation papers were filed.

[18]Salt Lake and Suburban Railway Co., 15 July 1902, Salt Lake County Foreign Corporate File 279, Series 4019, Utah State Archives. The company changed its name to Salt Lake Southern Railway Company on 19 January 1904.

[19]An outline and sequence of these claims and transactions may be seen in the abstracts for Township 1 South, Range 2 East, for this period. See Salt Lake County Recorder, Abstract Book D2, 89, 185–88.

[20]Salt Lake County Recorder, Deeds Book 11U, 171–72, Alvin V. Taylor to Oscar A. Kirkham, Trustee for Boy Scouts of America, 21 May 1919; *Salt Lake Tribune*, 19 September 1919; *Salt Lake Telegram*, 19 September 1919.

[21]Salt Lake County Recorder, Deeds Book 11V, 355, Third District Court, Boy Scouts of America vs. Salt Lake Southern Railway Company, 3 May 1924.

[22]*Deseret News*, 13 May 1920; *Salt Lake Tribune*, 15 May 1920.

[23]*Salt Lake Tribune*, 7 December 1923. Russel L. Tracy, Jr.'s death was reported in the *Deseret News*, 21 May 1921.

[24]Salt Lake County Court, Minute Book A, 26 March 1855.

[25]Geneva Lode, 30 September 1891, Big Cottonwood Mining District Book E, 412.

[26]Salt Lake County Court, Minute Book A, 8 December 1854.

[27]Case file: UP&L Church Fork, Water Right 57–18, Utah State Division of Water Resources.

[28]Salt Lake County Court, Minute Book B, 7 June 1859.

[29]Salt Lake County Court, Minute Book C, 4 May 1875.

[30]Salt Lake County Recorder, Deeds Book S, 130–37, tax sale, 21 December 1914; Deeds Book U, 136, tax sale, 15 January 1917.

[31]Salt Lake County Recorder, Deeds Book 11Q, 434, E. S. Hallock to L. F. Raines, 15 May 1923; Deeds Book 12K, 564–65, L. F. Rains to Ruth B. Rains, 23 July 1925; also see *Salt Lake Tribune,* 18 September 1994. The man's name was variously spelled as Rains or as Raines. He has been credited with being an architect but actually was a giant in the industrial world, having founded and operated several coal mining companies, organized and operated Utah's first large-scale steel plant at Ironton, and been involved in the founding of the Geneva steel plant in 1941; see *Salt Lake Tribune,* 12 September 1965.

[32]Salt Lake County Recorder, Deeds Book 194, 617–18, Ruth B. Rains and husband L. F. Rains to Zane W. Miller, 9 October 1936; Deeds Book 578, 370, Zane W. Miller to Gleed Miller, 2 May 1947; Deeds Book 2183, 124, Gleed Miller to Stanley W. Sprouse, 15 April 1964.

[33]Salt Lake County Court Minute Book A, 13 & 14 June 1853

[34]The Indian Pete legend was related by Asa Bowthorp, "History of Pioneer Sawmills and Local Canyons of Salt Lake Valley." The body was reported found in the *Salt Lake Tribune,* 19 June 1889. For the incorporation of the Indian Pete company see *Salt Lake Mining Review,* 15 August 1899. Claims filed by the company are in Big Cottonwood Mining District Book J between 1915 and 1917. The Wasatch Mountain Club hike was reported in "The Wasatch Rambler Yearbook, 1929," in Wasatch Mountain Club, "Papers."

[35]Salt Lake County Court, Minute Book A, 21 September 1853.

[36]In 1998 the Forest Service closed access to the Burch Hollow road. It constructed a new trail starting below Burch Hollow and going up many switchbacks to meet the Burch Hollow road where it doubled back to reach the pipeline trail.

[37]Salt Lake County Court, Minute Book A, 1 September 1856.

[38]Map of pipeline, upper and lower reservoirs, and power house based upon a survey completed on 12 June 1906. Utah Power & Light Co. "Upper Mill Creek Station."

[39]*Salt Lake Tribune,* 8 December 1936, 3 July 1937.

[40]Mammoth Coal Mine, 22 November 1865, Mountain Lakes Mining District Book A, 104.

[41]Boston and Last Chance Claims, 4 September 1903, Big Cottonwood Mining District Book I, 375, 377.

[42]Mammoth Coal Mine, 22 November 1865, Mountain Lakes Mining District Book A, 104; Gee Whiz, 25 July 1899, Big Cottonwood Mining District Book I, 63.

[43]Salt Lake County Court, Minute Book A, 3 March 1856.

[44]Salt Lake County Court, Minute Book A, 4 June 1860, 5 June 1860.

⁴⁵Thomas Fork was mentioned in the Klondyke Claim, 19 November 1898, Big Cottonwood Mining District Book I, 24. Norths Fork was mentioned in a water claim, 5 August 1898, Salt Lake County Recorder, Water Claims Book B, 107–8.

⁴⁶*Geology of Salt Lake County,* 65. The Peterson Basin name was used by A. V. Taylor, et al. in an 1898 water claim cited below.

⁴⁷Three water claims by A. V. Taylor, Joseph E. Taylor, and W. B. Stout, 3 August and 5 August 1898, Salt Lake County Recorder, Water Claims Book B. 107–9.

⁴⁸Delila Gardner Hughes, *Life of Archibald Gardner,* 54.

⁴⁹Mount Olympus Fruit and Livestock Farm Company, 3 June 1899, Salt Lake County Corporate File 2020, in Utah State Archives, series 3888.

⁵⁰Cottonwood Bell Claim, 5 March 1871, Big Cottonwood Mining District Book B, 91; Tiger Lode, 15 February 1895, Book F, 155; Howard Mine No. 1, 8 February 1895, Book F, 131.

⁵¹Henrietta Lode, 19 December 1870, Big Cottonwood Mining District Book B, 50.

⁵²Little Mrs. of the Mountains Lode, 16 March 1871, Big Cottonwood Mining District Book B, 97. The group of men filing this claim included Green Flake, one of two black slaves who were in the pioneer party arriving in the valley in July 1847.

⁵³C18 11.3, Map of Salt Lake County, 1 August 1908, on an aperture card found in Series 5947, Salt Lake County Surveyor Road Location Maps, Utah State Archives.

⁵⁴*Salt Lake Tribune,* 18 August 1876.

Notes to Chapter 13

¹*Salt Lake Tribune,* 1 September 1950, 17 February 1951.

²*Salt Lake Tribune,* 11 May 1956, 12 August 1957, 27 September 1957, 23 December 1957, 3 April 1959, 12 June 1959; *Deseret News,* 30 December 1957.

³Kenneth W. Baldridge, "Nine Years of Achievement: The Civilian Conservation Corps in Utah," 131, Appendix B. A photograph of CCC Camp F-38 appears in Alan K. Engen, *For the Love of Skiing: A Visual History,* 94–95.

⁴The lower end of this road was considerably altered in 1998–99 by contractors installing a new subsurface flume for the Granite Power Plant at the mouth of the canyon.

⁵Monitor Tunnel, located above Smoots Dugway, 30 December 1872, Big Cottonwood Mining District Book B, 615. A reference to Smoots Dugway was made as recently as 1913. Bell Claim, 13 February 1913, Big Cottonwood Mining District Book I, 580.

⁶John Henry Tew claims the highway was first paved, to a twelve-foot width, in 1928; see "Big Cottonwood Canyon and its Importance as a Water Source."

[7]For Mill C North Fork see Sir William Wallace Lode, 9 June 1880, and Cole Lode, 25 June 1880, Big Cottonwood Mining District Book D, 45, 59. Mule Cañon appeared in 1901 and Mule Gulch appears in several claims after 1905. See War Bonnet Claim, 10 June 1901, and Bluff Claim, 23 June 1905, Big Cottonwood Mining District Book I, 146, 536.

[8]The water claims were recorded in Big Cottonwood Mining District Book F, pp.303–4, 9 August 1895. Title for the land was through Homestead Certificate 8024, 11 May 1908, BLM office, Salt Lake City. For early Maxfield Lodge operations see *Salt Lake Tribune,* 11 August 1912, 4 September 1912, 27 June 1915, 1 July 1916. Also see Josie M. Reenders and Lois M. Recore, "Growing Up in Big Cottonwood Canyon" and "Maxfield History."

[9]*Salt Lake Tribune,* 4 March 1956, 14 May 1960, 6 July 1960, 20 November 1960, 3 May 1961.

[10]Elizabeth Lodes, 4 October 1886, Big Cottonwood Mining District Book D, 572, and 3 April 1895, Book F, 216. His obituary was in the *Deseret News,* 8 December 1936. Also see Reenders and Recore, "Growing Up in Big Cottonwood Canyon."

[11]Nelson Wheeler Whipple, "Journal, 1877 Jun–1878 August," entry for 22 July 1877; Brocks Fork Water Right, 21 December 1893, Salt Lake County Recorder, Water Claims Book A, 216; Mill B Electric Mill Site, water right appropriation and claim, 14 June 1894, Salt Lake County Recorder, Water Claims Book A, 231 (two entries).

[12]El Ray Nos. 1, 2, 3, and 5, 15 October 1908, Big Cottonwood Mining District Book I, 576–78; White Cap Placer, 20 January 1909, Big Cottonwood Mining District Book I, 579; see "Charles W. Hardy Collection."

[13]*Deseret News,* 30 July 1932; *Salt Lake Tribune,* 16 August 1932.

[14]Reenders and Recore, "Growing Up in Big Cottonwood Canyon."

[15]Articles of Agreement, 13 October 1855, "Brigham Young Letter Book," 2:427–28. This document may well have established the convention to name the Big Cottonwood tributaries by the closest sawmill.

[16]H. L. A. Culmer, "Mountain Scenery of Utah." The lakes were named sometime in the period from 1886 to 1891. Lambourne's daughter Lillian was born in September 1885, and Culmer published the lakes' names in February 1892.

[17]See Alfred Lambourne, "A Group of Wasatch Lakes," "Naming of the Wasatch Lakes," "Lake Lillian, Wasatch Mountains," *A Summer in the Wasatch.* See p. 30 of the latter for mention of the natural sundial.

[18]White Swan Claim, 7 August 1872, Big Cottonwood Mining District Book B, 527; minutes of miners' meeting, 22 February 1874, Argenta Mining District Book A, 26; *Utah Mining Gazette,* 21 March 1874.

[19]Oceola Lode, 6 December 1870, Big Cottonwood Mining District Book B, 48; Ellison Lode, 18 July 1872, Book B, 474; Bright Point Lode, 2 August 1872, Book B, 491, Hornet Lode, 2 August 1872, Book B, 522.

[20]Lone Pine Claim, 2 September 1899, Big Cottonwood Mining District Book I, 76.

[21]*Salt Lake Mining Review,* 16 June 1936.

[22]Silver Mountain Mine No. 2, 7 January 1871, Big Cottonwood Mining District Book B, 53; British Tunnel, 13 July 1880, Big Cottonwood Mining District Book D, 75. For the organization of the Silver Mountain Mining Company see the *Salt Lake Tribune,* 26 November 1879; *Salt Lake Herald,* 10 January 1880.

[23]*Salt Lake Tribune,* 1 January 1881, 1 January 1882, 1 January 1885, 1 January 1886. The road is reported in the *Salt Lake Tribune,* 13 July 1883, 24 July 1883, 15 September 1883.

[24]Meridian Quarry Lode, 5 July 1929, Big Cottonwood Mining District Book O, 587.

[25]Meridian Quarry Lode, 19 August 1936, Big Cottonwood Mining District Book Q, 260; Mineral Survey 7152, Meridian Quarry Lode, 1 August–3 September 1936, amended 10 February 1941, BLM office, Salt Lake City.

[26]Mineral Patent No. 1111576, Meridian Quarry and Meridian Quarry Lode, 9 June 1941, BLM office, Salt Lake City

[27]*Salt Lake Tribune,* 2 July 1943; *Polk's Ogden City Directory,* 1946.

[28]*Deseret News,* 31 October 1952.

[29]On 24 April 1866 D. H. Cahoon and Robert Maxfield proposed to haul lumber to "the Burnt Flat" for D. H. Wells; see letter, in D. H. Wells Collection, Incoming Letters, 1856–66.

[30]*Salt Lake Tribune,* 10 September 1875.

[31]Meridian Quarry Lode, amended notice. 19 August 1936. Salt Lake County Recorder, Location Book Q, 140–41.

[32]Peter Cooper Lode, 7 June 1878, Big Cottonwood Mining District Book C, 447; Green Back Lode, 13 May 1880, Big Cottonwood Mining District Book D, 38; Boadicia Mine, 9 June 1880, Book D, 46.

[33]William F. James called it White Butte when he recorded the Ninety Six Lode, 2 November 1896, Big Cottonwood Mining District Book F, 459. It was called Poison Dog Peak in Laurence P. James, *Geology, Ore Deposits and History of the Big Cottonwood Mining District, Salt Lake County, Utah,* plate 4.

[34]Silver Boy Lode, 1 January 1896, Big Cottonwood Mining District Book F, 368; Silver Girl Lode, 1 January 1896, Book F, 369.

[35]Afghan Claim, 2 November 1878, Big Cottonwood Mining District Book C, 498.

[36]Afghan Spring and Water Claim, 6 November 1878, Big Cottonwood Mining District Book C, 499.

[37]Jackson Hadley, Cottonwood Chief, Cottonwood Chief Tunnel and Paint Hill Tunnel Claims, 24 June 1872, 5 July 1872, 9 July 1872, and 31 July 1872, respectively, Big Cottonwood Mining District Book B, 396, 425, 458, 486.

[38]*Salt Lake Tribune,* 29 July 1872, 31 July 1872.

[39]*Salt Lake Tribune*, 19 August 1872, 24 August 1872; *Salt Lake Herald*, 25 August 1872.

[40]*Salt Lake Tribune*, 21 March 1880; *Salt Lake Herald*, 24 June 1881. Mrs. Graham was probably the wife of John D. Graham, who did considerable prospecting in Big Cottonwood in the 1880s.

[41]Salt Lake County Commission Minute Book M, 15 August 1904. For more on the Argenta community during this period see Charles L. Keller, "James T. Monk: The Snow King of the Wasatch."

[42]Lorenzo Brown, "Journal of Lorenzo Brown," entry for 26 January 1855.

[43]Nelson Wheeler Whipple, "Autobiography of Nelson Wheeler Whipple," 200.

[44]Kimbal Bros. Lode, 27 August 1870, Big Cottonwood Mining District Book A, 68; Reed and Benson Lode, 13 September 1870, Book A, 84; Pierless Lode, 17 October 1870, Book A, 131; Union Pacific Lode, 14 November 1870, Book B, 41.

[45]Reynolds's claim was Mill D Meadows No. 2, 31 December 1894, Big Cottonwood Mining District Book F, 19. The two claims referring to the cabin location were Dragoon and Dragoon No. 1, 13 July 1903, Book I, 351. Reynolds appeared in the press when one of his cowherders got into a shooting altercation with William F. James, superintendent of the Maxfield Mining Company; *Salt Lake Tribune*, 14 June 1897. He also received the attention of the county commissioners when some of his cattle died and were allowed to remain in the canyon near Smoots Dugway; Salt Lake County Commission Minute Book L, 7 May, 14 May 1900.

[46]*Salt Lake Tribune*, 21 August 1871.

[47]Savage Claim, 12 August 1870, Little Cottonwood Mining District Book A, 100.

[48]Bates & Selover Mining Company, 12 June 1871, Big Cottonwood Mining District Book B, 179. Bates was a party to at least nine claims in the Big and Little Cottonwood Mining Districts.

[49]*Salt Lake Herald* 2 June 1878. The post office name was cited in the *Salt Lake Herald*, 23 December 1879, and the *Salt Lake Tribune*, 23 December 1879. It is likely the post office was not actually moved from Argenta but simply received the new name. Batesville also appeared on a map published several years later. See "Rand McNally Indexed County and Township Map of Utah," c. 1883.

[50]*Salt Lake Tribune*, 25 February 1882, 26 February 1882.

[51]Will C. Higgins, "The South Fork of Big Cottonwood," *Salt Lake Mining Review*, 30 July 1911.

[52]*Salt Lake Tribune*, 15 March 1907; Reeds Peak Mining Company, 14 February 1907, Summit County Corporation File 128, Utah State Archives Series 7184; F. C. Calkins and B. S. Butler, *Geology and Ore Deposits of the Cotton-*

wood–American Fork Area, Utah, 111; McDougald Tunnel Claim, 16 December 1876, Big Cottonwood Mining District Book C, 288.

[53]McDougald Lode, 22 June 1871, Big Cottonwood Mining District Book B, 155. The trail was mentioned in the Kenilworth Claim, 4 October 1871, Book B, p. 306. The Kessler Peak North Trail is described in Veranth, *Hiking the Wasatch*, 113. Also see McDougald Mill Site, 25 October 1871, Book B, 344; Goodspeed Tunnel, 20 May 1872, Book B, 357; McDougald Tunnel, 16 December 1876, Book C, 288.

[54]McDougald Tunnel Claim, 6 February 1880, Big Cottonwood Mining District Book D, 10.

[55]Eagle Nest Lode, 21 May 1890, Big Cottonwood Mining District Book E, 214.

[56]Montreal Lode, 6 July 1871, Big Cottonwood Mining District Book B, 189; Quebec Tunnel Claim, 9 September 1871, Book B, 286; Walker Tunnel Claim, 27 May 1872, Book B, 357; Halifax Lode, 26 June 1874, Big Cottonwood Mining District Book B, 707.

[57]Vino Lode, 3 July 1871, Big Cottonwood Mining District Book B, 201; Vino Tunnel, 12 June 1872, Book B, 368; Vino Tunnel No. 2, 19 July 1875, Book C, 165. The Vino Flat name first appeared in the Pluto Tunnel Claim, 7 July 1876, Book C, 246. The Vina name appeared in the *Salt Lake Tribune*, 8 August 1878, and the Montreal No. 1 Claim, 25 October 1881, Big Cottonwood Mining District Book D, 238.

[58]An Act to Repeal Timber-Culture Laws, and for other purposes, 3 March 1891, U.S. Statutes at Large 26 (1891): 1095–1103.

[59]Samuel Trask Dana, *Forest and Range Policy, Its Development in the United States*, 100–1.

[60]Appropriations Act for Department of Agriculture, 4 March 1907, U.S. Statutes at Large 34 (1907): 1271.

[61]The Uintah reserve, at 875,520 acres (1,368 square miles), was designated on 22 February 1897. The Fish Lake reserve, at 67,840 acres (106 square miles), was designated on 10 February 1899. See Twentieth Annual Report of the U.S. Geological Survey, 1898–99, Part 5, Forest Reserves, and Twenty-first Annual Report of the U.S. Geological Survey, 1899–1900, Part 5, Forest Reserves.

[62]*Salt Lake Tribune*, 28 November 1900, 7 August 1901.

[63]*Salt Lake Tribune*, 25 February 1902. It was during the second half of this year that Albert F. Potter made his extensive survey of Utah's forest lands. See Potter, "Diary," and Charles S. Peterson, "Albert F. Potter's Wasatch Survey, 1902."

[64]*Salt Lake Tribune*, 25 February 1902, 20 March 1902, 23 March 1902; *Deseret News*, 26 February 1902. The Sevier Forest Reserve was reported in the *Salt Lake Tribune*, 21 August 1902.

[65]*Salt Lake Tribune*, 9 August 1908. For the origins of the Forest Service and the many forest reserves, see Jenks Cameron, *The Development of Govern-*

mental Forest Control in the United States; Milton Conover, *The General Land Office, Its History, Activities and Organization;* Marian Clawson, *The Bureau of Land Management;* Marian Clawson, *The Federal Lands Revisited;* Paul W. Gates, *History of Public Land Law Development;* Charles S. Peterson, *A History of the Wasatch-Cache National Forest;* and James T. Young, *The New American Government and its Work.*

[66]*Salt Lake Tribune,* 25 September 1905, 30 November 1905, 17 July 1906, 5 May 1910, 28 July 1911, 10 September 1913, 7 April 1915, 1 May 1918.

[67]*Salt Lake Tribune,* 28 August 1915, 23 September 1915, 16 July 1918.

[68]*Salt Lake Tribune,* 11 July 1919, 18 August 1920, 29 August 1920.

[69]*Salt Lake Tribune,* 10 June 1921, 5 July 1921, 10 July 1921, 7 July 1922, 8 July 1922; *Deseret News,* 23 June 1921, 18 July 1922.

[70]*Salt Lake Tribune,* 12 August 1923, 8 December 1936, 9 January 1937, 30 May 1937.

[71]Brilliant Lode, 14 September 1870, Big Cottonwood Mining District Book A, 89.

[72]John B. Wright, *Rocky Mountain Divide, Selling and Saving the West,* 168. In 1900 the patrolman in Parleys Canyon reported 1,700,000 sheep and 5,550 cattle driven over the designated stock trail in the spring and again in the fall; *Annual Reports of the Officers of Salt Lake City, Utah, for the year 1900,* 87.

[73]Walter P. Cottam and Fred A. Evans, "A Comparative Study of the Vegetation of Grazed and Ungrazed Canyons in the Wasatch Range," 145.

[74]*Salt Lake Tribune,* 31 August 1897.

[75]*Salt Lake Tribune,* 21 December 1915. It should be noted that the Wasatch Forest Reserve included much more than the mountains above Salt Lake County. The statement about prohibition of grazing was in the *Salt Lake Tribune,* 15 January 1924.

[76]J. A. Carollo, J. M. Montgomery, and N. T. Veatch, *Report on Investigation of Quality of Water Supplies, Metropolitan Water District of Salt Lake City,* 223. The report recommended the city acquire the grazing rights for the watershed canyons and allow the soil cover to return.

[77]*Deseret News,* 29 June 1954; *Salt Lake Tribune,* 29 June 1954.

[78]Sir Richard F. Burton, *The City of the Saints,* 347.

[79]Asa R. Bowthorpe, *History of Pioneer Sawmills and Local Canyons of Salt Lake Valley.* Seth E. Littleford died at the age of ninety-four on 15 May 1962. His year of birth was reported as 1858, but 1868 is consistent with his reported age. His place of birth was given as Silver Springs. *Salt Lake Tribune,* 16 May 1962.

[80]See Bowthorpe, *History of Pioneer Sawmills.*

[81]The Antelope Lode was described as situated at the head of the first south fork below the mill known as Mill F; 18 June 1870, Big Cottonwood Mining District Book A, 12. A few weeks later the Wellington Lode was described as being on the northeast side of Silver Fork; 6 July 1870, Book A, 21.

[82]Whipple, "Autobiography of Nelson Wheeler Whipple," entries for summer 1870, 230–32.

[83]Jim Light, conversations with the author.

[84]Fair Play Lode, 17 July 1888, Big Cottonwood Mining District Book E, 42. The Willow Heights name appeared in the Cactus Claim, 21 September 1903, Book I, 381.

[85]USGS Cottonwood Special Map, edition of March 1907; USGS Cottonwood Quadrangle, edition of 1939.

[86]Little Dollie Lode, 13 July 1894, Big Cottonwood Mining District Book E, 722; Little Dollie Nos. 2 to 8 Lodes, 26 July 1895, Book F, 272–78; Little Dollie Tunnel, 26 July 1895, Book F, 298; Giles Flat Lode, 23 March 1896, Book F, 401.

[87]Giles Mining & Milling Co., 24 September 1902, Salt Lake County Corporate File 3581.

[88]*Salt Lake Mining Review,* 30 July 1915; *Salt Lake Tribune,* 30 April 1916, 29 July 1917.

[89]*Deseret News,* 22 February 1924.

[90]USA to B. F. Redmond and Emmett G. Hunt, 22 August 1907, Salt Lake County Recorder Abstracts, Book D1, 165.

[91]Monarch Nos. 1 and 2 lodes, 26 June 1894, Big Cottonwood Mining District Book E, 704–5.

[92]*Salt Lake Tribune,* 22 May 1926, 23 May 1926, 24 May 1926.

[93]See William T. Larkins, "The Aircraft History of Western Air Lines."

[94]*Deseret News* 26 August 1914, 31 August 1914, 26 July 1915.

[95]*Salt Lake Telegram,* 8 July 1938, 9 July 1938; *Salt Lake Tribune,* 8 July 1938, 13 January 1945; *Deseret News,* 12 January 1945.

[96]*Salt Lake Tribune,* 9 December 1954.

[97]*Salt Lake Tribune,* 9 January 1955; Salt Lake County Commission Minute Book X, 28 December 1926.

[98]*Salt Lake Tribune,* 15 January 1955, 7 June 1956, 19 June 1957, 21 June 1957, 23 April 1959.

[99]Frederick Kesler, "Diary," entry for 5 August 1857; Eureka Lode, 20 July 1864, Mountain Lakes Mining District Book A, 7; Excelsior Lode, 20 July 1864, Book A, 10; Maraposa Lode, 29 August 1864, Book A, 29; El Dorado Lode, 14 June 1870, Big Cottonwood Mining District Book A, 11.

[100]Julie, 12 July 1871, Big Cottonwood Mining District Book B, 193; Big Cottonwood Tunnel, 10 July 1871, Book B, 203; Stella Brighton Nielsen, "William Stuart Brighton"; Mary Brighton Timmons, "Address, July, 1940." Both women were granddaughters of William S. Brighton.

[101]Desert Lode, 15 September 1971, Big Cottonwood Mining District Book B, 285; Janet Lode, 7 September 1872, Book B, 559.

[102]Evergreen Lode, 4 June 1870, Big Cottonwood Mining District Book A, 10.

[103]*Salt Lake Tribune,* 25 June 1872.

[104]Evergreen Consolidated Mining and Tunnel Co., 29 August 1874, Salt Lake County Corporate File 110.

[105]*Salt Lake Tribune,* 1 January 1880; *Salt Lake Herald,* 4 January 1880.

[106]Poor Mans Friend Lode, 16 May 1871, Big Cottonwood Mining District Book B, 115; Sailors Delight, 31 October 1871, Book B, 487.

[107]Julie, 12 July 1871, Big Cottonwood Mining District Book B, 193; Lee Lode, 11 October 1872, Book B, 598; Sailors Delight No. 2, 27 August 1877, Big Cottonwood Mining District Book C, 357.

[108]Zoah Tunnel, 24 June 1872, Big Cottonwood Mining District Book B, 398.

[109]Buttercup Claim, 27 March 1899, Big Cottonwood Mining District Book I, 36.

[110]Virginia State Lode, 10 October 1872, Big Cottonwood Mining District Book B, 592; North Virginia Mine, 28 July 1873, Book B 658; Granite Mountain Lode, 17 August 1890, Book E, 237.

[111]Amalric, "Mount Eyrie."

[112]See Nielsen "William Stuart Brighton," and Timmons, "Address, July, 1940."

[113]*Salt Lake Herald,* 24 June 1871, 20 October 1872, 13 March 1874; *Salt Lake Tribune,* 13 August 1872, 22 February 1878. Godbe's cottage was mentioned in the *Salt Lake Tribune,* 13 July 1883. For more on the Godbeite movement see Ronald W. Walker, *Wayward Saints: The Godbeites and Brigham Young.*

[114]*Salt Lake Herald,* 13 August 1899.

[115]*Salt Lake Tribune,* 2 August 1902; *Deseret News,* 3 August 1902. Charles P. Brooks died on 30 November 1918. *Salt Lake Tribune,* 1 December 1918.

[116]Wolverine Lode, 22 July 1870, Big Cottonwood Mining District Book A, 38; Wolverine, 1 September 1871, Little Cottonwood Mining District Book B, 50.

[117]*Salt Lake Herald,* 2 August 1896.

[118]Map of Salt Lake County, Utah, 1 August 1908, in Series 5947, Salt Lake County Surveyor Road Location Maps, Utah State Archives; *Salt Lake Tribune,* 9 August 1903.

[119]H. L. A. Culmer, "Mountain Scenery of Utah," 204–5; Alfred Lambourne, "Naming of the Wasatch Lakes," 1055; *Salt Lake Tribune,* 21 April 1888, 6 April 1890; Lake Phoebe Claim, 17 July 1888, Big Cottonwood Mining District Book E, 44.

[120]Culmer, "Mountain Scenery," 205–6; Lambourne, "Naming of the Wasatch Lakes," 1055.

[121]Culmer, "Mountain Scenery," 206; Lambourne, "Naming of the Wasatch Lakes," 1055; Dale L. Morgan, *The Great Salt Lake,* 338; *Salt Lake Tribune,* 17 March 1885.

[122]Col. Cooper Lode, 11 May 1866, Mountain Lakes Mining District Book A, 117.

¹²³Brighton Lode, 19 August 1870 Big Cottonwood Mining District Book A, 63; Catherine Lode, 6 September 1870, Book A, 78.

¹²⁴See Timmons, "Address, July, 1940"; and Nielsen, "William Stuart Brighton."

¹²⁵Sunbeam Lode, 24 February 1890, Big Cottonwood Mining District Book E, 195.

¹²⁶Allegany Lode, 6 September 1870, Big Cottonwood Mining District Book A, 91.

¹²⁷Culmer, "Mountain Scenery," 205; Lambourne, "Naming of the Wasatch Lakes," 1056.

¹²⁸*Emma Mine Investigation*, 44th Congress, 1st Session, House of Representatives Rept. 579, 25 May 1876, Serial 1711, p. 608.

¹²⁹*Salt Lake Tribune*, 13 August 1882. Clayton died on 3 July 1889 from injuries suffered in a stage accident near Coeur d'Alene, Idaho; see *Salt Lake Tribune*, 22 June 1889, 28 June 1889, 5 July 1889. See also *Salt Lake Herald*, 9 August 1896.

¹³⁰Scott Lode, 12 September 1870, Big Cottonwood Mining District Book A, 82. The Scott Hill name was used in the Majors Lode, 9 September 1871, Book B, 327.

NOTES TO CHAPTER 14

¹"The Ferguson Gold Mine," in Daughters of Utah Pioneers, *Chronicles of Courage*, v. 1, 420–21.

²*Salt Lake Tribune*, 16 June 1912.

³Salt Lake County Court, Minute Book C, 4 December 1876.

⁴Journal History, 23 March 1850; *Deseret News Weekly*, 15:56 (23 November 1865); Salt Lake County Court, Minute Book B, 4 December 1865, 4 June 1866.

⁵Salt Lake County Court, Minute Book C, 21 December 1875.

⁶*Salt Lake Herald*, 21 June 1871, 27 September 1879; Australia Lode, 26 June 1877, Big Cottonwood Mining District Book C, 311.

⁷*Salt Lake Tribune*, 21 August 1916, 24 August 1916, 28 August 1916, 20 July 1917.

⁸*Salt Lake Tribune*, 16 August 1920, 29 August 1920, 28 June 1921, 3 July 1921, 1 August 1922, 6 August 1923.

⁹*Salt Lake Tribune*, 22 September 1916; Donald C. McKay, "Ascent of Twin Peaks," in *Salt Lake Tribune*, 13 June 1920. While some hikers use the Little Willow or Lake Blanche routes to Twin Peaks today, the favored route is by way of Broads Fork. See John Varanth, *Hiking the Wasatch*, 94.

¹⁰Walter Prescott Webb, ed., *Handbook of Texas*, v. 2, 622.

¹¹Gold Hill and Hidden Treasure Claims, 24 May 1894, Little Cottonwood Mining District Book D, 436–37.

¹²Gold Hill Lode, 11 July 1894, Big Cottonwood Mining District Book E, 721.

¹³New State Mining and Milling Co., 1 December 1894, Salt Lake County Corporate File 1176, Utah State Archives, series 3888; *Salt Lake Tribune*, 20

January 1895. One of the principals in the New State company was Benjamin F. Redman, who was introduced in this volume's narration of the Redman Forest Camp in Big Cottonwood Canyon. Redman also appeared on three of Dalton's claims during 1894.

[14]*Salt Lake Tribune,* 22 January 1895, 23 January 1895.

[15]Cottonwood Gold Mining and Milling Company, 26 January 1895, Salt Lake County Corporate File 1180; *Salt Lake Tribune,* 26 January 1895, 27 January 1895.

[16]*Salt Lake Tribune,* 28 January 1895, 29 January 1895.

[17]Gold City Land and Townsite Company, 14 February 1895, Salt Lake County Corporate File 1192; *Salt Lake Tribune,* 4 February 1895, 7 February 1895, 10 February 1895.

[18]*Salt Lake Tribune,* 17 March 1895, 16 June 1895.

[19]*Park Record,* 6 July 1895; *Salt Lake Tribune,* 1 August 1896; Baby Ruth Claim, 4 June 1896, Little Cottonwood Mining District Book E, 33; *Salt Lake Mining Review,* 15 July 1901.

[20]*Salt Lake Tribune,* 30 December 1935, 1 January 1936; Model Steam Bakery Claim, 6 July 1894, Big Cottonwood Mining District Book E, 717. The Model Steam Bakery was at 555 South 900 East in Salt Lake City; Schmittroth was foreman there in 1892. See *Salt Lake City Directory, 1892–93.*

[21]Jefferson Gold and Copper Mining Company, 3 September 1899, Salt Lake County Corporate File 2072; *Salt Lake Tribune,* 7 September 1899; *Salt Lake Mining Review,* 15 September 1899.

[22]*Engineering and Mining Journal,* 13 June 1903; *Salt Lake Mining Review,* 15 December 1904, 15–17.

[23]*Engineering and Mining Journal,* 10 March 1906.

[24]Jefferson Extension Mining Company, 27 August 1908, Salt Lake County Corporate File 5050; Diana Gold and Copper Exploration Company, 2 December 1908, Salt Lake County Corporate File 5139; *Salt Lake Mining Review,* 15 September 1908, 15 December 1908, 15 March 1909.

[25]*Salt Lake Tribune,* 7 October 1908; *Engineering and Mining Journal,* 17 October 1908; *Salt Lake Mining Review,* 30 October 1908, 29 February 1916.

[26]Golden Phorphyry Mines Company, 12 December 1924, Salt Lake County Corporate File 10641; *Salt Lake Mining Review,* 15 February 1927, 30 March 1930, 30 June 1931, 27 December 1932.

[27]*Salt Lake Tribune,* 30 December 1935, 1 January 1936.

[28]Salt Lake County Commission Minute Book T, 2 January 1917, 23 February 1917.

[29]*Salt Lake Tribune,* 31 July 1912.

[30]*Salt Lake Tribune,* 30 April 1929, 28 June 1931; *Salt Lake Mining Review,* 6 October 1931.

[31]*Salt Lake Tribune,* 12 June 1983.

[32]Whitmore Oxygen Company, 20 September 1916, Salt Lake County Corporate File 8199.

[33]Water Right File 57–88, Whitmore Oxygen Company, Utah State Division of Water Rights.

[34]Journal History, 7 April 1851.

[35]Much confusion was caused by the use of the word foundation when describing the footing in early accounts of the temple's construction. As a result, there is widespread belief that when part of the temple foundations (footings) were taken up and relaid in 1862 the sandstone was replaced with granite, but that is not the case. The temple's granite foundation walls actually rest on footings of rough-hewn sandstone from the Red Butte quarries. See Paul C. Richards, "The Salt Lake Temple Infrastructure: Studying It Out in Their Minds."

[36]Journal History, 27 April 1855. The choice of materials was not publicly announced until 16 March 1856 when Brigham Young, speaking in the Salt Lake Tabernacle, said they intended to use granite; *Journal of Discourses*, 3:249.

[37]Wilford Woodruff, *Wilford Woodruff's Journal, 1833–1898*, 14 August 1847.

[38]Journal History, 21 August 1847.

[39]*Deseret News Weekly* 4:176 (1 February 1855), 5:56 (25 April 1855); 5:197 (29 April 1855); 7:21 (25 March 1857), 11:361 (14 May 1862).

[40]See "Cottonwood Quarry Book."

[41]Winslow Farr, "Diary, May 1856–September 1899"

[42]William Adams, "Autobiography January 1894"; Woodruff, *Journal,* 23 July 1857.

[43]*Deseret News Weekly,* 11:196 (18 December 1861).

[44]See William Dobbie Kuhn, "Recollections of Temple Quarry, Little Cottonwood Canyon and Old Granite (City)."

[45]"Cottonwood Quarry Book."

[46]*Salt Lake Herald,* 29 December 1871.

[47]*Salt Lake Tribune,* 12 November 1872.

[48]*Salt Lake Herald,* 11 March 1873.

[49]*Salt Lake Herald,* 5 August 1873.

[50]*Salt Lake Herald,* 12 March 1874.

[51]*Salt Lake Herald,* 11 May 1873.

[52]*Salt Lake Tribune,* 15 October 1899, 25 October 1899.

[53]*Salt Lake Tribune,* 30 June 1911, 17 August 1911, 10 June 1913; *Salt Lake Mining Review,* 30 June 1913.

[54]*Salt Lake Tribune,* 24 July 1912, 13 June 1913, 24 June 1913, 26 August 1913, 16 November 1913.

[55]*Salt Lake Tribune,* 24 July 1912; *Salt Lake Mining Review,* 30 November 1919.

[56]*Deseret News,* 3 July 1883; *Salt Lake Tribune,* 24 July 1886, 16 August 1887, 17 July 1892.

[57] *Salt Lake Herald,* 29 July 1894; *Salt Lake Tribune,* 16 July 1894, 1 June 1899.

[58] *Salt Lake Tribune,* 1 August 1920.

[59] *Salt Lake Herald,* 10 November 1870.

[60] *Salt Lake Tribune,* 14 July 1871.

[61] *Salt Lake Tribune,* 23 May 1871.

[62] *Salt Lake Herald,* 17 August 1872.

[63] Surveyor's notes R129, T3SR2E, BLM office, Salt Lake City.

[64] Maybird Gold, Silver and Lead Mining Co., 20 June 1874, Little Cottonwood Mining District Book C, 120; Snowdrop Claim, 22 August 1875, Little Cottonwood Mining District Book C, 250. J. B. White was killed in an avalanche at the Wellington Mine on 28 December 1876.

[65] *Salt Lake Tribune,* 16 December 1936 to 1 January 1937, 5–7 January 1937.

[66] *Salt Lake Tribune,* daily entries from 31 May 1937 to 7 June 1937.

[67] *Salt Lake Tribune,* entries from 8 June 1937 to 13 June 1937, 15 June 1937, 16 June 1937.

[68] *Salt Lake Tribune,* 17 June 1937, 3 July 1937, 4 July 1937, 6 July 1937, 7 July 1937, 16 July 1937, 17 July 1937, 4 August 1937, 5 August 1937, 12 August 1937.

[69] *Salt Lake Tribune,* 19 August 1937.

[70] Maybird Gold, Silver and Lead Mining Co., 20 June 1874, Little Cottonwood Mining District Book C, 120.

[71] El Paso Claim, 20 September 1874. Little Cottonwood Mining District Book C, 148; Stella Claim, 31 May 1882, Little Cottonwood Mining District Book D, 155; Mineral King Claim, 18 October 1901, Little Cottonwood Mining District Book G, 140.

[72] USGS 1:250,000 map, Salt Lake City, Utah; Wyoming, 1954.

[73] *Salt Lake Tribune,* 28 March 1939; Odell Peterson, conversations with the author.

[74] Ophir Claim, 12 August 1870, Little Cottonwood Mining District Book A, 101; Tanners Creek Claim, 30 September 1870, Book A, 120.

[75] Samuel A. Woolley, "Diaries, 1846–1899," entry for 26 April 1871; *Salt Lake Herald,* 17 May 1871, 21 June 1871; *Salt Lake Tribune,* 18 May 1871.

[76] *Salt Lake Herald,* 14 March 1871, 25 March 1871, 17 June 1871; *Salt Lake Tribune,* 26 April 1871.

[77] *Salt Lake Herald,* 26 August 1871, 16 January 1872; *Salt Lake Tribune,* 14 July 1871, 29 March 1872, 3 July 1872, 28 December 1872.

[78] *Salt Lake Herald,* 14 September 1872.

[79] *Salt Lake Herald,* 17 July 1873, 12 March 1874; *Salt Lake Tribune,* 30 May 1875.

[80] See Hickman Lode, 25 October 1864, Mountain Lakes Mining District Book A, 35; Alexina Lode, 2 July 1870, Mountain Lakes Mining District Book B, 76.

[81] Samuel A. Woolley, "Diaries," 29 May 1866. Red Pine Slide was first used

in the Sawyer Claim, 9 September 1870, Little Cottonwood Mining District Book A, 110. The Red Pine Canyon name was first in the Tanners Creek Claim, 30 September 1870, Book A, 120.

[82]Niagara First Easterly Extension, 26 May 1866, Mountain Lakes Mining District Book A, 121; Harald Claim, 3 January 1875, Little Cottonwood Mining District Book C, 176.

[83]Woolley, "Diaries," 4 November 1868. Woolley used the Crapo Creek name as late as 25 June 1874.

[84]Texas Ranger Claim, 11 September 1874, Little Cottonwood Mining District Book C, 145.

[85]Woolley, "Diaries," 29 May 1866. White Pine Slide was first used in the Furnace Site Claim, 6 May 1870, Little Cottonwood Mining District Book A, 52; White Pine Canyon was first used in the Silver Gray Claim, 26 January 1891, Book D, 364.

[86]Snowbird Claim, 20 October 1903, Little Cottonwood Mining District Book G, 240.

[87]Woolley, "Diaries," 2 June 1869.

[88]Peruvian Lode, 27 June 1870, Mountain Lakes Mining District Book B, 68.

[89]*Salt Lake Mining Review,* 15 August 1902.

[90]The first four were Iron Blosam Nos. 1–4, 13 September 1909, Little Cottonwood Mining District Book G, 459–60.

[91]Harper's Ferry Claim, 5 August 1872, Little Cottonwood Mining Distict Book B, 339; Harpers Ferry Tunnel, 20 August 1873, Book C, 79.

[92]Superior Lode, 16 September 1869, Mountain Lakes Mining District, Book B, 28.

[93]Hellgate, 2 January 1901, Little Cottonwood Mining District Book G, 103; Hellgate No. 2, 17 October 1901, Book G, 139; Hellgate No. 3, 16 June 1903, Book G, 203; Hellgate No. 4, 23 June 1904, Book G, 279.

[94]The Rettichs's mining activities at Alta are described in Charles L. Keller, "Tales of Four Alta Miners," 108–11.

[95]Desoto Claim, 24 June 1872, Little Cottonwood Mining District Book B, 237; Baker No. 6, 24 November 1902, Book G, 181; Toledo Claim, 10 October 1871, Little Cottonwood Mining District Book B, 114. Toledo Hill appears in Toledo Tunnel Claim, 29 April 1876, Book C, 309; Baker No. 10, 24 November 1902, Book G, 183.

[96]Emily Claim, 28 June 1871, Little Cottonwood Mining District Book A, 320. Emily Hill appears in Queen of the Hill Claim, 17 August 1880, Book D, 85; Skylark Claim, 29 August 1885, Book D, 239.

[97]Davenport claim, 19 July 1870, Little Cottonwood Mining District Book A, p.88. The Davenport Hill name appeared in the Dexter claim, 2 October 1891, Book B, p.105. It still appears on the USGS Brighton, Utah 7–1/2 minute quadrangle, 1955

[98]Flagstaff Claim, 9 April 1870, Little Cottonwood Mining District Book A, 43. Flagstaff Hill appears in the *Salt Lake Herald,* 2 September 1871; Sleeper Claim, 4 January 1908, Little Cottonwood Mining District Book M, 74.

[99]*Salt Lake Mining Review,* 15 July 1926, 30 October 1926; Jay W. Jacobson interview, 25 December 1964, in Laurence P. James, "Little Cottonwood Canyon Collection." Jay Jacobson was a son of A. O. Jacobson, manager of the Mineral Veins mines.

[100]Lexington Claim, 31 October 1871, Little Cottonwood Mining District Book B, 138. Lexington Hill name was used in O'Connell Claim, 16 June 1872, Book B, 212, and most recently in Sells Tunnel Claim, 17 August 1896, Book E, 45. Lexington Tunnel claims were on 5 September 1874, Book C, 144, and 5 October 1880, Book D, 88. Also see *Salt Lake Tribune,* 31 January 1873, 12 December 1873; *Salt Lake Herald,* 14 December 1873.

[101]Alpha Lode, 20 August 1870, Mountain Lakes Mining District Book B, 89. Alpha Hill was used in Pepperbox Claim, 4 August 1873, Little Cottonwood Mining District Book C, 70, and Fair Play, 26 September 1885, Book D, 244. Peruvian Hill was used in the Lucy Long Claim, 9 June 1872, Book B, 200, and Water Right notice, 15 October 1891, Book D, 376.

[102]Flora Claim, 26 July 1871, Little Cottonwood Mining District Book A, 386.

[103]Rustler Claim, 15 August 1893, Little Cottonwood Mining District Book D, 413.

[104]The Iris Claim notice has not been found, but the hill was known as Iris Hill when the Iris Tunnel Claim was recorded, 7 September 1872, Little Cottonwood Mining District Book B, 387. It was still in use in 1904; see Happy Go Lucky claim, 2 January 1904, Book G, 251. Emerald Claim, 31 July 1871, Book A, 394. Emerald Hill was used in the Homestake Claim, 25 December 1871, Book B, 153, and was still in use for the Cobra Claim, 5 January 1903, Book G, 187.

[105]Germania Claim, 21 October 1905, Little Cottonwood Mining District Book M, 18; Alta Germania Mining Company, 4 September 1903, Salt Lake County Corporate File 3799. The Germania ski lift was constructed in 1954; see *Salt Lake Tribune,* 15 August 1954, 10 October 1954, 23 November 1954.

[106]Greeley Claim, 1 September 1884, Little Cottonwood Mining District Book D, 210; *Salt Lake Tribune,* 7 November 1893; *Salt Lake Mining Review,* 15 March 1925.

[107]Neversweat, Neversweat No. 1, and Neversweat No. 2 claims, 31 July 1901, 5 January 1903, and 4 December 1903, respectively, Little Cottonwood Mining District Book G, 124, 187, 246.

[108]Patsey Marley Claim, 23 May 1870, Little Cottonwood Mining District Book A, 62.

[109]*Salt Lake Herald,* 13 July 1870.

[110]Restraining Order, 28 March 1871, Little Cottonwood Mining District Book A, 210.

[111]*Deseret News,* 12 July 1870.

[112]*Salt Lake Tribune,* 5 March 1873.

[113]*Salt Lake Herald,* 14 October 1873.

[114]*Salt Lake Tribune,* 7 February 1874.

[115]*Salt Lake Tribune,* 2 July 1872.

[116]*Salt Lake Tribune,* 15 December 1916, 23 December 1916; Keller, "Tales of Four Alta Miners," 106–8.

[117]Wellington Lode, 5 August 1870, Little Cottonwood Mining District Book A, 96; *Salt Lake Herald,* 2 February 1872; *Salt Lake Tribune,* 28 December 1872.

[118]*Salt Lake Herald,* 13 April 1872, 14 April 1872; *Salt Lake Tribune,* 13 April 1872.

[119]*Salt Lake Herald,* 3 January 1877; *Salt Lake Tribune,* 3 January 1877, 5 January 1877, 9 January 1877, 24 July 1877.

[120]Albion Claim, 3 January 1880, Little Cottonwood Mining District Book C, 573.

[121]*Salt Lake Mining Review,* 28 February 1919; *Engineering and Mining Journal,* 15 March 1919.

[122]*Salt Lake Mining Review,* 30 October 1921; *Engineering and Mining Journal,* 5 November 1921; South Hecla Mines Company, 19 October 1921, Salt Lake County Corporate File 9889.

[123]Mountain Lake Lode, 27 August 1866, Mountain Lakes Mining District Book A, 151; Flora Claim, 26 July 1871, Little Cottonwood Mining District Book A, 386; Flora No. 2 Claim, 9 October 1871, Book B, 111; M. E. Pat. 4266, 14 October 1880, Mineral Survey 111, BLM Office, Salt Lake City.

[124]Katherina Claim, 26 August 1885, Little Cottonwood Mining District Book D, 238; Salt Lake County map, 15 June 1937, on aperture cards in series 5947, Salt Lake County Surveyor, Utah State Archives.

[125]Lambourne, "Naming of the Wasatch Lakes," 1056.

[126]The known paintings are *Lake Minnie—Wasatch Mountains 1879; Lake Minnie, Little Cottonwood Canyon 1887; Lake Minnie, Wasatch Mountains 1889;* and *Lake Minnie, Wasatch Mountains 1896;* see Alfred Lambourne, "Catalogue Raisonne." The books are *Jo, A Christmas Tale of the Wasatch,* and *Plet, A Christmas Tale of the Wasatch.*

[127]North Pole Claim, 8, January 1878, Little Cottonwood Mining District Book C, 460.

[128]For Cases Cabin see Katherina Claim, 26 August 1885, Little Cottonwood Mining District Book D, 238. For Cases Lake see Releigh and Rathbone Claims, 15 December 1890, Book D, 357, and Star of the West Claim, 16 September 1895, Book D, 554.

[129]Secret Mining and Milling Company, 7 April 1906, Salt Lake County Corporate File 4327, Utah State Archives.

[130]Cecret Claim, 16 September 1905, Little Cottonwood Mining District Book G, 322. Cecret Nos. 2 and 3 claims were filed the same day, recorded in Book G, 322–23. Cecret No. 4 was filed 20 October 1905, Book G, 328.

[131]Cecret Nos. 1 to 3 amended claims, 1 October 1907, Little Cottonwood Mining District Book M, 69–70.

[132]Cole and Cecret Nos. 1–4, M. E. Pat. 205917, 12 June 1911, Mineral Survey 5803, to Secret Mining and Milling Co., BLM *Office, Salt Lake City.*

[133]*Salt Lake Tribune,* 30 December 1916; *Salt Lake Mining Review,* 30 May 1919, 30 June 1919.

[134]Devils Slide Claim, 19 December 1879, Little Cottonwood Mining District Book C, 569.

[135]Alfred Lambourne, "The Wasatch Story of Plet," *Salt Lake Tribune,* 23 December 1912.

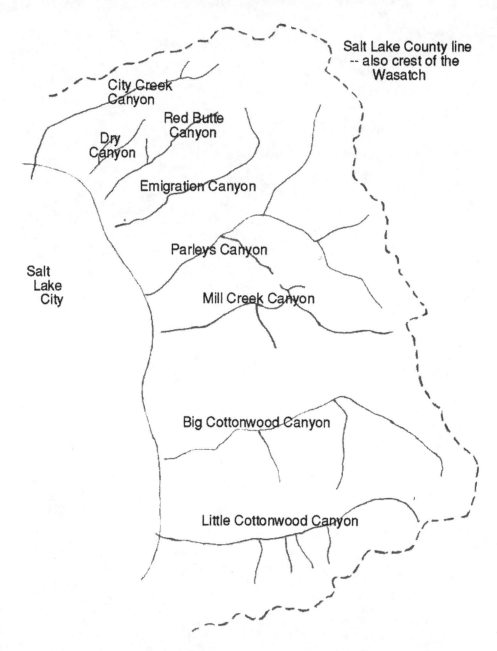

Wasatch Mountain Canyons in Salt Lake County

Maps:
Above: Wasatch Mountain Canyons in Salt Lake County
Mill Creek and Neffs Canyons (4 panels)
Big Cottonwood Canyon (4 panels)
Little Cottonwood Canyon (4 panels)

    The last three maps are composites of the Sugar House, Draper, Mount Aire, Dromedary Peak, Park City West, and Brighton USGS 7.5-minute series maps. They have captions for milepost markers (MP), tributaries, and other features mentioned in the text. The individual maps were stitched together through the courtesy of Mark Silver of iGage in Salt Lake City.

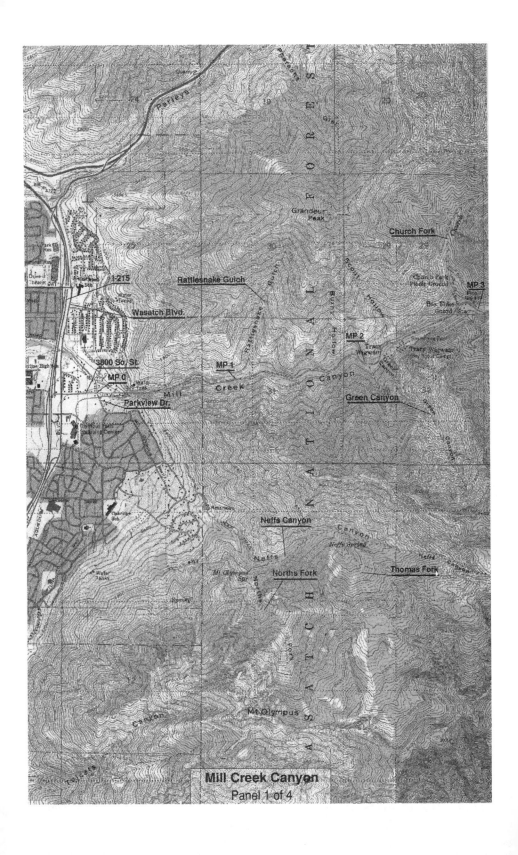

Mill Creek Canyon
Panel 1 of 4

Mount Aire

Canyon

Elbow Fork

"The Elbow"

MP 6

Burch Hollow

W   A   S

Church Fork

MP 5

Church Park
Picnic Ground

MP 4

Maple Grove
Picnic Ground

Creek   Creek

Evergreen
Picnic Ground

Box Elder
Ground Sta

MP 3

Main Box Elder
Picnic Ground

Maple Grove
Campground

Terrace
Campground

Bowman Fork

Thayne Canyon

Porter Fork

Green Canyon

N   A   T   I

Neffs Canyon

Mount Raymond

Mill Creek Canyon
Panel 2 of 4

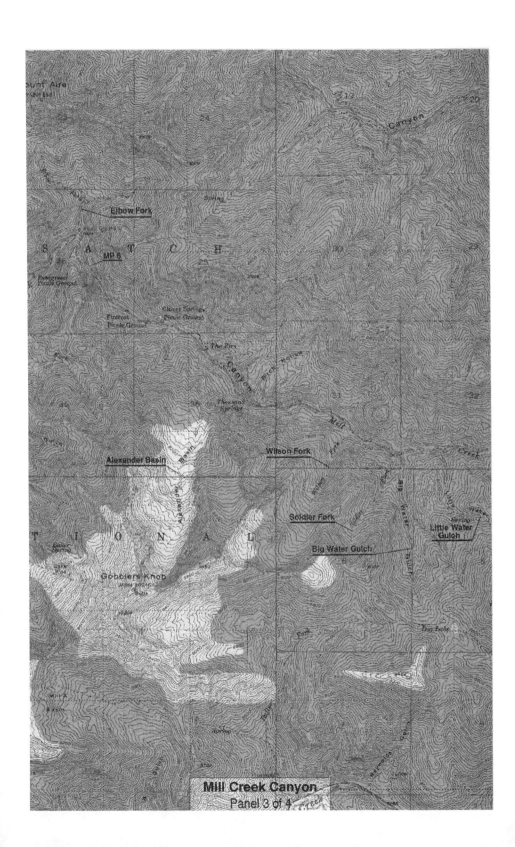

Mill Creek Canyon
Panel 3 of 4

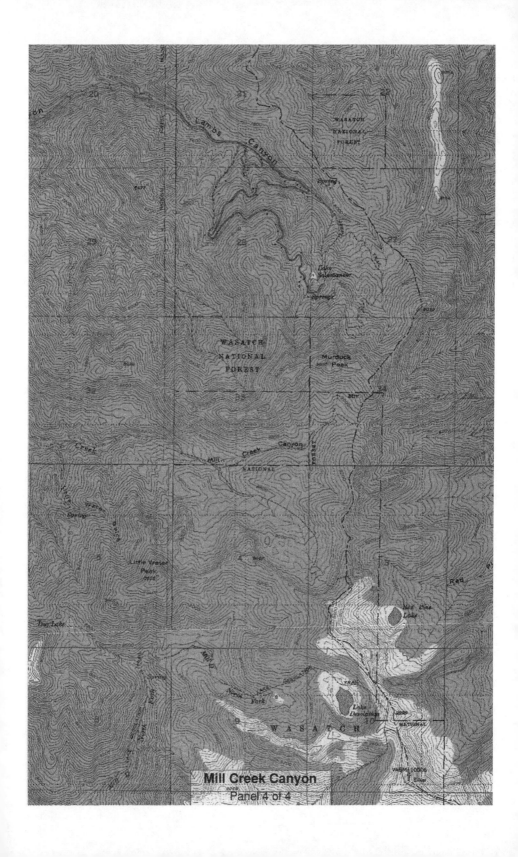

Mill Creek Canyon
Panel 4 of 4

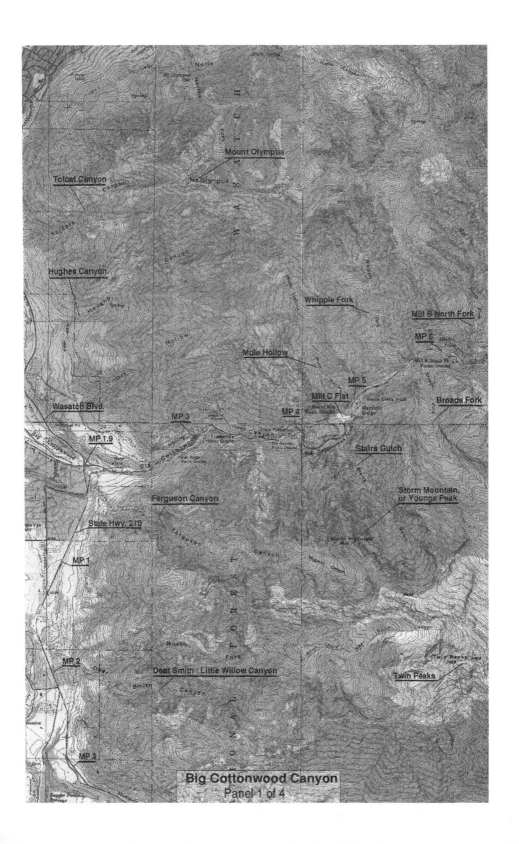

Mount Olympus

Tolcat Canyon

Whipple Fork

Mill B North Fork

MP 6

Hughes Canyon

Mule Hollow

MP 5

Mill C Flat

Broads Fork

Wasatch Blvd.

MP 3

MP 4

MP 1.9

Stairs Gulch

Ferguson Canyon

Storm Mountain,
or Youngs Peak

State Hwy. 210

MP 1

MP 2

Twin Peaks

Deaf Smith / Little Willow Canyon

MP 3

Big Cottonwood Canyon
Panel 1 of 4

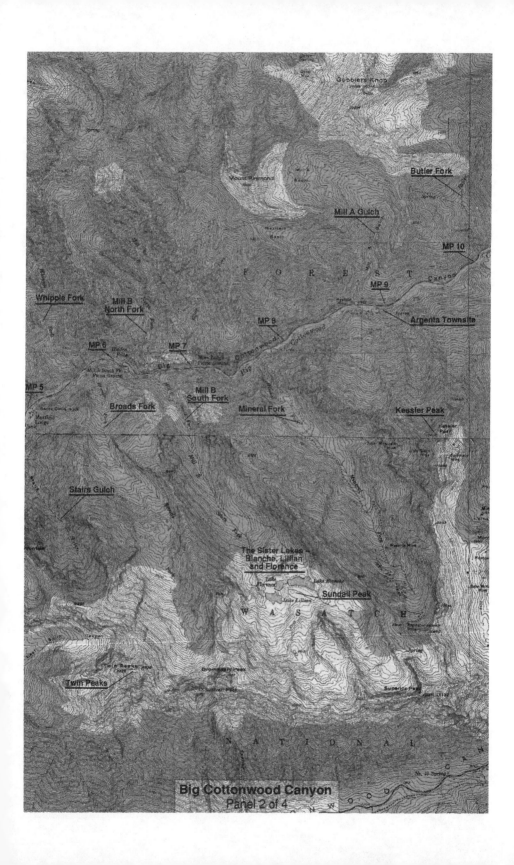

Gobblers Knob

Butler Fork

Mill A Gulch

Mount Raymond

MP 10

FOREST

MP 9

Whipple Fork

Mill B
North Fork

Argenta Townsite

MP 8

MP 6

MP 7

MP 5

Broads Fork

Mill B
South Fork

Mineral Fork

Kessler Peak

Stairs Gulch

The Sister Lakes—
Blanche, Lillian
and Florence

Sundail Peak

W   A   S   A   T   C   H

Twin Peaks

Superior Peak

N   A   T   I   O   N   A   L

Big Cottonwood Canyon
Panel 2 of 4

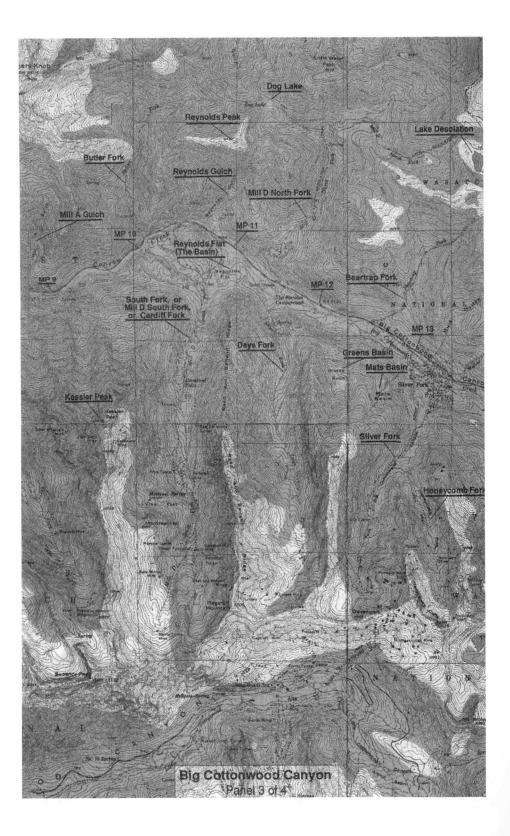

Dog Lake

Reynolds Peak

Lake Desolation

Butler Fork

Reynolds Gulch

Mill D North Fork

Mill A Gulch

MP 11

MP 10

Reynolds Flat
(The Basin)

Beartrap Fork

MP 9

MP 12

MP 13

South Fork, or
Mill D South Fork,
or Cardiff Fork

Days Fork

Greens Basin

Mats Basin

Silver Fork

Kessler Peak

Silver Fork

Honeycomb Fork

Big Cottonwood Canyon
Panel 3 of 4

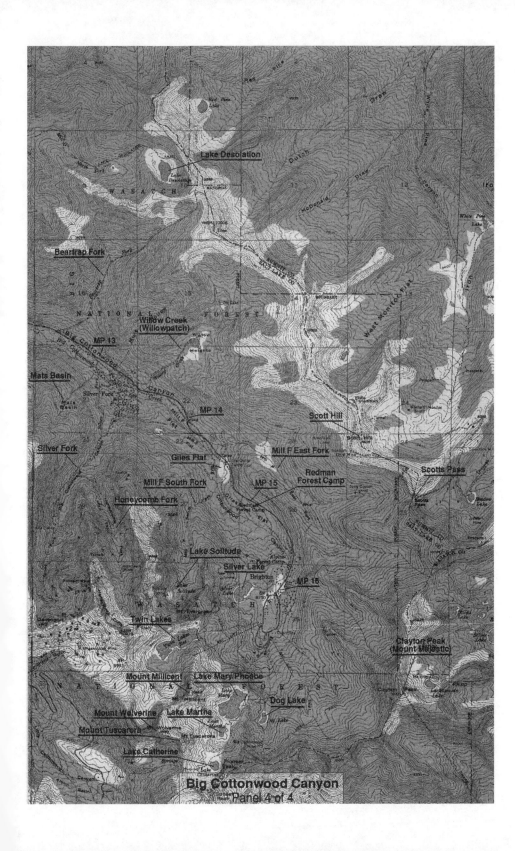

Lake Desolation

Beartrap Fork

Willow Creek
(Willowpatch)

MP 13

Mats Basin

MP 14

Scott Hill

Silver Fork

Giles Flat

Mill F East Fork

Scotts Pass

Mill F South Fork

MP 15

Redman
Forest Camp

Honeycomb Fork

Lake Solitude

Silver Lake

MP 16

Twin Lakes

Clayton Peak
(Mount Majestic)

Mount Millicent

Lake Mary/Phoebe

Mount Wolverine

Lake Martha

Dog Lake

Mount Tuscarora

Lake Catherine

Big Cottonwood Canyon
Panel 4 of 4

Little Cottonwood Canyon
Panel 1 of 4

Little Cottonwood Canyon
Panel 2 of 4

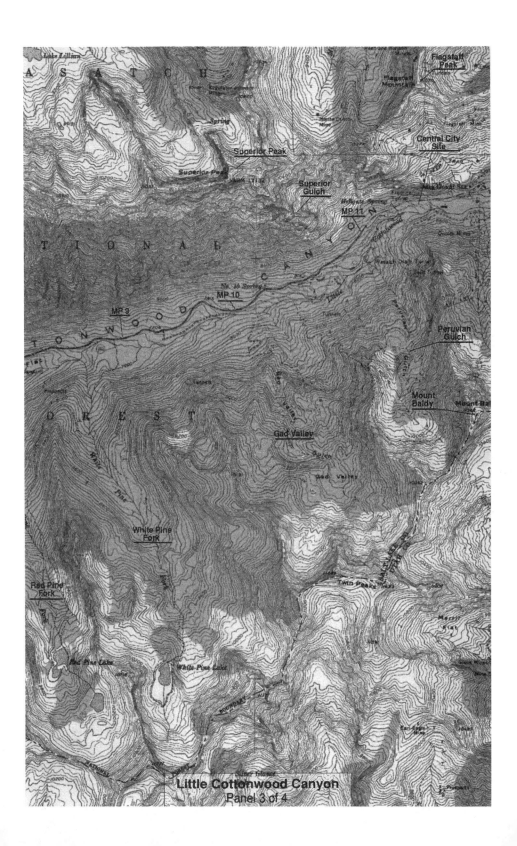

Little Cottonwood Canyon
Panel 3 of 4

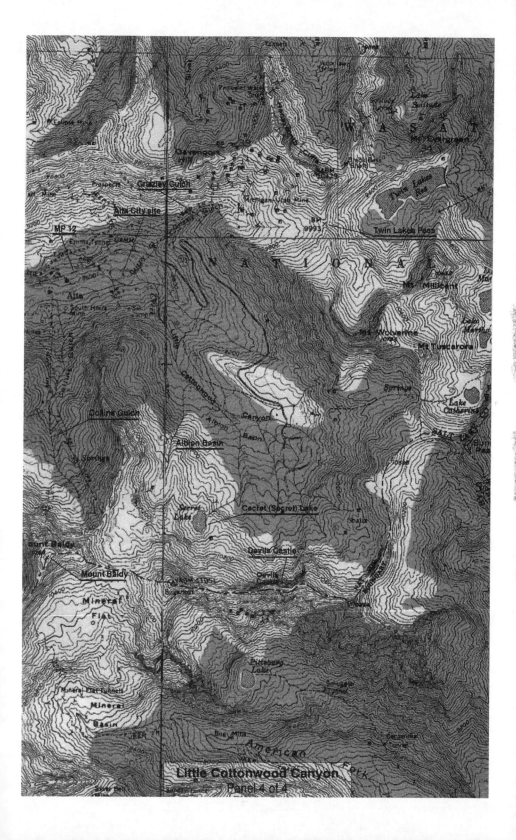

Little Cottonwood Canyon
Panel 4 of 4

Bibliography and Sources

In this bibliography the following abbreviations for sources are used:

BLM—Bureau of Land Management. The BLM maintains an office in Salt Lake City where it has many records pertaining to its operations as well as those of the Land Office that preceded it. Included are mineral surveys, mining and various other land patents, and surveyor's notes from original surveys when range and township boundaries were established. All records are on microform, but hard copies of many original maps are still on file.

CHOA—Church Historian's Office Archive. The Church of Jesus Christ of Latter-day Saints maintains an excellent historical archive. It is located in the Church Office Building in Salt Lake City. While its focus is on the history of the church and its members, early Utah history and church history are virtually one and the same. The archive contains many early diaries, journals, letterbooks, and other period documentation, as well as many photographs. Most documents have been placed on microfilm, but the copying is usually well done and the archive reading room has many well-maintained microfilm readers.

CHOL—Church Historian's Office Library. The library, like the archive, is located in the Church Office Building. It has a large assortment of books, directories, periodicals, newspaper files for papers published in Utah, and other documents. A special reading room is used for rare books that are not stored on the open shelves.

UHS—Utah State Historical Society. The Historical Society maintains a library containing a large collection of published and unpublished material relating to Utah history. It has a large collection of maps, directories, newspapers, periodicals, photographs, and other documents. A well-kept card catalog allows material to be located with more ease than is possible with many computer-based catalogs. The Society publishes the *Utah Historical Quarterly,* which in 2001 was in its sixty-ninth year.

UofU—University of Utah Marriott Library. Sometimes it seems that this library has more material than could be explored in a lifetime. Its extensive holdings include government documents and publications; maps; newspapers and periodicals, both in hard copy and on microfilm, microfiche, and microform; theses from its own student body as well as those of other institutions; and an assortment of published volumes located in countless rows of open stacks. The library also has a Special Collections Department that maintains numerous collections, like that of Western Americana. The computerized catalog allows material to be quickly located within the large five-story building, and it is accessible on the Internet.

Other sources, not cited by abbreviations, include the following:

Family History Library—This library, maintained in Salt Lake City by the Church of Jesus Christ of Latter-day Saints, has genealogy as its primary focus. Its collections contain material from all over the world. While it has many bound volumes in open stacks, the bulk of the collection is on microfilm. In research for this volume the library was especially useful for city directories, phone directories, and newspapers for cities outside of Utah.

Salt Lake County Recorder's Office—The county recorder's records are a treasure-trove for research into local history. In the archives are found most of the original mining district recorder books, many of them tattered and water-stained from their years of storage in the mountain cabins of the mining recorders. Also found in the county recorder's office are deed records, some of them going back to the earliest days of Salt Lake County, long before the Land Office opened in the territory; water-right records; and records of various legal documents, such as liens, mortgages, and agreements.

Utah State Archive—This state agency is charged with the storage and maintenance of inactive records of the various divisions of state government. Records still active are maintained by the agencies to which they belong. The state archives also hold copies of records of many local governmental agencies, such as county courts, county commissions, and city commissions. It stores many judicial court records and corporation files, at both the county and state levels. Most records are on microfilm, and recent emphasis has been on microfilming all records and discarding the originals. Unfortunately, inattention to focus and use of 16-mm film cause many filmed records to be much less legible than the originals, and some are not legible. The archive reading room has very few microfilm readers, and they are poorly maintained or totally inoperative. Hard-copy records are stored in an off-site warehouse, so they must be requested in advance of use.

BIBLIOGRAPHY

Acts, Resolutions and Memorials, passed at the several annual sessions of the legislative assembly of the territory of Utah [1850–1855] . . . published by virtue of an act approved Jan. 19, 1855. Great Salt Lake City: Joseph Cain, 1855.

Adams, William. "Autobiography Jan. 1894." (42 pp. typescript) CHOA, MS 8039.

Amalric. "Mount Eyrie," *The Contributor* 2 (1880–81): 308–10.

Annual Reports of the Officers of Salt Lake City, Utah, for the year 1900. (An annual report was prepared, printed and bound each year, starting as early as 1893, possibly earlier, and continued at least through 1915. Early in the twentieth century, at least by 1909, the city engineer prepared his own bound report. Copies of many of the reports were found in the office of the city engineer.)

Armstrong, F. W. "F. W. Armstrong Collection." UHS B-175. (Includes material on the Armstrong and Bagley Lumber Company.)

Arrington, Leonard J. "Abundance from the Earth: The Beginnings of Commercial Mining in Utah." *Utah Historical Quarterly* 31 (1963): 192–219.

———. "Banking Enterprises in Utah, 1847–1880," *Business History Review* 29 (1955): 312–34.

———. *From Quaker to Latter-day Saint: Bishop Edwin D. Woolley.* Salt Lake City: Deseret Book Co., 1976.

Baldridge, Kenneth W. "Nine Years of Achievement: The Civilian Conservation Corps in Utah." Ph.D. diss., Brigham Young University, 1971.

Beadle, John H. *The Undeveloped West.* Philadelphia: National Publishing Company, [1873].

Big Cottonwood Lumber Company. "Balance Sheet, March 1859" (rl. 141, bx. 113, fd. 1); "Ledger A (part 1), Nov. 1854—March 1863" (rl. 141, bx. 113, fd. 3); "Ledger A (part 2), Nov. 1854–March 1863" (rl. 141, bx. 113, fd. 4); "Misc. Papers, 1855, 1862" (rl. 142, bx. 114, fd. 1). CHOA, MS 1234.

Big Cottonwood Mining District Recorder's Books. Deposited in the Salt Lake County Recorder Archives.

Black, Susan Lindsay Ward. *Membership of the Church of Jesus Christ of Latter day Saints, 1830–1848.* 50 vols. Compiled by Susan Ward Easton. Provo: 1984–88. Microfilm copy is found in CHOA.

Bliss, Jonathan. *Merchants and Miners in Utah: The Walker Brothers and Their Bank.* Salt Lake City: Western Epics, 1983.

Bliss, Robert S. "The Robert S. Bliss Journal." Edited by Everett L. Cooley. *Utah Historical Quarterly* 27 (1959): 381–404.

Bowman, Henry Eyring "Autobiography of Henry Eyring Bowman." (9 pp. typescript.) CHOA, MS 9548, item 13.

Bowthorpe, Asa R. "History of Pioneer Sawmills and Local Canyons of Salt Lake Valley." (Copy of 30 pp. typescript) 1961. UHS, MSS A2243.

"Brighton LDS Camp History." (6 pp. typescript) ca. 1980. CHOA, MS 9323.

Brown, Craig Jay. "The Allocation of Timber Resources Within the Wasatch Mountains of Salt Lake County Utah, 1847–70." Master's thesis, University of Utah, 1982.

Brown, John. *Autobiography of Pioneer John Brown, 1820–1896.* Arranged and published by his son, John Zimmerman Brown. Salt Lake City: 1941.

Brown, Lorenzo. "Diaries, 1853–1859." CHOA, MS 1563. Also included in *Journal of Lorenzo Brown, 1823–1900.*

———. *The Journal of Lorenzo Brown, 1823–1900.* Heritage Press, 198?.

Burton, Sir Richard F. *The City of the Saints; and Across the Rocky Mountains to California.* 1862. Reprint, Niwot, CO: University of Colorado Press, 1990.

Butler, B. S., G. F. Loughlin, V. C. Heikes et al., *The Ore Deposits of Utah.* Washington, DC: U.S. Dept. of the Interior, 1920. USGS Professional Paper 111. 1920.

Butters, Judith B. *History of the Butlerville Area.* Cottonwood Heights, UT: Brighton Stake Relief Society, 1997.

Calkins, Frank C., and B. S. Butler. *Geology and Ore Deposits of the Cottonwood-American Fork Area, Utah.* Washington: Government Printing Office, 1943. USGS Professional Paper 201.

Cameron, Jenks. *The Development of Governmental Forest Control in the United States.* Baltimore: Johns Hopkins Press, 1928.

Carnahan, C. T., and C. T. Wright. *Report Covering Public Health Aspects of Salt Lake City's Water Supply, With Reference to Watershed Developments.* Salt Lake City: Utah State Department of Health, 1947.

Carollo, J. A., J. M. Montgomery, and N. T. Veatch. *Report on Investigation of Quality of Water Supplies.* Salt Lake City: Metro Water District, 1950.

Carson, Kit. *Kit Carson's Own Story of His Life.* Edited by Blanche C. Grant. 1926. Reprint. Lincoln: University of Nebraska Press, 1966.

Clapp, Elisha Drown. "Notebook, 1880–1883." CHOA, MS 3934 3.

Clawson, Marion. *The Bureau of Land Management.* New York: Praeger Publishers, 1971.

———. *The Federal Lands Revisited.* Washington, DC: Resources for the Future, 1983.

Clayton, William. *William Clayton's Journal. . . .* Salt Lake City: Deseret News, 1921.

Clyman, James. *James Clyman, Frontiersman....* Edited by Charles L. Camp. Portland, OR: Champoeg Press, 1960.

Conover, Milton. *The General Land Office, Its History, Activities and Organization.* Baltimore: Johns Hopkins Press, 1923.

Cook, Phineas Wolcott. "Reminiscences and Journal, 1844–1886." CHOA, MS 6288.

Cottam, Walter P., and Fred A. Evans. "A Comparative Study of the Vegetation of Grazed and Ungrazed Canyons in the Wasatch Range." *Ecology* 26, no. 2 (1945): 145.

Cottonwood Observer. Alta City, Utah. 1873.

"Cottonwood Quarry Book." (rl. 7, bx. 11, fd. 6 no. 1) CHOA, CR 5 7. (A small account and time book kept by the foreman of the stonecutters at the first granite quarries in 1857.)

Culmer, H. L. A. "Mountain Scenery of Utah." *The Contributor* 13 (1892): 173–78 ff.

Culmer, Kennett Alley. "Some Memories of the Life of H. L. A. Culmer—Artist." (11 pp. typescript) UHS, MSS A 1012.

Cumming, Elizabeth. "The Governor's Lady, A Letter from Camp Scott, 1857." *Utah Historical Quarterly* 22 (1955): 163–73.

Daily Union Vedette (1864–1867). Also issued as the *Union Vedette* (1863) and *Daily Vedette* (1864).

Dana, Samuel Trask. *Forest and Range Policy, Its Development in the United States.* New York: McGraw-Hill, 1956.

Daughters of Utah Pioneers. *An Enduring Legacy.* Vols. 1–12. Salt Lake City: 1978–1989.

———. *Chronicles of Courage.* Vols. 1–4. Salt Lake City: 1990–1993.

———. *Heart Throbs of the West.* Vols. 1–12. Edited by Kate B. Carter. Salt Lake City: 1939–1951.

———. *Our Pioneer Heritage.* Vols. 1–20. Edited by Kate B. Carter. Salt Lake City: 1958–1977.

———. *Tales of a Triumphant People: A History of Salt Lake County, Utah, 1847–1900.* Salt Lake City: Stevens & Wallis Press, 1947.

———. *Treasures of Pioneer History.* Vols. 1–6. Edited by Kate B. Carter. Salt Lake City: 1952–1957.

(With the more than fifty volumes in the five series of annual publications, the Daughters of Utah Pioneers have undertaken a monumental effort. The books have been especially useful to students who use them as source material for writing assignments and to those who enjoy reading short stories of former times. While they contain some excellent, well-documented articles, journals, and diaries, their primary content has been personal or family histories, folk or family lore, and recollections. As with oral histories, they contain the seeds of truth but often are distorted by time and the frailties of human memory. As a result, these volumes have been a time bomb in the font of historical authenticity, for there are those who accept such material as indisputable fact and repeat it as genuine history without attempting a second source verification, thereby propagating the inaccuracies it contained. Read these volumes and enjoy them, but always ask: Is this really history?)

Day-Holm, Kimberly. "The Importance of Frederick Kesler to the Early Economic History of Utah, 1851–1865." Master's thesis, University of Utah, 1980.

Deseret News (1867–).

Deseret News Weekly (1850–).

Despain, Robert Earl. *Robert Henry Despain Family, 1875 to 1967.* Lovell, WY: Mountain States Printing, ca. 1978.

Despain, Solomon Joseph. "Autobiography." ca. 1868. CHOA, MS 8318.

Dwyer, Robert Joseph. *The Gentile Comes to Utah.* Salt Lake City: Western Epics, 1971.

"Early History of Utah Power and Light Co." (380 pp. typescript) 1940. CHOA, MS 6291.

"Electrifying Zion: The Big Cottonwood Power Transmission, Salt Lake City, Utah." *Electrical Engineer* 20, no. 384 (1895): 245–49.

Ellsworth, Edmund. "Account Book, 1850–1877." (47 pp.) CHOA, MS 14935.

———. "Autobiography." (35 pp. handwritten) ca. 1892. CHOA, MS 9334.

Emma Mine Investigation. United States Congress. House Report No. 579. Washington: Government Printing Office, 1876.

Engen, Alan K. *For the Love of Skiing: A Visual History.* Salt Lake City: Gibbs Smith, 1998.

Engineering and Mining Journal (New York, 1869–1922).

Esshom, Frank. *Pioneers and Prominent Men of Utah.* Salt Lake City: Utah Pioneers Book Publishing Co., 1913.

Farr, Winslow. "Diary, May 1856–Sept. 1899." (252 pp. typescript) CHOA, MS 1743.

Foote, Warren. "Autobiography and Journals, 1837–1903." CHOA, 1123 1–3.

Frémont, John Charles. *Memoirs of My Life.* Chicago: Balford, Clark & Co., 1887.

Froiseth, B. A. M. *Froiseth's Map of Little Cottonwood Mining District and Vicinity, Utah.* Salt Lake City: Froiseth, 1873. (Copy found in University of Utah Special Collections.)

Gardner, Clarence. "Autobiography and Journals, 1895–1957." CHOA, MS 5509.

Gardner, Robert. *Robert Gardner, Jr. 1819–1906, Utah Pioneer 1847. . . .* Cedar City, UT: n.p., 1973.

Gates, Paul W. *History of Public Land Law Development.* Washington, DC: Government Printing Office, 1968.

Gazetteer of Utah and Salt Lake City Directory, 1874. Compiled and edited by Edward L. Sloan. Salt Lake City: Salt Lake Herald Publishing Co., 1874.

Geology of Salt Lake County. Bulletin 69, Utah Geological and Mineral Survey, 1964.

Gilchrist, Charlotte E. "Pen Picture from Silver Lake Glen, Big Cottonwood Canyon, Utah." (150 pp. typescript) 1881, 1884. UHS, MS A1931.

Greeley, Horace. *An Overland Journey from New York to San Francisco in the Summer of 1859.* Reprint. New York: Knopf, 1964.

Greenland, Helen. "Williams, Pugh and North Family Biographies." Item 13: Spencer, Lilly North. "History of Levi North." 1950. CHOA, MS 8096 1.

Groo, Isaac. "Autobiography." (3 pp. typescript) CHOA, MS 1003 2.

Hardesty, W. P. "The Water Power and Electric Transmission Plant of the Big Cottonwood Power Co." *Engineering News* 36 (1895): 220–22.

———. "The Water Supply System of Salt Lake City, Utah." *Engineering News* 36 (1896): 258–60.

Hardy, Charles W. "Charles W. Hardy Collection." Brigham Young University Library, MSS 997.

Hardy, Clyde Brian. "The Historical Development of Wasatch Trails in Salt Lake County." Master's thesis, Brigham Young University, 1975.

Harlan, Jacob Wright. *California '46 TO '88.* San Francisco: Bancroft Co., 1888.

Hastings, Lansford W. *The Emigrant's Guide to Oregon and California.* New York: DaCapo Press, 1969. Reprint of the 1845 edition.

Haycock, Obed C. "Electric Power Comes to Utah." *Utah Historical Quarterly* 45 (1977): 173–87.

Higgins, Will C. "The South Fork of Big Cottonwood." *Salt Lake Mining Review* 13, no. 8 (30 July 1911): 13–18.

Hiking the Wasatch. The Official Wasatch Mountain Club Trail Map for the Tri-Canyon Area. Salt Lake City: University of Utah Press, 1994.

"Historical Department Journals 1844–1890." CHOA, CR 100 1.

"History of the Church." (This document was formerly called the Manuscript History of Brigham Young. Reels 9–11: 1848–60, 1839–ca. 1882. CHOA, CR 100 102.

History of the Early Settlement and the Spokane Falls of Today. Spokane Falls, WA: 1889.

Hot Springs Mining District Recorder's books. Salt Lake County Recorder Archives.

Hughes, Delila Gardner. *Life of Archibald Gardner.* West Jordan, UT: Archibald Gardner Family Genealogical Association, 1939.

Hunter, Louis C. *A History of Industrial Power in the United States, 1870–1930.* Volume 1: *Waterpower.* Charlottesville: University Press of Virginia, 1979.

Jackman, Levi. "Letter from Salt Lake City, 1850." (4 pp. handwritten) CHOA, MS 7888.

Jackson, W. Turrentine. "British Impact on the Utah Mining Industry." *Utah Historical Quarterly* 31 (1963): 347–75.

————. "The Infamous Emma Mine: A British Interest in the Little Cottonwood District, Utah Territory." *Utah Historical Quarterly* 23 (1955): 339–62.

Jacobs, Zebulon Wm. "Reminiscences and Diaries, 1861–1914." CHOA, MS 1715.

James, Laurence P. *Geology, Ore Deposits, and History of the Big Cottonwood Mining District, Salt Lake County, Utah.* Bulletin 114, Utah Geological and Mineral Survey, 1979.

————. "Little Cottonwood Canyon Collection." UofU, MS 632.

James, Laurence P., and Henry O. Whiteside. "Promoting the Alta Tunnel: The Rise and Fall of F. V. Bodfish." *Journal of the West* 20 no. 2 (1981): 89–102.

Jenson, Andrew. *Latter-day Saint Biographical Encyclopedia.* 4 vols., 1914. Reprint. Salt Lake City: Western Epics, 1971.

"Journal History of the Church of Jesus Christ of Latter-day Saints." Church Historian's Office, Salt Lake City.

Journal of Discourses. Reprint of 1855 edition. Salt Lake City: Deseret Book, 1966.

Kearl, J. R., Clayne L. Pope, and Larry Wimmer, comps. *Index to the 1850, 1860 & 1870 Census of Utah: Heads of Household.* Baltimore: Genealogical Publishing Co, 1981.

Keller, Charles L. "James T. Monk: The Snow King of the Wasatch," *Utah Historical Quarterly* 66 (1988): 139–58.

————. "Tales of Four Alta Miners." *Utah Historical Quarterly* 68 (2000): 100–111.

Kelly, Charles. *Salt Desert Trails.* Salt Lake City: Western Epics, 1969.

Kelner, Alexis. *Skiing in Utah, A History.* Salt Lake City: Kelner, 1980.

Kesler, Frederick. "Frederick Kesler Papers." UofU, MS49.

Koch, Michael. *The Shay Locomotive, Titan of the Timber.* Denver: World Press, 1971.

Korns, J. Roderic. "West from Fort Bridger: The Pioneering of the Immigrant Trails across Utah, 1846–1850." *Utah Historical Quarterly* 19 (1951): 1–297. Revised, updated, and published under the same title by Will Bagley and Harold Schindler. Logan: Utah State University Press, 1994.

Kuhn, Wm. Dobbie. "Recollections of Temple Quarry, Little Cottonwood Canyon and Old Granite (City)." (11 pp. typescript) CHOA, MS 13729.

Lambourne, Alfred. "Catalogue Raisonne." Springville Art Museum, Springville, Utah. (A list of 600 of Lambourne's works, located in the research library at the art museum.)

————. "A Group of Wasatch Lakes." *Improvement Era* 27 (1923): 1041–46.

————. *Jo, A Christmas Tale of the Wasatch.* Chicago: Belford-Clark Co., 1891.

————. "Lake Lillian, Wasatch Mountains." *Improvement Era* 19 (1915): 97.

————. "Naming of the Wasatch Lakes." *Improvement Era* 28 (1924): 1055–57.

————. *Plet, A Christmas Tale of the Wasatch.* Salt Lake City: Deseret News, 1909. (Previously published by S. E. Cassino, Boston, 1895, and Dodge Publishing Co., New York, 1906.)

————. *A Summer in the Wasatch.* Boston: Samuel E. Cassino, 1895.

Larkins, William T. "The Aircraft History of Western Air Lines." *American Aviation Historical Society Journal* 21 (1976): 9–24.

Laub, George. "Reminiscences and Journals, Jan. 1845–Apr. 1857." CHOA, MS 9628.

Little Cottonwood Mining District Recorder's books. Salt Lake County Recorder Archives.

Lockerbie, C. W. "Sidelight Stories of Romantic Alta." *News Bulletin of the Mineralogical Society of Utah* 4 (1943): 26–31.

"Maxfield History." Compiled by Alice M. Johnson. 1976. UHS.

Miller, David E. *Utah History Atlas*. Salt Lake City: D. E. Miller, ca. 1968. (Reprinted many times, presumably for classroom notes at the University of Utah.)

Mining and Scientific Press (San Francisco, 1860–1869). Became *Engineering and Mining Journal* in 1869.

Morgan, Dale L. *The Great Salt Lake*. Salt Lake City: University of Utah Press, 1995.

———. "The State of Deseret—(1847–1849)." *Utah Historical Quarterly* 8 (1940): 64–239.

Morgan, Nicholas Groesbeck. *Our Groesbeck Ancestors in America*. Salt Lake City: author, 1963. CHOL.

Mountain Lakes Mining District Recorder's books. Salt Lake County Recorder Archives.

Mountaineer. (Salt Lake City, 1859).

Murphy, John R. *The Mineral Resources of the Territory of Utah, with Mining Statistics and Maps*. Salt Lake City: James Dwyer, 1872.

Nielsen, Stella Brighton. "William Stuart Brighton." (20 pp. typescript) UHS, MS A769-2. (This document has no date, but the Utah Historical Society's copy of the acceptance letter is dated March 1963.)

"One of a Family, Two of a Nation: Three Pioneer Families Who Helped Build the Great Basin Kingdom—The Brightons, The Thornleys and The Timmins of Utah." (249 pp., ca. 1979) CHOL.

Orton, Richard H. *Records of California Men in the War of the Rebellion, 1861–1867*. Sacramento State Office, 1890.

Park Record. (Park City, 1884–1960).

Parkinson, Preston W. *The Utah Woolley Family*. Salt Lake City: 1967.

Peterson, Charles S. "Albert F. Potter's Wasatch Survey, 1902: A Beginning for Public Management of Natural Resources in Utah." *Utah Historical Quarterly* 39 (1971): 238–53.

———. *A History of the Wasatch-Cache National Forest*. Logan: Utah State University, 1980.

Peterson, Lloyd Leon. "The Pioneer Ancestors of Lloyd L. Peterson." CHOL.

Poll, Richard D., et al., eds. *Utah's History*. Provo: Brigham Young University Press, 1978.

Potter, Albert F. "Diary of Albert F. Potter, July 1, 1902 to November 22, 1902."
     UHS, A-625.
"Rand McNally Indexed County and Township Map of Utah, ca. 1883." UHS,
     PAM 11181.
Rawlins, Joseph Lafayette. *The Unfavored Few: The Autobiography of Joseph L.
     Rawlins.* Carmel, CA: n.p., ca. 1956.
Raymer, Robert G. "Early Mining in Utah." *Pacific Historical Review* 8 (1939):
     81–88.
Reenders, Josie, and Lois Recore. "Maxfield." (5 pp. typescript) UHS, MS A832.
     Also published as "Growing Up in Big Cottonwood Canyon." *Beehive His-
     tory* 19 (1993): 19–23.
Rich, Roxie N. *The History and People of Early Sandy.* Privately published, no date.
Richards, Paul C. "The Salt Lake Temple Infrastructure: Studying It Out in
     Their Minds." *BYU Studies* 36, no. 2 (1996–97): 203–25.
Richardson, Arthur. *The Life and Ministry of John Morgan.* [Salt Lake City?]:
     Nicholas G. Morgan, ca. 1965.
Roberts, Allen D. "The Chase Mill." *An Enduring Legacy* 7 (1984): 137–92.
———. "Pioneer Mills and Milling." *An Enduring Legacy* 7 (1984): 85–136.
Rogers, Fred B. *Soldiers of the Overland.* San Francisco: Grabhorn Press, 1938.
Romney, Edyth. "Typescript Collections." CHOA, MS 2737. (This collection
     consists of typescripts of many original documents in the Church Histo-
     rian Office Archives.)
Russell, Osborne. *Osborne Russell's Journal of a Trapper.* Edited by Aubrey L.
     Haines. Portland: Oregon Historical Society, 1955.
Salt Lake County Court Minute Books. Utah State Archives. (After Utah be-
     came a state in 1896 the county court was replaced by a county commis-
     sion, whose minute books continued the letter designation established
     years earlier.)
Salt Lake City Directory, 1867. Microfilm copy at CHOL.
Salt Lake City Directory and Business Guide for 1869. Salt Lake City: E. L. Sloan
     & Co., 1869.
Salt Lake Daily Telegraph (1868–1870).
Salt Lake Mining Review (1899–1926). (Became the *Mining Review* in 1927,
     and the *Mining and Contract Review* in 1935.)
"Salt Lake Stake High Council Minutes 1847–1848." (41 pp. handwritten, 4
     October 1847 to 22 July 1848) CHOA, MS 3426.
Salt Lake Telegram (1915–1952).
Salt Lake Tribune (1871–   ).
Sinclair, Peter. "Diary 1856 Jan.–1862 Mar." CHOA, MS 12483.
Smith, Elias. "Elias Smith; Journal of a Pioneer Editor, March 6,1859–Septem-
     ber 23, 1863." Edited by A. R. Mortensen. *Utah Historical Quarterly* 21
     (1953): 1–24, 137–68, 237–66, 331–60.
Smith, Jedediah S. *The Southwest Expedition of Jedediah S. Smith, His Personal*

Account of the Journey to California 1826–1827. Edited by George R. Brooks. Glendale, CA: A. H. Clark Co., 1977.

Spence, Clark C. *British Investments and the American Mining Frontier, 1860–1901.* Ithaca: Cornell University Press, 1958.

Stenhouse, T. B. H. [Thomas Brown Holmes]. *The Rocky Mountain Saints.* London: Ward, Lock, and Tyler [1874].

Stevenson, Joseph Grant, ed. *Porter Family History,* vol. 1. Delta and Provo, UT: n.p., 1957.

Stewart, Watt. *Keith and Costa Rica: A Biographical Study of Minor Cooper Keith.* Albuquerque: University of New Mexico Press, 1964.

Stout, Hosea. *On the Mormon Frontier, The Diary of Hosea Stout.* Edited by Juanita Brooks. Salt Lake City: University of Utah Press, 1964.

Tanner, Mary Jane Mount. *A Fragment: The Autobiography of Mary Jane Mount Tanner.* Salt Lake City: University of Utah Library, 1980.

Taylor, George H. "The Autobiography of George H. Taylor." (72 pp. typescript) 1949. CHOL.

Tew, John Henry. "Big Cottonwood Canyon and its Importance as a Water Source." Master's thesis, University of Utah, 1971.

Timmons, Mary Brighton. "Address." (8 pp. typescript) July 1940. CHO, MS 615.

Tracy, Albert. "The Utah War, Journal of Albert Tracy, 1858–1860." *Utah Historical Quarterly* 13 (1945): 1–128.

Tullidge, Edward Wheelock. *The History of Salt Lake City and its Founders, ...* Salt Lake City: E. W. Tullidge [1886].

Tullidge's Quarterly Magazine 1–3 (1880–1885).

Tuttle, Daniel Sylvester. *Missionary to the Mountain West: Reminiscences of Episcopal Bishop Daniel S. Tuttle, 1866–1886.* Salt Lake City: University of Utah Press, 1987. (Originally published in 1906 as *Reminiscences of a Missionary Bishop.*)

Utah Directory and Gazetteer for 1879–80. Salt Lake City: J. C. Graham & Co., 1879.

Utah Gazetteer and Directory of Logan, Provo, Ogden and Salt Lake City, 1884. Salt Lake City: Herald Printing & Publishing Co., 1884.

Utah Light & Traction Co. "Utah Light and Traction Company: History of Origin and Development." (Prepared in connection with Federal Power Commission request, order dated 11 May 1937.)

Utah Mining Gazette (Salt Lake City, 1873–1874).

Utah Power & Light Co. "Instruction to Employees. Power Department." 1925. UofU, MS 206, Bx. 2, Bk. 8.

———. "Upper Mill Creek Station." Folder with blueprints. Ca. 1911. UofU, MS 206, Bx. 4.

———. "Utah Power & Light Company: History of Origin and Development." UHS. (Prepared in connection with Federal Power Commission request, order dated 11 May 1937.)

"Utah Territory Census Schedules 1880." Microfilm copy, CHOA, MS 9207.

Van Cott, John W. *Utah Place Names.* Salt Lake City: University of Utah Press, 1990.

Veranth, John. *Hiking the Wasatch: A Hiking and Natural History Guide to the Central Wasatch.* Salt Lake City: Wasatch Mountain Club, 1988.

Walker, Ronald W. *Wayward Saints: The Godbeites and Brigham Young.* Urbana: University of Illinois Press, 1998.

Wall, C. James. "Gold Dust and Greenbacks." *Montana, The Magazine of Western History* 7 (1957): 24–31.

Wasatch Mountain Club. "Papers." UofU, MS 283.

Watson & Co., George H. "Market Letter, The Story of Alta." (4 pp. flier) [1931?] UHS, PAM 6589.

Webb, Walter Prescott, ed. *Handbook of Texas.* 3 vols. Austin: Texas State Historical Association, 1952–76.

Wells, Daniel H. "Collection." CHOA, MS 1344.

Wendel, C. H. *The Circular Sawmill.* Lancaster, PA: Stemgas Publishing Co., 1994.

Whipple, Nelson Wheeler. "Autobiography and Journal of Nelson Wheeler Whipple, 1819–1885." CHOA, MS 5348.

———. "Journal, 1877 Jun–1878 Aug." CHOA, MS 9995, fd. 1.

Woodruff, Wilford. *Wilford Woodruff's Journal, 1833–1898.* 9 vols. Edited by Scott G. Kenney. Midvale, UT: Signature Books, 1983–1985.

Woolley, John Mills. "Diaries, 1851–1859." CHOA, MS 4147, fd. 1–4.

———. "Diaries, 1854–1859." CHOA, MS 4720.

Woolley, Samuel Amos. "Diaries 1846–1899." CHOA, MS 1556.

Wright, John B. *Rocky Mountain Divide: Selling and Saving the West.* Austin: University of Texas Press, 1993.

Wrigley, Irene Black. "Life of Samuel Thompson." 1992. CHOA, MS 13475.

Young, Brigham. "Collection." CHOA, MS 1234.

———. "Letterbooks." Originals: Young, Brigham. "Papers 1832–1876 Letterbooks." CHOA, MS 1234. See index at CHOA. Typescript transcriptions: Romney, Edyth Jenkins. "Typescript transcriptions of portions of the Brigham Young Collection." CHOA, MS 2736.

———. "Letter 29 Sept. 1855." [To son-in-law Edmund Ellsworth] (4 pp., handwritten) CHOA, MS 4823.

Young, James T. *The New American Government and its Work.* New York: Macmillan, 1923.

Young, John R. "Reminiscences of John R. Young." *Utah Historical Quarterly* 3 (1930): 83–86.

Young, Lorenzo D. "Diary of Lorenzo Dow Young." *Utah Historical Quarterly* 14 (1946): 133–70.

Index